Hydrodynamic Design of Ships

Hydrodynamic Design of Ships

Editor

Gregory Grigoropoulos

 Basel • Beijing • Wuhan • Barcelona • Belgrade • Novi Sad • Cluj • Manchester

Editor
Gregory Grigoropoulos
National Technical University
of Athens
Athens
Greece

Editorial Office
MDPI AG
Grosspeteranlage 5
4052 Basel, Switzerland

This is a reprint of articles from the Special Issue published online in the open access journal *Journal of Marine Science and Engineering* (ISSN 2077-1312) (available at: https://www.mdpi.com/journal/jmse/special_issues/design_ships).

For citation purposes, cite each article independently as indicated on the article page online and as indicated below:

Lastname, A.A.; Lastname, B.B. Article Title. *Journal Name* **Year**, *Volume Number*, Page Range.

ISBN 978-3-7258-1605-7 (Hbk)
ISBN 978-3-7258-1606-4 (PDF)
doi.org/10.3390/books978-3-7258-1606-4

© 2024 by the authors. Articles in this book are Open Access and distributed under the Creative Commons Attribution (CC BY) license. The book as a whole is distributed by MDPI under the terms and conditions of the Creative Commons Attribution-NonCommercial-NoDerivs (CC BY-NC-ND) license.

Contents

About the Editor vii

Preface ix

Gregory J. Grigoropoulos
Hydrodynamic Design of Ships
Reprinted from: *J. Mar. Sci. Eng.* **2022**, *10*, 512, doi:10.3390/jmse10040512 1

Guangyu Shi, Alexandros Priftis, Yan Xing-Kaeding, Evangelos Boulougouris, Apostolos D. Papanikolaou, Haibin Wang, et al.
Numerical Investigation of the Resistance of a Zero-Emission Full-Scale Fast Catamaran in Shallow Water
Reprinted from: *J. Mar. Sci. Eng.* **2021**, *9*, 563, doi:10.3390/jmse9060563 5

Rui Deng, Shigang Wang, Yuxiao Hu, Yuquan Wang and Tiecheng Wu
The Effect of Hull Form Parameters on the Hydrodynamic Performance of a Bulk Carrier
Reprinted from: *J. Mar. Sci. Eng.* **2021**, *9*, 373, doi:10.3390/jmse9040373 29

Gregory J. Grigoropoulos, Christos Bakirtzoglou, George Papadakis and Dimitrios Ntouras
Mixed-Fidelity Design Optimization of Hull Form Using CFD and Potential Flow Solvers
Reprinted from: *J. Mar. Sci. Eng.* **2021**, *9*, 1234, doi:10.3390/jmse9111234 45

Kai Xu, Gang Wang, Luyao Zhang, Liquan Wang, Feihong Yun, Wenhao Sun, et al.
Multi-Objective Optimization of Jet Pump Based on RBF Neural Network Model
Reprinted from: *J. Mar. Sci. Eng.* **2021**, *9*, 236, doi:10.3390/jmse9020236 60

Lawrence Doctors
Reanalysis of the Sydney Harbor RiverCat Ferry
Reprinted from: *J. Mar. Sci. Eng.* **2021**, *9*, 215, doi:10.3390/jmse9020215 78

Miles P. Wheeler, Konstantin I. Matveev and Tao Xing
Numerical Study of Hydrodynamics of Heavily Loaded Hard-Chine Hulls in Calm Water
Reprinted from: *J. Mar. Sci. Eng.* **2021**, *9*, 184, doi:10.3390/jmse9020184 94

Wei Wang, Guobin Cai, Yongjie Pang, Chunyu Guo, Yang Han and Guangli Zhou
Bubble Sweep-Down of Research Vessels Based on the Coupled Eulerian-Lagrangian Method
Reprinted from: *J. Mar. Sci. Eng.* **2020**, *8*, 1040, doi:10.3390/jmse8121040 112

Daniele Peri
Direct Tracking of the Pareto Front of a Multi-Objective Optimization Problem
Reprinted from: *J. Mar. Sci. Eng.* **2020**, *8*, 699, doi:10.3390/jmse8090699 129

Apostolos Papanikolaou, Yan Xing-Kaeding, Johannes Strobel, Aphrodite Kanellopoulou, George Zaraphonitis and Edmund Tolo
Numerical and Experimental Optimization Study on a Fast, Zero Emission Catamaran
Reprinted from: *J. Mar. Sci. Eng.* **2020**, *8*, 657, doi:10.3390/jmse8090657 141

Igor Nesteruk, Srecko Krile and Zarko Koboevic
Electrical Swath Ships with Underwater Hulls Preventing the Boundary Layer Separation
Reprinted from: *J. Mar. Sci. Eng.* **2020**, *8*, 652, doi:10.3390/jmse8090652 159

Stefan Harries and Sebastian Uharek
Application of Radial Basis Functions for Partially-Parametric Modeling and Principal Component Analysis for Faster Hydrodynamic Optimization of a Catamaran
Reprinted from: *J. Mar. Sci. Eng.* **2021**, *9*, 1069, doi:10.3390/jmse9101069 . . . . **168**

Duck Young Yoon and Jeong Hee Park
Adaptive Polynomials for the Vibration Analysis of an L-Type Beam Structure with a Free End
Reprinted from: *J. Mar. Sci. Eng.* **2021**, *9*, 300, doi:10.3390/jmse9030300 . . . . **192**

About the Editor

Gregory Grigoropoulos

Gregory Grigoropoulos is the a Professor of Ship and Marine Hydrodynamics at the School of Naval Architecture and Marine Engineering of the National technical University of Athens, Greece. He had published more than 90 original papers with major scientific areas of focus, including the following: ship hydrodynamics, optimization using deterministic and stochastic methods, experimental techniques in hydrodynamics and high-speed craft.

Preface

During the last two decades, the process of designing a ship has encompassed and incorporated its hydrodynamic performance in calm water and in waves, as well as that of the propulsion units, as a major aspect of its merit in service. The computational tools developed during the second half of the 20th century for the evaluation of a ship's resistance, propulsion, and seakeeping, including added resistance in waves, are exploited in the preliminary design process. In most cases, formal optimization strategies are implemented. They are based on biomimetic methods, i.e., genetic algorithms, evolutionary strategies, and artificial neural networks, which are capable of handling multi-objective problems, and are implemented on an existing (parent) hull form or generic design. A variety of techniques used to accelerate the execution time of these methods, taking into account the available hardware and resources, without a significant reduction in their accuracy and robustness, are proposed in this Special Issue. In these cases, the revealed properties of the objectives under consideration are exploited to reduce randomness in the search, and to drive the procedure to reach the optimum faster, with reduced variant evaluations. Both potential and viscous flow environments, with a wide range of grid densities and fidelities, and in various mixtures, are used in the optimization process. The validity of the final outcome is evaluated using model tests.

During the last decade, international organizations, the European Union, and national authorities have focused their interest on environmental protection. In this respect, the CO_2, greenhouse gases (GHG), and carbon particles emitted by means of transportation, including waterborne transportation, using carbon-based fossil fuel should be radically reduced until 2050. The issued guidelines for the shipping industry indicate a strong target towards the optimization of the operation of ships, which is mainly feasible via the optimization of their hydrodynamic performance and improvements in the performance of their main engines and propulsion characteristics, as well as their dynamic responses and additional power requirements in actual seaways. Aiming to reduce the operating expenses, the main target is to minimize the fuel consumption in all sailing conditions. The outcome of the optimization of ship design will also be applicable in the case of alternative fuels (LNG, methanol, ammonia, or hydrogen) and all electric propulsions.

The high-quality papers published in this Special Issue are directly related to most of the aforementioned aspects of the hydrodynamic performance of a ship, including novel techniques, in ship hydrodynamics, emission reduction, optimization strategies, hull form optimization, seakeeping, resistance, propulsion, maneuvering, as well as some interesting case studies. Monohulls, catamarans, and SWATHs, in both deep and shallow water, are considered.

Enjoy reading this Special Issue on the "Hydrodynamic Design of Ships".

Gregory Grigoropoulos
Editor

Editorial

Hydrodynamic Design of Ships

Gregory J. Grigoropoulos

Laboratory of Ship and Marine Hydrodynamics, School of Naval Architecture and Marine Engineering, National Technical University of Athens, 15780 Athens, Greece; gregory@central.ntua.gr

Citation: Grigoropoulos, G.J. Hydrodynamic Design of Ships. *J. Mar. Sci. Eng.* **2022**, *10*, 512. https://doi.org/10.3390/jmse10040512

Received: 26 January 2022
Accepted: 31 March 2022
Published: 7 April 2022

Publisher's Note: MDPI stays neutral with regard to jurisdictional claims in published maps and institutional affiliations.

Copyright: © 2022 by the author. Licensee MDPI, Basel, Switzerland. This article is an open access article distributed under the terms and conditions of the Creative Commons Attribution (CC BY) license (https://creativecommons.org/licenses/by/4.0/).

During the last two decades, the process of designing a ship has encompassed and incorporated its hydrodynamic performance in calm water and in waves, as well as that of the propulsion units, as a major aspect of its merit in service. The computational tools developed during the second half of the 20th century, for the evaluation of a ship's resistance, propulsion and seakeeping, including added resistance in waves, are exploited in the preliminary design process. In most cases, formal optimization strategies are implemented. They are based on biomimetic methods, i.e., genetic algorithms, evolutionary strategies and artificial neural networks, which are capable of handling multi-objective problems, and are implemented on an existing (parent) hull form or generic design. A variety of techniques to accelerate the execution time of these methods, taking into account the available hardware, resources, without a significant reduction in their accuracy and robustness, are proposed in this Special Issue. In these cases, the revealed properties of the objectives under consideration are exploited to reduce randomness in the search, and to drive the procedure to reach the optimum faster, with reduced variant evaluations. Both potential and viscous flow environments, with a wide range of grid densities and fidelities, and in various mixtures, are used in the optimization process. The validity of the final outcome is evaluated by model tests.

During the last decade, international organizations, the European Union and national authorities have focused their interest on environmental protection. In this respect, the CO_2, greenhouse gases (GHG) and carbon particles emitted by transportation means, including waterborne transportation, using carbon-based fossil fuel, should be radically reduced until 2050. The issued guidelines for the shipping industry indicate a strong target towards the optimization of the operation of ships, which is mainly feasible via the optimization of their hydrodynamic performance, and improvement in the performance of their main engines and propulsion characteristics, as well as their dynamic responses and additional power requirements in actual seaways. Aiming to reduce the operating expenses, the main target is to minimize the fuel consumption in all sailing conditions. The outcome of the optimization of ship design will also be applicable in the case of alternative fuels (LNG, methanol, ammonia or hydrogen) and all electric propulsion.

The high-quality papers published in this Special Issue are directly related to most of the aforementioned aspects of the hydrodynamic performance of a ship, including novel techniques, in ship hydrodynamics, emission reduction, optimization strategies, hull form optimization, seakeeping, resistance, propulsion, maneuvering, as well as some interesting case studies. Monohulls, catamarans and SWATHs, in deep and shallow water, are considered. To be more specific, the following contributions are included in this Special Issue:

Shi et al. [1] numerically evaluated the resistance at full scale of a zero-emission, highspeed catamaran, in both deep and shallow water, for a Froude number (Fn) ranging from 0.2 to 0.8. The numerical methods are validated by the available model and a blind validation, using two different flow solvers. The total resistance is highly affected by the pressure component, which is maximized at Fn = 0.58 in deep water and at Fn = 0.30 in shallow water, when the secondary trough is created at the stern, leading to the largest trim angle. The vessel witnesses a hump near the critical speed (Fn = 0.30) in shallow water, due to the interaction between the wave systems created by the demi-hulls.

Deng et al. [2] study the principal dimensions and the hull form of a bulk carrier, to optimize its hydrodynamic performance. They considered ship resistance and seakeeping, while maneuverability was estimated by empirical methods. A new parent ship was chosen from 496 sets of hulls, after comprehensive consideration. A further hull form optimization was performed on the new parent ship, according to the minimum wave-making resistance. He concluded that optimization with respect to the principal dimensions provides a high-quality parent ship, which can be further optimized for both the principal dimensions and the hull form parameters.

Zoon and Park [3] use the component mode method to carry out vibration analyses when they design local structures on ships. The method provides natural mode functions and, eventually, reasonable natural frequencies. In their study, they use adaptive polynomials as additional flexible model functions, or a purely mathematical approach, with very good numerical results.

Xu et al. [4] used an RBF (radial basis function) neural network and NSGA-II (non-dominated sorting genetic algorithm) to optimize the hydraulic performance of the annular jet pump applied in submarine trenching and dredging. The suction angle, diffusion angle, area ratio and flow ratio were selected as the design variables. On the basis of CFD numerical simulation, an RBF neural network approximation model was established. Finally, the NSGA-II algorithm was selected to carry out multi-objective optimization and obtain the optimal design variable combination. The results show that both optimization criteria, the jet pump efficiency and the head ratio, were accurately modelled via the RBF neural network, while the optimization resulted in a 30% increase in the head ratio and a slight improvement in the efficiency.

Doctors [5] revisited the hydrodynamics supporting the design and development of the RiverCat class of catamaran ferries, which have operated in Sydney Harbor since 1991. They used more advanced software to account for the hydrodynamics of the transom demi-sterns that experience partial or full ventilation, depending on the vessel speed, which gives rise to hydrostatic drag. On the other hand, the transom creates hollowness in the water, causing effective hydrodynamic lengthening of the vessel, leading to a reduction in the wave resistance. The associated detailed analysis quite accurately predicts the phenomena and allows for the optimization of the vessel using affine transformations of the hull geometry, including the size of the transom.

Wheeler et al. [6] deal with heavily loaded hard-chine boats, which are usually intended for high-speed regimes in planing mode and relatively light displacement. They present the results for the steady-state hydrodynamic performance of these boats at nominal weight and when overloaded in calm water using the CFD solver program STAR-CCM+. The resistance and attitude values of a constant-deadrise reference hull and its modifications, with more pronounced bows of concave and convex shapes, are obtained. On average, 40% heavier hulls showed about 30% larger drag over the speed range from the displacement to planing mode. The hull with a concave bow is found to have 5–12% lower resistance than the other hulls in the semi-displacement regime and heavy loadings, and 2–10% lower drag in the displacement regime and nominal loadings, while this hull is also capable of achieving fast planning speeds at the nominal weight, with the typical available thrust. Selected near-hull wave patterns and hull pressure distributions are also presented and discussed.

Wang et al. [7] explored the bubble sweep-down phenomenon of research vessels and its effect on the position of the stern sonar of a research vessel; the use of a fairing was investigated as a defoaming appendage. The separation vortex turbulence model was selected for simulation, and the coupled Eulerian–Lagrangian method was adopted to study the characteristics of the bubble sweep-down motion, captured using a discrete element model. The interactions between the bubbles, water, air, and hull were defined via a multiphase interaction method. The bubble point position and bubble layer were calculated separately. The spatial movement characteristics of the bubbles were extracted from the bubble trajectories. It was demonstrated that the bubble sweep-down phenomenon is closely related to the distribution of the bow pressure field, and that the bubble motion characteristics are related to the speed and initial bubble position. When the initial bubble position is between the water surface and the ship bottom, the impact on the middle of the ship bottom is greater, and increases further with increasing speed. A deflector forces the bubbles to both sides through physical shielding, strengthening the local vortex structure and keeping the bubbles away from the middle of the ship bottom.

Peri [8] applied some methodologies aimed at the identification of the Pareto front of a multi-objective optimization problem. He presented the following three different approaches: local sampling, Pareto front resampling and normal boundary intersection (NBI). The first approximation of the Pareto front is obtained by regular sampling of the design space, and then the Pareto front is improved and enriched using the other two abovementioned techniques. A detailed Pareto front is obtained for an optimization problem where algebraic objective functions are applied and also compared with standard techniques. Encouraging results are obtained for two different ship design problems. The use of algebraic functions allows for a comparison with the real Pareto front, correctly detected. The variety of ship design problems allows for the applicability of the methodology to be generalized.

Papanikolaou et al. [9] focused on the hydrodynamic hull form optimization of a zero-emission, battery-driven, fast catamaran vessel. A two-stage optimization procedure was implemented to identify, in the first stage (global optimization), the optimum combination of a ship's main dimensions and, later on, in the second stage (local optimization), the optimal ship hull form, minimizing the required propulsion power for the set operational specifications and design constraints. The numerical results of the speed-power performance for a prototype catamaran, intended for operation in the Stavanger area (Norway), were verified by model experiments at Hamburgische Schiffbau Versuchsanstalt (HSVA), proving the feasibility of this innovative, zero-emission, waterborne urban transportation concept.

Nesteruk et al. [10] studied the body shapes of aquatic animals, which ensure laminar flow, without boundary layer separation, at rather high Reynolds numbers. The commercial efficiencies (drag-to-weight ratio) of similar hulls were estimated. Examples of neutrally buoyant vehicles, with high commercial efficiency, were proposed. It was shown that such hulls can be effectively used in both water and air. The authors discussed their application in SWATH (small-water-area twin hulls) vehicles, where the seakeeping characteristics of such ships can be improved, due to the use of underwater hulls. In addition, the special shape of these hulls allows the total drag to be reduced, as well as the energetic needs and pollution. The presented estimations show that a weight-to-drag ratio of 165 can be achieved for a yacht with such specially shaped underwater hulls, permitting the use of electrical engines only, and solar cells to charge the batteries.

Harries and Uharek [11] applied a flexible approach of partially parametric modelling, on the basis of radial basis functions (RBF), for the modification of an existing hull form (baseline). Contrary to other similar approaches, RBF functions allow sources that lie on the baseline and targets that define the intended new shape to be identified. Sources and targets can be corresponding sets of points, curves and surfaces, used to derive a transformation field that subsequently modifies those parts of the geometry that shall be subjected to variation, making the approach intuitive and quick to set up. Since the RBF approach may potentially introduce quite a few degrees of freedom, a principal component analysis (PCA) is utilized to reduce the dimensionality of the design space. PCA allows the deliberate sacrifice of variability, in order to define variations of interest with fewer variables, denoted as principal parameters. The aim of combining RBFs and PCA is to make simulation-driven design (SDD) easier and faster to use. Ideally, the turn-around time within which to achieve noticeable improvements should be 24 h, including the time needed to set up both the CAD model and the CFD simulation, as well as to run the first optimization campaign. The methodology was implemented on an electric catamaran, using a potential (SHIPFLOW) and a viscous (NEPTUNO) solver in the environment of CAESES, a versatile process integration and design optimization software. The combination of RBF and PCA proved quite efficient, resulting in meaningful reductions in total resistance and, hence, improvements in energy efficiency within very few simulations. For a deterministic search strategy via a one-stop steepest descent, 10 to 12 CFD runs are needed to identify better hulls, within one working day and a night for CFD runs.

Grigoropoulos et al. [12] proposed a new mixed-fidelity method to optimize the shape of ships using genetic algorithms (GA) and potential flow codes, to evaluate the hydrodynamics of variant hull forms, enhanced by a surrogate model, based on an artificial neural network (ANN), to account for viscous effects. The performance of the variant hull forms generated by the GA is evaluated for calm water resistance using potential flow methods, which are quite fast when they are run on modern computers. However, these methods do not take into account the viscous effects, which are dominant in the stern region of the ship. Solvers of the Reynolds-averaged Navier–Stokes equations (RANS) should be used in this respect, which, however, are too time consuming to be used for the evaluation of some hundreds of variants within the GA search. In this study, a RANS solver is used prior to the execution of the GA, to train the ANN to model the effect of stern design geometrical parameters only. The potential flow results, accounting for the geometrical design parameters of the rest of the hull, are combined with the aforementioned trained meta-model for evaluation of the final hull form. This work concentrates on the provision of a more reliable framework for the evaluation of hull form performance in calm water, without a significant increase in the computing time.

Enjoy reading this Special Issue on "Hydrodynamic Design of Ships".

Funding: This research received no external funding.

Conflicts of Interest: The author declares no conflict of interest.

References

1. Shi, G.; Priftis, A.; Xing-Kaeding, Y.; Boulougouris, E.; Papanikolaou, A.D.; Wang, H.; Symonds, G. Numerical Investigation of the Resistance of a Zero-Emission Full-Scale Fast Catamaran in Shallow Water. *J. Mar. Sci. Eng.* **2021**, *9*, 563. [CrossRef]
2. Deng, R.; Wang, S.; Hu, Y.; Wang, Y.; Wu, T. The Effect of Hull Form Parameters on the Hydrodynamic Performance of a Bulk Carrier. *J. Mar. Sci. Eng.* **2021**, *9*, 373. [CrossRef]
3. Yoon, D.Y.; Park, J.H. Adaptive Polynomials for the Vibration Analysis of an L-Type Beam Structure with a Free End. *J. Mar. Sci. Eng.* **2021**, *9*, 300. [CrossRef]
4. Xu, K.; Wang, G.; Zhang, L.; Wang, L.; Yun, F.; Sun, W.; Wang, X.; Chen, X. Multi-Objective Optimization of Jet Pump Based on RBF Neural Network Model. *J. Mar. Sci. Eng.* **2021**, *9*, 236. [CrossRef]
5. Doctors, J.D. Reanalysis of the Sydney Harbor RiverCat Ferry, JMSE Special Issue Hydrodynamic Ship Design. *J. Mar. Sci. Eng.* **2021**, *9*, 215. [CrossRef]
6. Wheeler, M.P.; Matveev, K.I.; Xing, T. Numerical Study of Hydrodynamics of Heavily Loaded Hard-Chine Hulls in Calm Water. *J. Mar. Sci. Eng.* **2021**, *9*, 184. [CrossRef]
7. Wang, W.; Cai, G.; Pang, Y.; Guo, C.; Han, Y.; Zhou, G. Bubble Sweep-Down of Research Vessels Based on the Coupled Eulerian-Lagrangian Method. *J. Mar. Sci. Eng.* **2020**, *8*, 1040. [CrossRef]
8. Peri, D. Direct Tracking of the Pareto Front of a Multi-Objective Optimization Problem. *J. Mar. Sci. Eng.* **2020**, *8*, 699. [CrossRef]
9. Papanikolaou, A.D.; Xing-Kaeding, Y.; Strobe, J.; Kanellopoulou, A.; Zaraphonitis, G.; Tolo, E. Numerical and Experimental Optimization Study on a Fast, Zero Emission Catamaran. *J. Mar. Sci. Eng.* **2020**, *8*, 657. [CrossRef]
10. Nesteruk, I.; Krile, S.; Koboevic, Z. Electrical Swath Ships with Underwater Hulls Preventing the Boundary Layer Separation. *J. Mar. Sci. Eng.* **2020**, *8*, 652. [CrossRef]
11. Harries, S.; Uharek, S. Radial Basis Functions for Partially-Parametric Modeling and Principal Component Analysis for Faster Hydrodynamic Optimization of a Catamaran. *J. Mar. Sci. Eng.* **2021**, *9*, 1069. [CrossRef]
12. Grigoropoulos, G.J.; Bakirtzoglou, C.; Papadakis, G.; Ntouras, D. Mixed-Fidelity Design Optimization of Hull Form Using CFD and Potential Flow Solvers. *J. Mar. Sci. Eng.* **2021**, *9*, 1234. [CrossRef]

Article

Numerical Investigation of the Resistance of a Zero-Emission Full-Scale Fast Catamaran in Shallow Water

Guangyu Shi [1], Alexandros Priftis [1], Yan Xing-Kaeding [2], Evangelos Boulougouris [1,*], Apostolos D. Papanikolaou [3], Haibin Wang [1] and Geoff Symonds [4]

1. Maritime Safety Research Centre (MSRC), Department of Naval Architecture, Ocean and Marine Engineering, University of Strathclyde, Glasgow G4 0LZ, UK; Guangyu.Shi@strath.ac.uk (G.S.); Alexandros.Priftis@strath.ac.uk (A.P.); haibin.wang.100@strath.ac.uk (H.W.)
2. Computational Fluid Dynamics Department, Hamburg Ship Model Basin (HSVA), 22305 Hamburg, Germany; Xing-Kaeding@hsva.de
3. Ship Design Laboratory, National Technical University of Athens (NTUA), 15780 Athens, Greece; papa@deslab.ntua.gr
4. Uber Boat by Thames Clippers, London E14 0JY, UK; geoff.symonds@thamesclippers.com
* Correspondence: evangelos.boulougouris@strath.ac.uk

Citation: Shi, G.; Priftis, A.; Xing-Kaeding, Y.; Boulougouris, E.; Papanikolaou, A.D.; Wang, H.; Symonds, G. Numerical Investigation of the Resistance of a Zero-Emission Full-Scale Fast Catamaran in Shallow Water. *J. Mar. Sci. Eng.* **2021**, *9*, 563. https://doi.org/10.3390/jmse9060563

Academic Editors: Javad Mehr and Abbas Dashtimanesh

Received: 24 April 2021
Accepted: 19 May 2021
Published: 23 May 2021

Publisher's Note: MDPI stays neutral with regard to jurisdictional claims in published maps and institutional affiliations.

Copyright: © 2021 by the authors. Licensee MDPI, Basel, Switzerland. This article is an open access article distributed under the terms and conditions of the Creative Commons Attribution (CC BY) license (https://creativecommons.org/licenses/by/4.0/).

Abstract: This paper numerically investigates the resistance at full-scale of a zero-emission, high-speed catamaran in both deep and shallow water, with the Froude number ranging from 0.2 to 0.8. The numerical methods are validated by two means: (a) Comparison with available model tests; (b) a blind validation using two different flow solvers. The resistance, sinkage, and trim of the catamaran, as well as the wave pattern, longitudinal wave cuts and crossflow fields, are examined. The total resistance curve in deep water shows a continuous increase with the Froude number, while in shallow water, a hump is witnessed near the critical speed. This difference is mainly caused by the pressure component of total resistance, which is significantly affected by the interaction between the wave systems created by the demihulls. The pressure resistance in deep water is maximised at a Froude number around 0.58, whereas the peak in shallow water is achieved near the critical speed (Froude number ≈ 0.3). Insight into the underlying physics is obtained by analysing the wave creation between the demihulls. Profoundly different wave patterns within the inner region are observed in deep and shallow water. Specifically, in deep water, both crests and troughs are generated and moved astern as the increase of the Froude number. The maximum pressure resistance is accomplished when the secondary trough is created at the stern, leading to the largest trim angle. In contrast, the catamaran generates a critical wave normal to the advance direction in shallow water, which significantly elevates the bow and creates the highest trim angle, as well as pressure resistance. Moreover, significant wave elevations are observed between the demihulls at supercritical speeds in shallow water, which may affect the decision for the location of the wet deck.

Keywords: fast catamaran; shallow water resistance; full-scale CFD

1. Introduction

Low-carbon, environmentally-friendly maritime transport is playing an important role in reducing the emission of greenhouse gases and building a sustainable future. The need for technological innovations in the design of zero-emission ships is posing challenges for the maritime industry in the coming decades. The research presented herein was conducted in the European Commission (EC) funded research project TrAM (Transport: Advanced and Modular, https://tramproject.eu/ (accessed on 23 April 2021)), which aims at designing and manufacturing battery-powered fast catamarans operating in coastal areas and inland waterways by implementing modular design and production methods. Given the significant lower specific energy content of batteries compared to conventional fuels [1], the design of zero-emission high-speed marine vehicles poses unique challenges and limitations which are tackled within the TrAM project. These include the selection of the appropriate battery technology and specification, safety considerations, and of course, multi-objective hull form optimisation in the presence of shallow water effects [1–4]. The present study is focused on the battery-driven, zero-emission 'TrAM London Demonstrator', designed for The Thames River. It examines the hydrodynamic performance of the preliminary design of this high-speed catamaran in shallow water as it affects directly the rate by which the vessel consumes the stored energy. Therefore, it verifies and validates the computational methods employed in the hydrodynamic optimisation of the hull form.

Catamarans, due to their favourable performance in efficiency and stability at high speeds, have been widely studied experimentally, theoretically and numerically over the past decades [5–7]. A series of model tests were carried out by Insel and Molland [8] and Molland et al. [9] investigating the calm water resistance of fast catamarans with symmetrical demihulls, whereas Zaraphonitis et al. [10] have studied asymmetrical demihulls. Their studies emphasised the effects of demihull dimensions and separation distance on the resistances and motions of the catamarans over a wide range of Froude numbers ($0.2 \leq Fn \leq 1.0$). van't Veer [11] also experimentally investigated the resistance and dynamic motion characteristics using Delft 372 catamaran, which has been used as a benchmark for numerical simulations. Later experimental studies with Delft 372 catamaran were concentrated on the hydrodynamic interference between demihulls [12,13] and seakeeping [14–16]. Broglia et al. [13] carried out experimental work examining the interference effects between the demihulls of a catamaran. It was found that positive inference only occurred within a narrow range of testing conditions, and the interaction between demihulls could increase the total resistance by up to 30%. The interference effects were less strong at very low and very high Froude numbers ($Fn < 0.3$ or $Fn > 0.7$). Zaraphonitis et al. [17] studied the optimisation of the hull shape with regards to powering and wash for a high speed catamaran. Souto-Iglesias et al. [18] also experimentally investigated the interference phenomenon of a catamaran and compared the wave systems created by the catamaran and the corresponding monohull. They concluded that the non-centred inner wave cuts are also important evidence for the analysis of wave interference. Later, Souto-Iglesias et al. [19] further studied the influence of demihull separation and testing condition on the interference resistance of a Series 60 catamaran and found that the free sinkage-trim condition enhanced both the favourable and unfavourable interference effects compared with fixed condition cases. Danışman [20] found that the wave interference resistance between the demihulls could be considerably reduced by placing an optimised Centrebulb, which led to a favourable secondary wave interaction.

With the fast development of computer science and numerical methods, computational fluid dynamics (CFD) has become a feasible approach with sufficient accuracy to investigate ship hydrodynamics [21]. Various CFD solvers have been applied to examine the calm water resistance and seakeeping of both monohulls [22–24] and multihulls [25–29]. A combined experimental and numerical study was carried out by Zaghi et al. [30] to analyse the interference effects between the demihulls and the dependency on the separation of a high-speed catamaran. Two humps were found in the total resistance coefficient curves, and the second one was much higher, corresponding to a stronger interference. Besides, a smaller separation distance led to a stronger interaction and a larger speed where the peak occurred. Broglia et al. [31] conducted a numerical analysis on the interference phenomena between the demihulls of the catamaran with emphasis on the validation of the CFD code and the Reynolds number effect. It was found that the numerical results agreed very well with the experiment in terms of resistance and wave cuts, and the dependency on the scale effect was rather weak. He et al. [32] computationally investigated the effects of Froude number, and demihull separation distance on the resistance and motion of the catamaran. They found that the resistance coefficient became higher at smaller separation distances, indicating stronger interference between the demihulls. Besides, the strongest demihull interaction occurred when Froude number is between 0.45 and 0.65 ($0.45 < Fn < 0.65$). When the Froude number is below 0.45 or above 0.65, the variation of the separation distance had a negligible effect on the resistance, as well as the sinkage and trim of the catamaran. Haase et al. [33] proposed a novel CFD-based method for predicting full-scale ship resistance, which relied on the results of the model test experiment and CFD simulation at both model-scale and full-scale Reynold number. Farkas et al. [34] carried out a numerical study on the interference of resistance components for a Series 60 catamaran at medium Froude numbers, where the interference factor was decomposed into viscous interference and wave interference. They found that the form factor of the catamaran was independent of the Froude number, but decreased to the value of the monohull when the separation distance became larger. It was also observed that the viscous interference factor was independent of the Froude number, but relied on the separation ratio of the catamaran.

The shallow water effects must be considered when designing ships for restricted waterways (e.g., inland rivers, canals). Previous studies regarding shallow water effects for monohulls [35–39] revealed that the depth Froude number ($Fn_H = U/\sqrt{gH}$, where U is the ship speed, g is gravity acceleration and H is the water depth) is playing a key role in determining the performance of the vessel. A ship moving near the critical depth Froude number ($Fn_H = 1.0$) will experience a surge in total resistance coefficient and drastic changes in motions and wave patterns, which should be taken into account when passing through shallow water areas. In terms of catamarans operating in shallow water, several experimental and numerical studies are also available [40–43]. Molland et al. [44,45] experimentally investigated the resistance of a series of fast displacement catamarans in shallow water. Similar to monohulls, the catamarans experienced large increases in total resistance and wave elevation, and significant changes in sinkage and trim near the critical depth Froude number. The resistance increase was higher for the smaller water depth. Gourlay [46] theoretically predicted the sinkage and trim of various catamaran configurations in shallow water. It was found that the maximum sinkage and trim occurred at the trans-critical speed range. Lee et al. [47] designed and tested the shallow water behaviours of a small catamaran and further investigated the influence of the separation ratio between the demihulls on the resistance characteristics. The residual resistance coefficient surged near the critical depth Froude number and the sinkage and trim also varied significantly in the critical region. Castiglione et al. [48] studied the interference effects between the demihulls of a high-speed catamaran in shallow water using a CFD method. They concluded that for all separation ratios, the total resistance coefficients were significantly increased, due to shallow water effects, with peaks achieved near the critical depth Froude number. However, at extreme subcritical and supercritical speeds, the total resistance coefficients in shallow water became smaller than the values of corresponding deepwater cases. It was also found that the interference factor reached its peak values around the critical speed and increased for smaller separation distances. Moreover, the sinkage and trim were also increased compared with deep water values and maximised at the critical speed.

Despite the extensive studies on the calm water resistance and interference of high-speed catamarans, CFD simulations on full-scale fast catamarans in shallow water are still rare. As aforementioned, the work in the present paper is part of the ongoing TrAM Project (https://tramproject.eu/ (accessed on 23 April 2021)), and the objectives of the current study are twofold: (1) Validate the numerical methods and setups that will be employed in the hull optimisation stage, (2) investigate the shallow water effect on the calm water resistance, sinkage, trim and wave creation of the full-scale London Demonstrator catamaran using a CFD method. The rest of this paper is organised as follows: In Section 2, the geometry of the London Demonstrator and parameters used for analysis are presented. The computational methods are introduced in Section 3, and they are validated in Section 3.4. In the Section 4, the numerical results are given. The conclusions are drawn in the Section 5.

2. Geometry and Parameters
2.1. Catamaran Geometry and Dimensions

The London Demonstrator catamaran investigated in the present work is designed by the Maritime Safety Research Centre (MSRC) at the University of Strathclyde, which is a partner at the ongoing EU funded project TrAM (https://tramproject.eu/ (accessed on 23 April 2021)). The London Demonstrator is designed for The Thames River as a battery-driven, zero-emission passenger ferry. As the catamaran is still at the initial design stage, the geometry illustrated in Figure 1 is selected as a showcase validating the numerical methods and examining the shallow water effect. Some main dimensions of the London Demonstrator are summarised in Table 1, where L_{pp} is the length between perpendiculars. The vertical and longitudinal centres of gravity are measured as the distances below the waterline and ahead of the aft perpendicular of the catamaran, respectively. The gyration radii used to calculate the moment of inertia for pitch motion is 0.25 L_{pp}.

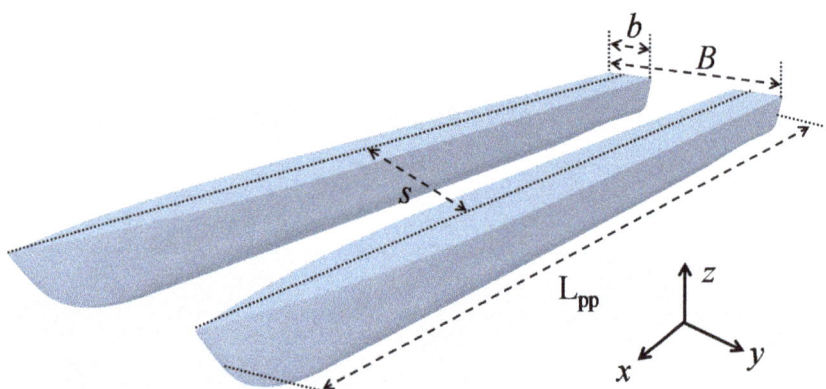

Figure 1. The geometry of the London demonstrator.

Table 1. Main dimensions of the TrAM London Demonstrator catamaran.

Dimension	Symbol	Value
Demihull breadth	b/L_{pp}	0.068
Separation	s/L_{pp}	0.187
Draught	T/L_{pp}	0.033
Depth/draught	H/T	2.0
Vertical centre of gravity	VCG/L_{pp}	0.012
Longitudinal centre of gravity	LCG/L_{pp}	0.447

2.2. Parameters for Analysis

In ship hydrodynamics, Froude number is an important non-dimensional parameter measuring the speed of the vessel, which is defined as:

$$Fn = \frac{U}{\sqrt{gL_{pp}}} \quad (1)$$

where U is the ship speed relative to the incoming flow, g is gravity acceleration. For ships advancing in shallow waterways, the Froude number defined based on the water depth is playing a significant role in determining the ship hydrodynamics:

$$Fn_H = \frac{U}{\sqrt{gH}} \quad (2)$$

The hydrodynamic performance of the TrAM London Demonstrator is analysed by examining its total resistance (R_T), sinkage (σ) and trim (θ). The total resistance is decomposed into the frictional component (R_F) and pressure component (R_P), i.e., $R_T = R_F + R_P$. The sinkage and trim are measured based on the centre of mass of the catamaran, and they defined as positive when the catamaran goes down and the bow moves up, respectively. Moreover, the resistance coefficients are also used for analysis. The total resistance coefficient is defined as

$$C_T = \frac{R_T}{0.5\rho U^2 S} \tag{3}$$

Similarly, the frictional and pressure resistance coefficients can be formulated as

$$\begin{aligned} C_F &= \frac{R_F}{0.5\rho U^2 S} \\ C_P &= \frac{R_P}{0.5\rho U^2 S} \end{aligned} \tag{4}$$

where ρ is the water density, U is the moving speed of the catamaran and $S \in (S_{sw}, S_{dw})$ is the wetted surface area, where S_{sw} and S_{dw} are the static and dynamic wetted surface areas, respectively. The frictional resistance coefficient can also be estimated by the ITTC 1957 correlation line formula

$$C_{F,ITTC} = \frac{0.075}{[log_{10}(Re) - 2]^2} \tag{5}$$

3. Methodology

3.1. Computational Methods

3.1.1. Flow Simulation

The unsteady incompressible turbulent flow in the present study is simulated by solving the unsteady Reynolds-Averaged Navier-Stokes (URANS) equations. The corresponding continuity and momentum equations can be formulated as [49]

$$\frac{\partial(\rho \overline{u_i})}{\partial x_i} = 0 \tag{6}$$

$$\frac{\partial(\rho \overline{u_i})}{\partial t} + \frac{\partial}{\partial x_j}\left(\rho \overline{u_i u_j} + \rho \overline{u'_i u'_j}\right) = -\frac{\partial \overline{p}}{\partial x_i} + \frac{\partial \overline{\tau_{ij}}}{\partial x_j} = 0 \tag{7}$$

where ρ is the fluid density, x_i and x_i are the components of the position vector in Cartesian coordinate, $\overline{u_i}$ and $\overline{u_j}$ are the components of the mean velocity vector, $\overline{u'_i u'_j}$ is the Reynolds stresses and $\overline{p}$ is the mean pressure. $\overline{\tau_{ij}}$ are the components of the mean viscous stress tensor, which can be written as

$$\overline{\tau_{ij}} = \mu\left(\frac{\partial \overline{u_i}}{\partial x_j} + \frac{\partial \overline{u_j}}{\partial x_i}\right) \tag{8}$$

where μ is the fluid dynamic viscosity.

The $k - \omega$ SST turbulence model, which has been widely used for marine hydrodynamics [21,29,32], is employed as the closure for Equations (6) and (7). Here, the flow governing equations, including the turbulent model, are solved using a finite volume method implemented in the commercial code Star CCM+ 14.06. The Semi-Implicit Method for Pressure-Linked Equations (SIMPLE) was used as the solution procedure, where the continuity and momentum equations are solved sequentially and then coupled via a predictor-corrector approach. The spatial discretisation was achieved using a second-order scheme, while a first-order scheme was employed for temporal discretisation, since we are only focused on the final converged equilibrium state.

3.1.2. Free Surface Capturing

For marine hydrodynamics, the appropriate capture of the free surface is of great importance to accurately predict the wave height. In the present work, the volume of fluid (VOF) method in combination with the High-Resolution Interface Capturing (HRIC) scheme was adopted to calculate the wave elevation induced by the motion of the catamaran. To avoid the wave's reflection at the boundaries of the computational domain, a wave forcing method was used at relevant boundaries to guarantee that the wave is completely damped out when it reaches the domain boundary. The wave forcing length and relevant boundaries are illustrated in Figure 2.

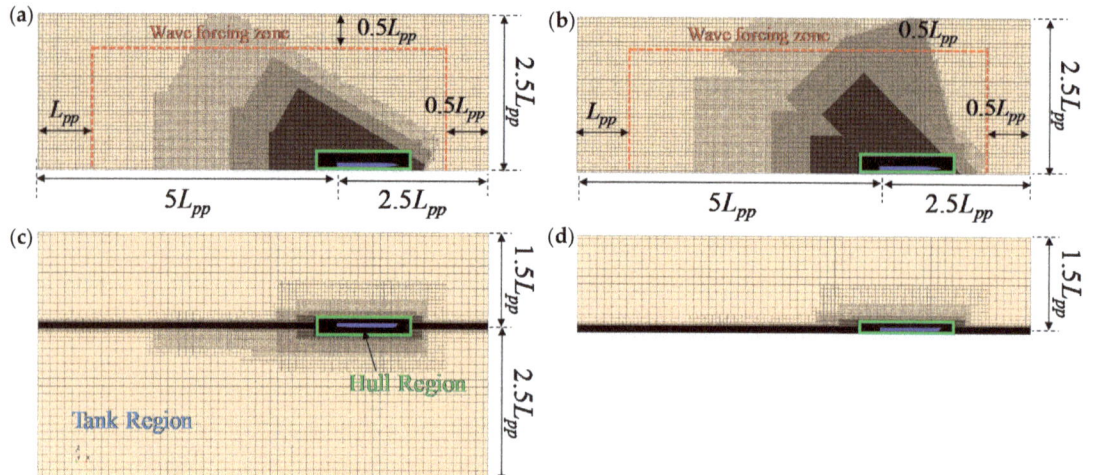

Figure 2. Computational meshes and domain dimensions used for deep water (**a**,**c**) and shallow water cases (**b**,**d**).

3.1.3. Dynamic Trim and Sinkage

As the catamaran is advancing in the water, the surface of the hull will interact with the surrounding water, leading to a fluid-body interaction problem. In the present study, only the heave and pitch motions were allowed, while the rest degrees of freedom were fixed. The Dynamic Fluid-Body Interaction (DFBI) method provided in Star CCM+ 14.06 package was employed to calculate the sinkage and trim of the catamaran according to the fluid forces and moments acting on the hull surface. As the overset grid strategy was used in the present study, the DFBI method was only applied to the demi-hull and its associated region (see Section 3.2).

3.1.4. Coordinate System

In the present simulation, two different coordinate systems are used: An earth-fixed (global) system and a ship-fixed (local) system. The flow simulation was carried out within the earth-fixed coordinate system, and the computed forces and moments were then transformed to the ship-fixed coordinate system whose origin was located at the centre of mass of the catamaran. The equations of the motion were solved based on the latest forces and moments using the DFBI method. The new position and velocity of the hull were then converted back to the earth-fixed system as the boundary condition for the flow simulation. After updating the position of the hull, the connectivity between the two sub-domains in the overset grid method was re-calculated accordingly.

3.2. Computational Domain and Boundary Conditions

As the catamaran is geometrically symmetrical about its mid-plane, only one demi-hull was used for CFD simulation to reduce computational cost. Besides, the overset grid method was employed in the present study, i.e., the entire computational domain was decomposed into two regions: An inner region around the demi-hull (Hull Region) and an outer region forming the virtual tank (Tank Region). Figure 3 shows the two regions and corresponding boundary conditions. The flow variables between the two flow regions were exchanged at the overlapping boundary via linear interpolation. For deep water scenarios, as demonstrated in Figure 2a,b, the Tank Region was extended 1.5 L_{pp} in front of the hull and 5 L_{pp} behind it. The lower and upper boundaries were 2.5 L_{pp} and 1.5 L_{pp} away from the undisturbed water level, respectively. The side boundary of the Tank Region was 2.5 L_{pp} away from the symmetry plane of the catamaran. The velocity inlet condition was applied at the inlet, top, bottom and side boundaries. The pressure outlet condition was used for the outlet boundary. The demi-hull surface was considered as the no-slip wall. To avoid wave refection, a wave forcing method was applied to the regions near the inlet, outlet and side boundaries, as shown in Figure 2a,b. For shallow water scenarios, the size of the Tank Region remained the same as the one used for deep water cases except that the bottom surface was 2.15 m below the water level, where the slip wall boundary condition was applied. The size of the Hull Region is determined by guaranteeing there are sufficient cells (at least five cell layers) in the overlapping area between the two regions. Besides, the cell size in the overlapping area should be comparable. In the present work, the Hull Region was 0.1 L_{pp} in front of the forward perpendicular and 0.15 L_{pp} behind the aft perpendicular. The lower and upper boundaries were 0.05 L_{pp} away from the waterline, and the side boundaries were 0.05 L_{pp} away from the mid-plane of the demihull.

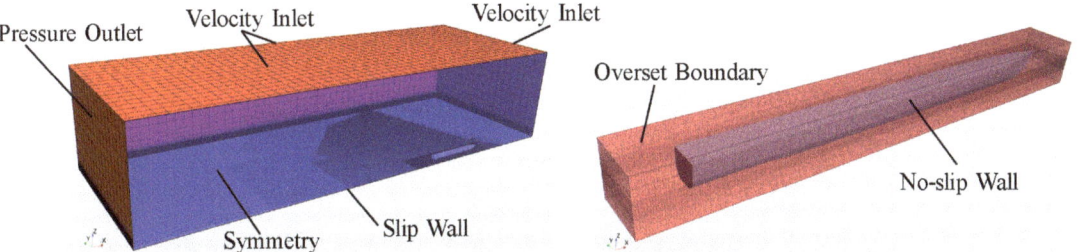

Figure 3. Computational domains (**left**: Tank Region; **right**: Hull Region) and boundary conditions.

3.3. Mesh Generation

The CFD mesh used in the present study was generated using the automated meshing functionality in Star CCM+ 14.06, which was comprised of prism cells around the hull and the hexahedral cells in the rest region. The meshes for the Tank Region and Hull Region were generated separately, and an overset grid interface was created between the two regions. Anisotropic mesh refinements were performed in various areas to appropriately capture the flow features. Specifically, three volumetric mesh controls in three different levels were created around the hull. Similarly, such volumetric mesh controls were also generated to capture the Kelvin waves, the flow wakes behind the hull and the free surface, as demonstrated in Figure 2. All mesh refinements were done by setting a target mesh size relative to a base size specified by the user in the automated meshing tool of Star CCM+ 14.06. The mesh density can also be controlled by varying the value of this base size. Additionally, special attention was given to the overlapping area of the Tank Region and Hull Region when generating the volume mesh. First, the mesh cells within the overlapping region were of similar size. Besides, the overlapping zone was comprised of at least five cell layers in both regions to ensure an accurate and conservative interpolation.

To properly resolve the flow boundary layer, prism mesh layers should be used in the vicinity of the hull. Here, the thickness of the turbulent boundary layer was estimated by

$$\delta = 0.37 L_{pp} / Re^{0.2} \qquad (9)$$

where Re is the Reynolds number based on L_{pp}. Ten layers of prism cells were placed within the boundary layer. As the wall function was used in the turbulent model, the distance of the first prism layer to the hull surface was targeted at $y+ = 100$. Figure 4 demonstrates the computed $y+$ distribution on the hull surface, and it can be observed that for both Froude numbers, the $y+$ values are within the range ($30 < y+ < 300$) that the wall function can be appropriately applied.

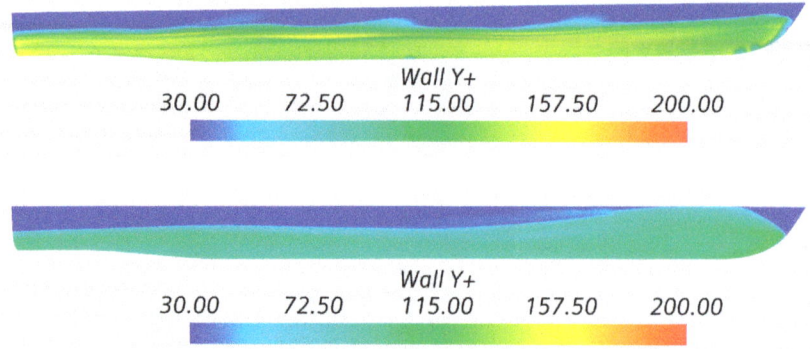

Figure 4. Computed $y+$ distribution on the demihull surface at $Fn = 0.287$ (**upper**) and 0.805 (**below**) in shallow water.

3.4. Numerical Validation and Verification

3.4.1. NPL 4a02 Catamaran

The first case used to validate the computational methods used in the present study was the NPL 4a02 catamaran from a series of model tests carried out by Molland et al. [9]. Table 2 gives the main particulars of this catamaran. The same computational methods and mesh generation strategies presented in Section 3 were also applied here. The total number of mesh cells used for this validation case was around 4.7 million. Figure 5 demonstrates the total resistance coefficients, sinkage-to-draught ratios and trim angles as functions of the Froude number, from which it is observed that the computed results are in good agreement with the experimental data.

Table 2. Main dimensions of the NPL 4a02 catamaran.

Dimension	Symbol	Value
Demihull breadth	b/L_{pp}	0.096
Separation	s/L_{pp}	0.200
Draught	T/L_{pp}	0.064
Vertical centre of gravity	VCG/L_{pp}	0.020
Longitudinal centre of gravity	LCG/L_{pp}	0.436

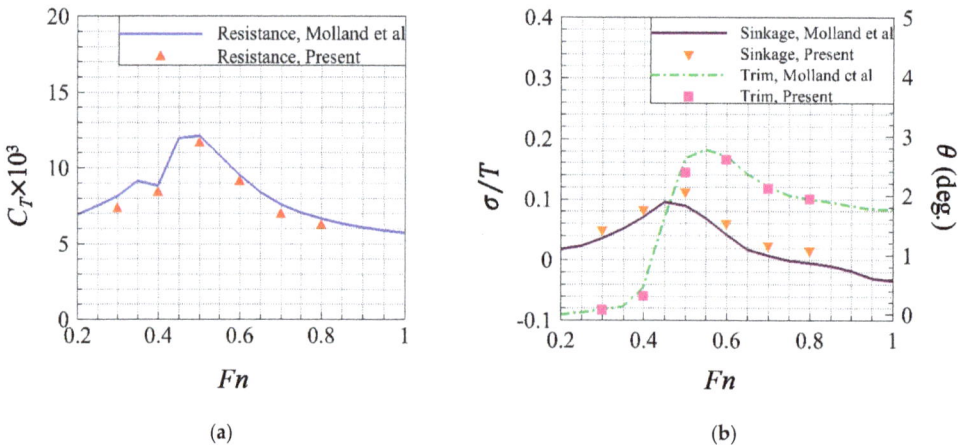

Figure 5. Total resistance coefficients (**a**), sinkage and trim (**b**) of NPL 4a02 model.

3.4.2. Stavanger Demonstrator

The computational methods presented in Section 3 were further validated against the experimental data of the Stavanger demonstrator [2,3] measured in the Hamburg Ship Model Basin (HSVA). The main dimensions are demonstrated in Table 3. The geometry of the Stavanger demonstrator is illustrated in Figure 6a, and the mesh used for simulation is demonstrated in Figure 6b, which was consisted of about 11.4 million cells. The computational domain, boundary conditions and mesh system were generated in similar manners to those presented in Section 3. Table 4 compares the total resistance coefficients of the Stavanger demonstrator obtained from CFD simulations with that from physical model tests. It is seen that for the four speeds considered here, the difference between the present numerical result and the experimental data is within 1.5%.

Table 3. Main dimensions of the Stavanger Demonstrator catamaran.

Dimension	Symbol	Value
Demihull breadth	b/L_{pp}	0.074
Separation	s/L_{pp}	0.227
Draught	T/L_{pp}	0.045
Vertical centre of gravity	VCG/L_{pp}	0.016
Longitudinal centre of gravity	LCG/L_{pp}	0.450

(a)   (b)

Figure 6. The geometry of the Stavanger demonstrator (**a**) and CFD mesh used for simulation (**b**).

Table 4. Total resistance coefficient of Stavanger demonstrator obtained from model tests and CFD simulation.

Fn	$C_{T,CFD} \times 10^3$	$C_{T,Exp} \times 10^3$	Error
0.57	5.476	5.520	−0.79%
0.63	4.844	4.899	−1.11%
0.69	4.404	4.437	−0.74%
0.75	4.098	4.157	−1.42%

3.4.3. Mesh Convergence Study for the London Demonstrator

To justify the mesh size used in the simulation of the London Demonstrator and quantify the uncertainties, due to spatial discretisation, a mesh convergence study was carried out. In the present paper, the numerical convergence and uncertainty, due to grid density are evaluated using the grid convergence index (GCI) method described in Stern et al. [50]. The convergence ratio ($\overline{R}$) is used to assess the convergence condition, which is calculated as

$$\overline{R} = \frac{\varphi_2 - \varphi_1}{\varphi_3 - \varphi_2} \qquad (10)$$

where φ_1, φ_2 and φ_3 correspond to the solutions (total resistance coefficient) with the fine, medium and coarse grids. Based on the value of $\overline{R}$, the four resulting convergence conditions are: (1) Monotonic convergence ($0 < \overline{R} < 1$); (2) oscillatory convergence ($\overline{R} < 0$, $|\overline{R}| < 1$); (3) monotonic divergence ($\overline{R} > 1$); and (4) oscillatory divergence ($\overline{R} < 0$, $|\overline{R}| > 1$).

For convergence conditions, the numerical errors can be predicted as follows. For a constant refinement ratio ($\bar{r} = \sqrt[3]{N_1/N_2} = \sqrt[3]{N_2/N_3}$), where N_1, N_2 and N_3 are the number of cells in millions for fine, medium and coarse grids, respectively, the order of accuracy (q) can be calculated as

$$q = \frac{\ln[(\varphi_3 - \varphi_2)/(\varphi_2 - \varphi_1)]}{\ln(\bar{r})} \tag{11}$$

The extrapolated values can be obtained by

$$\varphi_{ext}^{21} = \frac{\bar{r}^q \varphi_1 - \varphi_2}{\bar{r}^q - 1} \tag{12}$$

The approximate relative error and the extrapolated relative error can be computed using the following formulas

$$E_{a,21} = \left| \frac{\varphi_1 - \varphi_2}{\varphi_1} \right| \tag{13}$$

$$E_{ext,21} = \left| \frac{\varphi_{ext}^{21} - \varphi_1}{\varphi_{ext}^{21}} \right| \tag{14}$$

Finally, the fine-grid convergence index can be predicted as

$$GCI_{fine,21} = \frac{1.25 E_{a,21}}{\bar{r}^q - 1} \tag{15}$$

Table 5 summarises the results for grid convergence study at the design speed ($Fn = 0.805$) in shallow water, from which we can observe that the present simulation achieved a monotonic convergence and the uncertainty, due to the spatial discretisation for the fine grid is around 1.5%. As the medium grid is used in the present work, the uncertainty, including the difference between the fine and medium grids, is approximately 2.3%.

Table 5. Results of the mesh convergence study for the London Demonstrator.

$\bar{r}$	N_1	N_2	N_3	φ_1	φ_2	φ_3	$\bar{R}$	$E_{a,21}$	$E_{ext,21}$	$GCI_{fine,21}$
1.2	10.35	5.99	3.47	2.537	2.557	2.589	0.607	0.78%	1.22%	1.50%

4. Results and Discussion

4.1. Resistance, Sinkage and Trim

As the present paper aims to validate the numerical methods adopted for simulation, a blind validation study was carried out by MSRC and HSVA, where the commercial solver Star CCM+ 14.06 was used by MSRC, whereas an in-house code FreSCo+ [51] was employed by HSVA. The number of mesh cells for both simulations was approximately 6 million. Figure 7 shows the resistances and motions of the full-scale London Demonstrator in deep water, from which we can observe that very good agreement is accomplished between the present results (Strath) and those from HSVA. It is seen from Figure 7a that the total resistance (R_T) rises monotonously as the speed of the catamaran increases. The relation between R_T and Fn is almost linear when $Fn < 0.4$. A continuous change in the slope of the total resistance curve can be observed when $0.4 < Fn < 0.6$, which is also reported by Zaghi et al. [30] in the same Froude number range and indicates the experience of unfavourable interferences. The frictional component (R_F) also rises monotonously with increased speed, while the pressure component (R_P) experiences a peak of $Fn = 0.575$. Besides, R_P is the larger component at lower speeds whilst it becomes smaller when Fn is greater than 0.65.

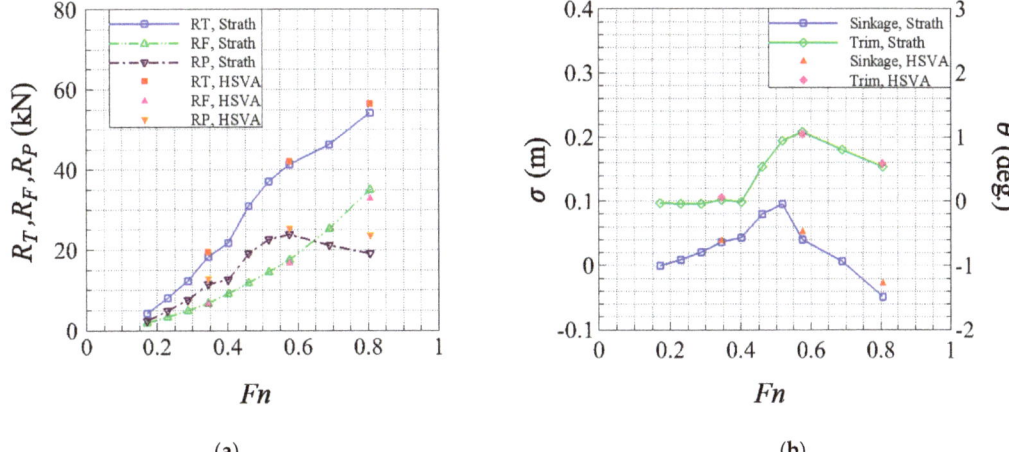

Figure 7. Resistances (**a**) and motions (**b**) of the London Demonstrator in deep water.

The sinkage and trim are demonstrated in Figure 7b, and it is observed that the trim angle of the catamaran is always positive, i.e., the stern goes down for all speeds considered here. At lower speeds ($Fn < 0.4$), the trim angle of the London Demonstrator remains almost zero. When Fn becomes higher than 0.4, it rises significantly and reaches its peak of $Fn = 0.575$ where R_P also achieves its maximum value. In terms of the sinkage of the catamaran, it keeps positive (the hull moves downwards) until the Fn is higher than 0.7. The largest sinkage is experienced at $Fn = 0.517$, which is slightly smaller than the Froude number where the trim maximum is accomplished. It should also be noted that the significant changes in trim and sinkage occur when $0.4 < Fn < 0.6$, corresponding to the range where the total resistance curve varies. It will be shown in the following sections that these behaviours of resistance and motion are closely associated with the position and strength of the crests and troughs at the central plane of the catamaran.

Figure 8 compares the resistances and motions of the London Demonstrator in shallow water obtained from the present calculation with those computed by HSVA using FreSCo+. The results from both solvers also agree very well with each other for shallow water scenarios. It is interesting to observe from Figure 8a that R_T experiences a hump at $Fn = 0.287$, corresponding to a depth Froude number ($Fn_H = 1.12$) around the critical value. It has been widely acknowledged that fast catamarans will experience a dramatic surge in total resistance coefficient near the critical speed in shallow water [45,48]. However, the existence of such a hump in total resistance rather than the coefficient near the critical depth Froude number is rarely reported in previous studies. R_T rises monotonously after the hump (when $Fn > 0.35$) as the continuous increase of the frictional resistance. An inspection of R_P and R_F curves reveal that the hump comes from the pressure component of the resistance, indicating it is the consequence of wave interference between the demihulls. Unlike the total resistance, R_P declines after the hump and the frictional resistance exceeds R_P and becomes the larger part of the total resistance when $Fn > 0.55$. The existence of such a hump in the R_T curve should be carefully considered in the design of the catamaran to guarantee that the installed power is sufficient to overcome the hump resistance in the process of accelerating the vessel to the designed speed. It is observed from Figure 8b that the sinkage and motion of the catamaran change significantly near the critical speed, which agrees with previous studies on high speed catamarans [45,48].

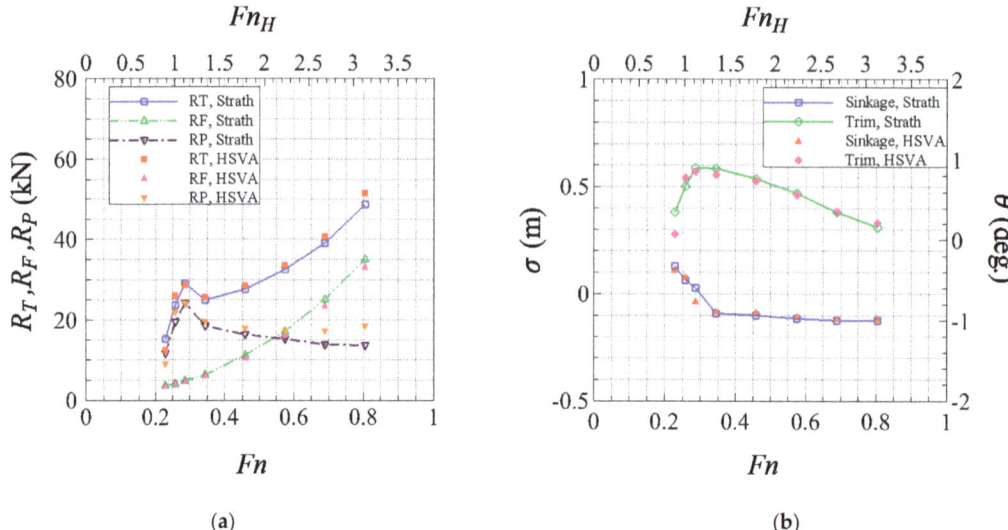

Figure 8. Resistances (**a**) and motions (**b**) of the London Demonstrator in shallow water (H = 2.15 m).

Figure 9 compares the resistances and motions of the London Demonstrator in deep and shallow water. Hereafter, only the results computed using Star CCM+ 14.06 are used for further analysis. The total resistance in shallow water is higher than that in deep water at smaller Froude numbers ($Fn < 0.45$) because of the hump near the critical speed. When Fn further increases, R_T in shallow water becomes lower, due to the reduction of pressure resistance R_P. The frictional resistances in deep and shallow water are almost the same, i.e., the difference between the total resistance in deep and shallow water results from significantly different wave patterns and interferences, which will be demonstrated in the following sections. By comparing the motions of deep and shallow water cases, it is found that the maximum trim angles accomplished in deep and shallow water are close to each other ($\approx$1.0 degree). However, the maximum of trim in shallow water is reached near the critical speed ($Fn = 0.287$), whereas the peak in deep water is achieved at $Fn = 0.575$. Similar to deep water cases, the maximum value of shallow water trim is also achieved at the Froude number where the pressure resistance peaks ($Fn = 0.287$). The sinkage of the catamaran in shallow water is larger in sub- and trans-critical ranges ($Fn < 0.3$), whilst in the supercritical region, the sinkage in shallow water becomes smaller than that in deep water, which leads to a considerable reduction in pressure drag, as observed from Figure 9a. Furthermore, the sinkage of the catamaran in shallow water is positive at subcritical speeds. With the further increase of Froude number, the catamaran's centre of mass starts to move upward and when $Fn > 0.35$, the change rate of sinkage becomes less significant.

The resistance coefficients of the London Demonstrator in deep and shallow water are illustrated in Figures 10 and 11, respectively. The total resistance coefficients (C_T) are normalised using both static and dynamic areas and the differences are small for both deep and shallow water cases. Generally, the coefficients calculated based on the dynamic wetted area are slightly smaller, and the difference only becomes noticeable for the highest speed ($Fn \approx 0.8$). The frictional resistance coefficients (C_F) of the catamaran in both deep and shallow water agree well with those predicted using the ITTC 1957 correlation line formula, indicating the frictional resistance is not significantly affected by shallow water. Moreover, for deep water cases, shown in Figure 10, C_T and C_P experience multiple peaks as the increase of Froude number. The peaks at lower Froude numbers ($Fn < 0.4$) are higher than that at $Fn = 0.46$. The total resistance coefficient drops significantly with the further increase of the advance speed. The present C_T curve differs from those observed in some previous studies, where the humps at smaller Froude numbers were usually lower [9,13,30].

This may be associated with the exact hull form and configuration of the catamaran, which leads to a different wave interference between the demihulls.

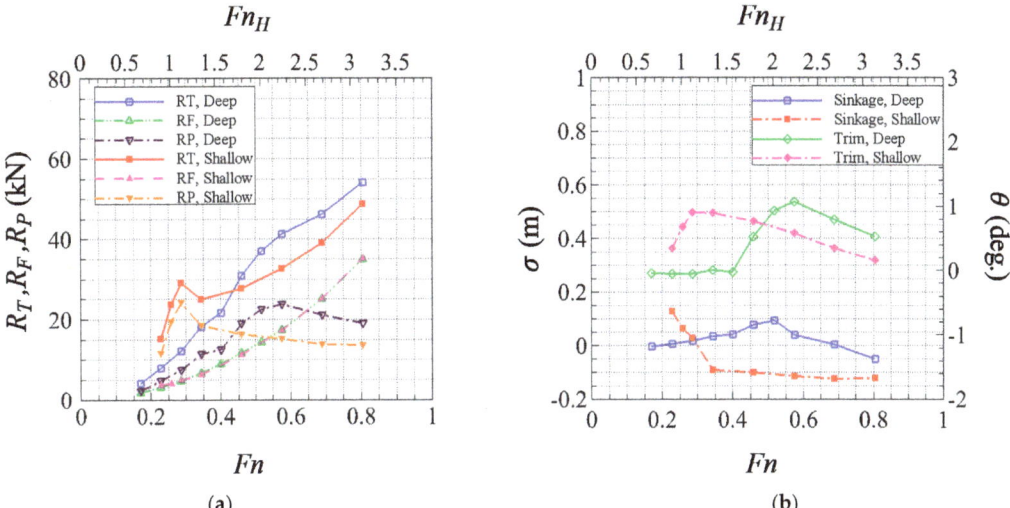

Figure 9. Comparison of resistances (**a**) and motions (**b**) of London Demonstrator in deep and shallow water ($H = 2.15$ m).

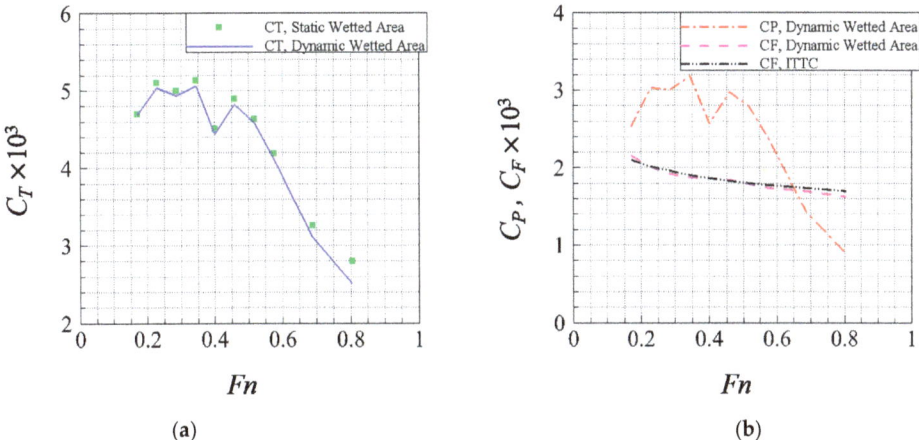

Figure 10. Total resistance coefficient (**a**), and pressure and frictional resistance coefficients (**b**) of the London Demonstrator in deep water.

For the shallow water scenario (Figure 11), the resistance coefficient of the catamaran reaches its peak value around the critical depth Froude number and then declines dramatically as the moving speed increases. The maximum C_T value in shallow water is approximately 2.4 times higher than that created in deep water. This ratio is smaller than the value obtained by Castiglione et al. [48] for a similar catamaran configuration, where the C_T peak in shallow water is about 4.2 times larger than that in deep water. Unlike the hump of the R_T curve in shallow water, as shown in Figure 9a, which is not commonly seen in previous papers, the dramatic increase of C_T near the critical speed has been widely observed in both model tests and numerical simulations [45,48]. It is worth noting that the maximum total resistance coefficient does not correspond to the maxima of the total resistance, according to which the propulsion power should be installed. For the London

Demonstrator examined here, the maximum total resistance is accomplished at the highest speed considered here (see Figure 9a), where C_T reaches its minimum value.

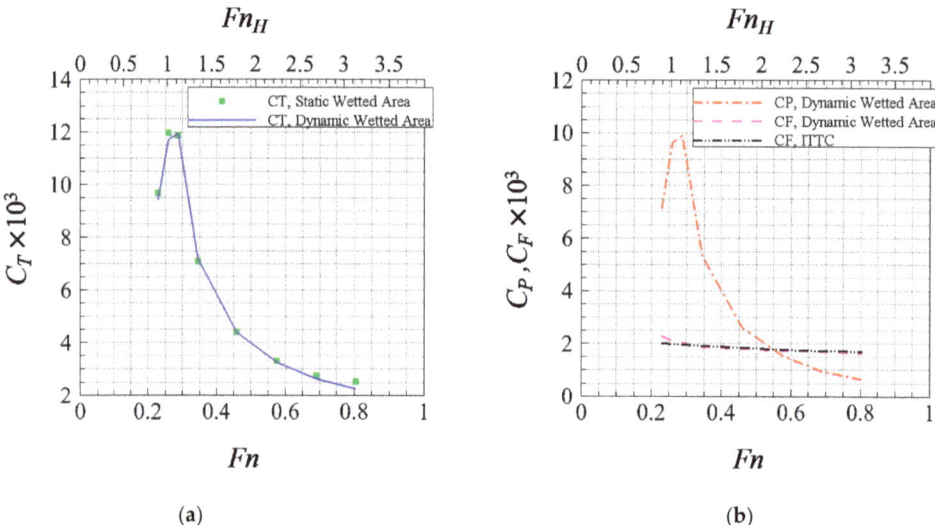

(a)   (b)

Figure 11. Total resistance coefficient (**a**), and pressure and frictional resistance coefficients (**b**) of the London Demonstrator in shallow water.

4.2. Wave Patterns

The wave patterns created by the London Demonstrator at various speeds in deep water are demonstrated in Figure 12. The catamaran generates typical Kelvin wave patterns at lower speeds, which comprise both transverse and divergent waves. As the increase of the Froude number, the amplitude and length of the induced wave also increase, while the Kelvin wave angle becomes smaller. Besides, the divergent waves become dominant in the wave pattern at $Fn = 0.805$. Figure 13 demonstrates the wave elevations of the catamaran in shallow water, which are profoundly different from those shown in Figure 12. As expected, when the depth Froude number is near its critical value ($Fn_H = 1.0$), the Kelvin wave angle is close to 90 degrees, and the critical wave is created at $Fn_H = 1.12$, which is located right in front of the catamaran. The critical wave is normal to the advance direction of the vessel, and its attitude is significantly elevated, which leads to the hump observed in the R_T curve in Figure 8a and the remarkable C_T peak, shown in Figure 11b. Besides, the critical wave significantly elevates the bow, creating the trim maxima observed from Figure 8b. Behind the stern of the vessel, divergent waves are generated. As the moving speed increases to the supercritical range, the critical wave disappears, and divergent waves are created near both the bow and stern of the hull. The further increase of the Froude number reduces the angles of the divergent waves. However, the overall wave patterns are not significantly changed. In both deep and shallow water, the decrease of the Kelvin wave angle leads the intersection point of the bow waves created by the two demihulls to move astern, which will be more clearly observed from Figures 14 and 15, as well as the wave cuts demonstrated in the next section.

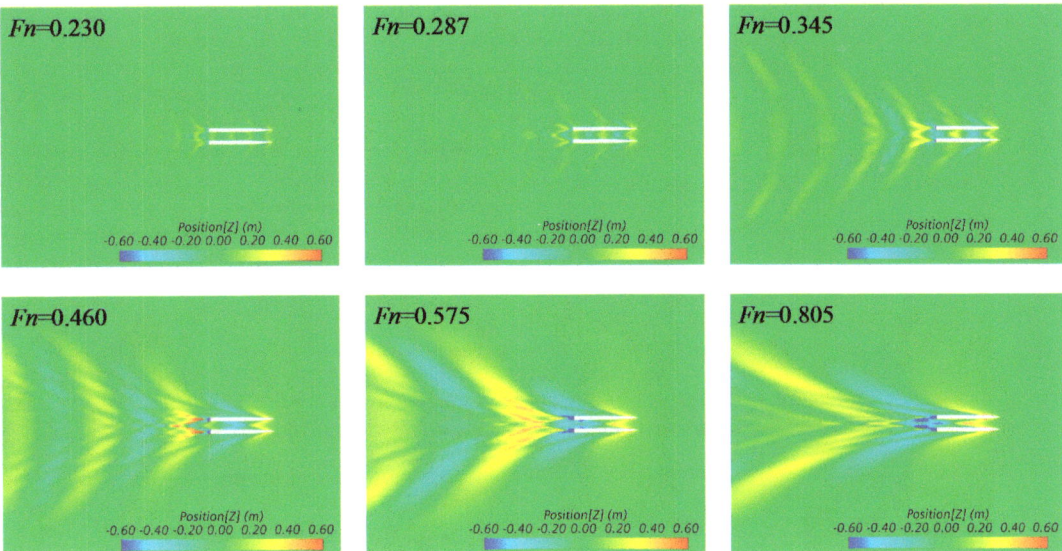

Figure 12. Wave patterns created by the London Demonstrator in deep water.

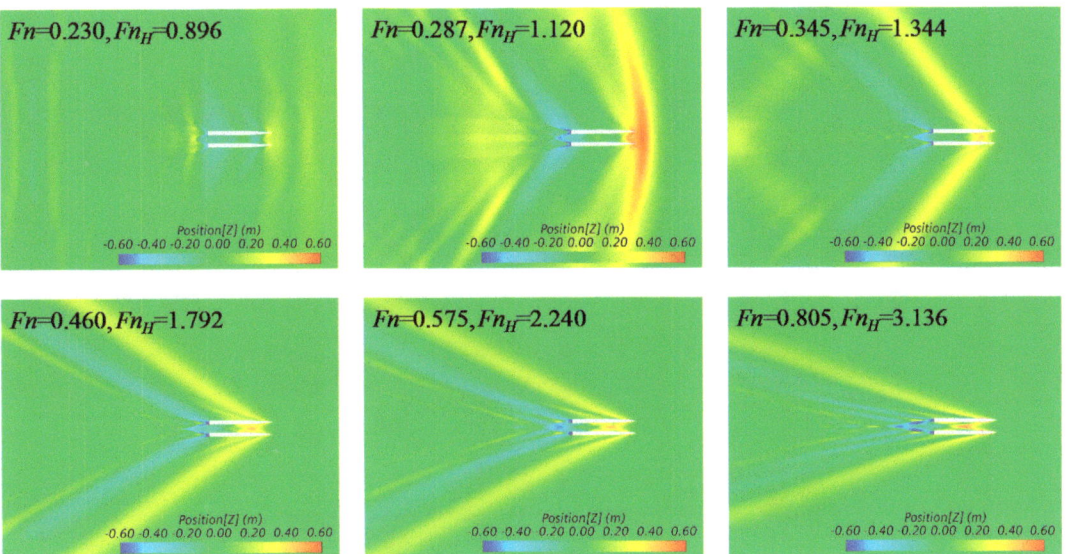

Figure 13. Wave patterns created by the London Demonstrator in shallow water.

The behaviours of the resistance, trim and sinkage discussed in the previous section can be better understood by analysing the interaction between the wave systems generated by the demihulls. Figure 14 shows a closer inspection of the wave interference between the demihulls in deep water. We can observe that at smaller Froude numbers (e.g., when $Fn < 0.3$), multiple crests and troughs exist within the inner region between the two hulls. Enhanced crests and troughs become pronounced when $Fn = 0.345$ at the symmetry plane of the catamaran, where the waves meet and strengthen each other. At $Fn = 0.46$, another two troughs are generated on each side of the symmetry plane apart from the one created at the central plane, indicating a significant secondary wave interference. At this

Froude number, the secondary troughs are located slightly behind midship. As the Froude number increases to 0.575, the crest and troughs between the demihulls are moved further downstream, which has also been reported in previous studies [13,30]. In particular, the secondary wave troughs are generated near the stern with higher amplitudes, which leads to a larger sinkage at the stern, thereby creating the peak of trim, as shown in Figure 7b. Moreover, as discussed in Figure 9a, the pressure resistance R_P reaches its maximum value at $Fn = 0.575$, implying the wave interference is the strongest at this Froude number. When $Fn = 0.805$, the wave troughs created, due to the secondary wave interaction are moved behind the aft of the catamaran (see Figure 12), which leads to a decrease in the trim as the secondary troughs are closer to the hull surface, thereby having a more direct impact on the motion of the demihull. Another observation from the wave pattern at $Fn = 0.805$ is that the first crest in the inner region is produced near midship, which results in the reduction of the moment causing the pitch motion, leading to the decrease in trim angle. On the other hand, with the first crest further strengthened and moved near the catamaran's centre of mass, this crest will lift the entire catamaran instead of the bow. Therefore, the sinkage becomes negative (the hull moves upward) at higher Froude numbers.

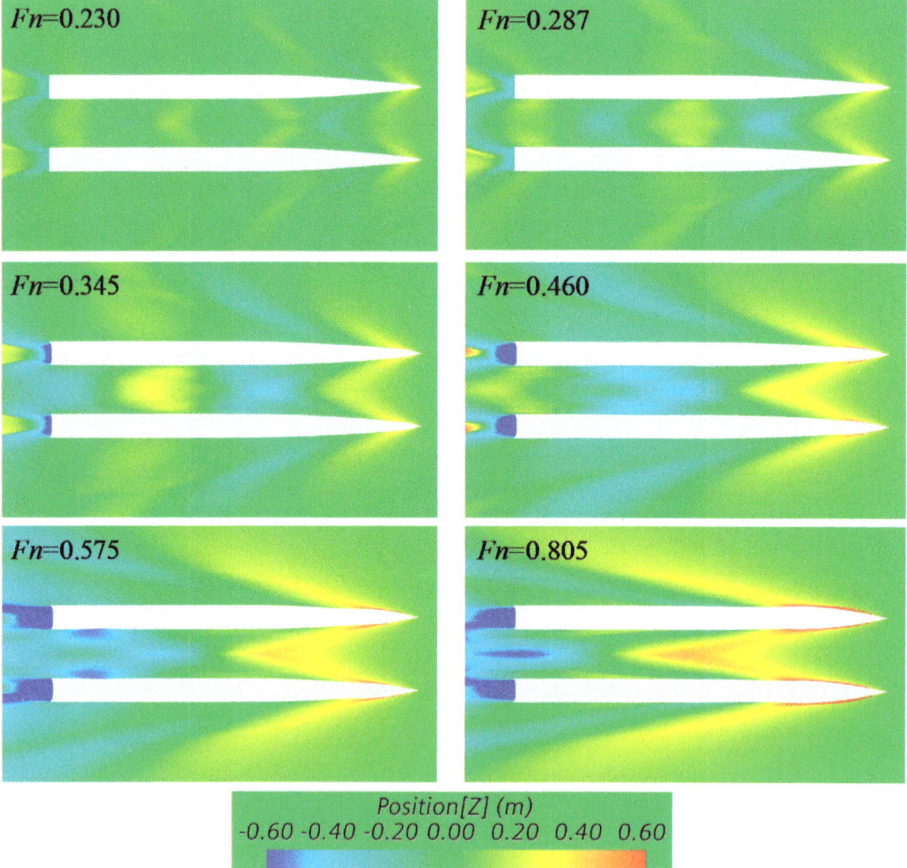

Figure 14. The wave interaction between demihulls in deep water.

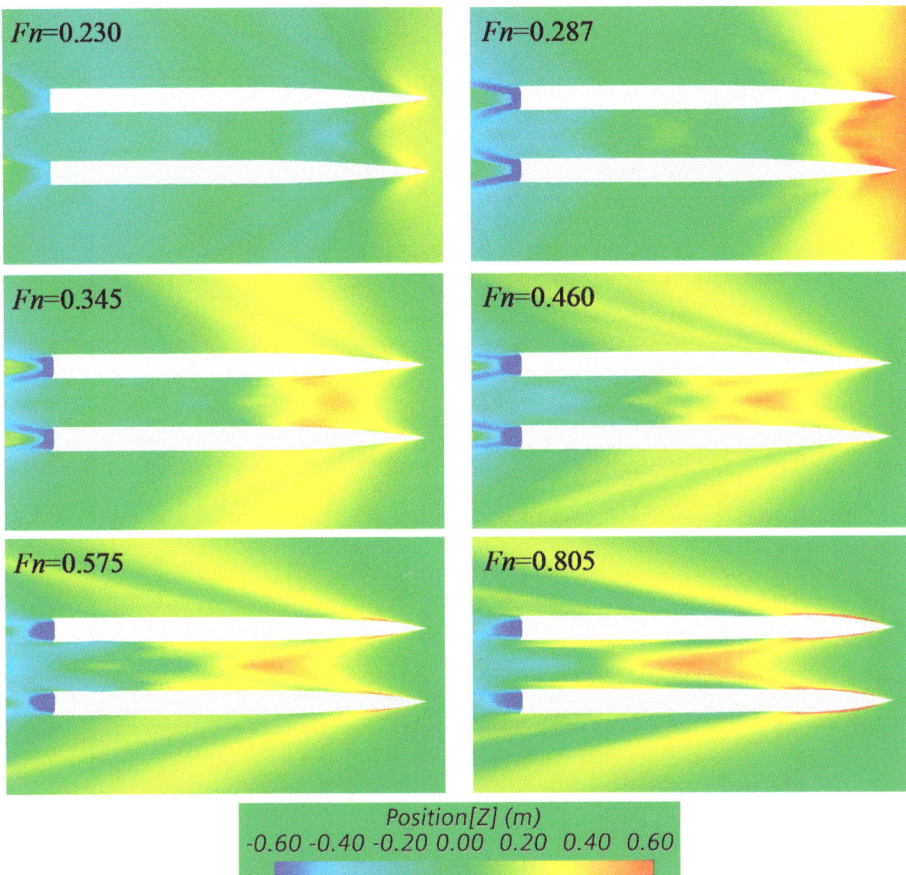

Figure 15. The wave interaction between demihulls in shallow water.

The wave interferences between demihulls in shallow water are demonstrated in Figure 15. Several significant differences from those in deep water can be observed. First, at trans-critical speeds (Fn = 0.23 and 0.287), wave interactions between the demihulls seem to be suppressed, due to creating the critical wave in front of the catamaran (see Figure 13), i.e., the phenomenon of existing multiple crests and troughs within the inner region disappears. At supercritical speeds ($Fn > 0.345$), the three troughs observed in deep water (e.g., in Figure 14 when Fn = 0.46) are not seen in shallow water cases. Instead, another two secondary crests are generated apart from the primary one at the catamaran's central plane. As the Froude number increases, the wave crests are stretched and moved towards the stern. As previously discussed, both trim and sinkage will be decreased with the first crest moving midship. This trend will be further enhanced, due to creating the secondary crests, i.e., at higher speeds, both the trim and sinkage in shallow water are smaller, as seen from Figure 9b.

4.3. Longitudinal Wave Cuts

The wave propagation within the inner region can be better understood by analysing the longitudinal wave cuts at the central plane of the catamaran, as demonstrated in Figure 16. It is seen that the wave starts to come into being at the forward perpendicular (FP) for all cases except those at trans-critical speeds (Fn_H = 0.896 and 1.12) in shallow water, where the water is elevated at least 0.5 L_{pp} ahead of the catamaran and reaches the maximum height near the FP. In deep water, both the wave height and wave length increase as the Froude number rises, confirming the observations from Figure 14. The increase of the wave length leads to a reduction in the number of waves between FP and aft perpendicular (AP). For example, there are approximately three waves between FP and AP when $Fn = 0.23$, while the number becomes less than one when Fn increases to 0.805. It is interesting to observe that at $Fn = 0.575$, the wave number between FP and AP is approximately unity and this Froude number corresponds to the maximum value of the pressure component of total resistance (see Figure 9a). In shallow water, the first wave crest behind the bow is always higher than that created in deep water, especially near the critical speed. The difference is considered small only when the Froude number is greater than 0.575. Moreover, no noteworthy wave troughs are generated between FP and AP in shallow water, which significantly differs from those in deep water. Furthermore, the catamaran generates higher wave crests behind the stern in deep water, while creating deeper wave troughs in shallow water.

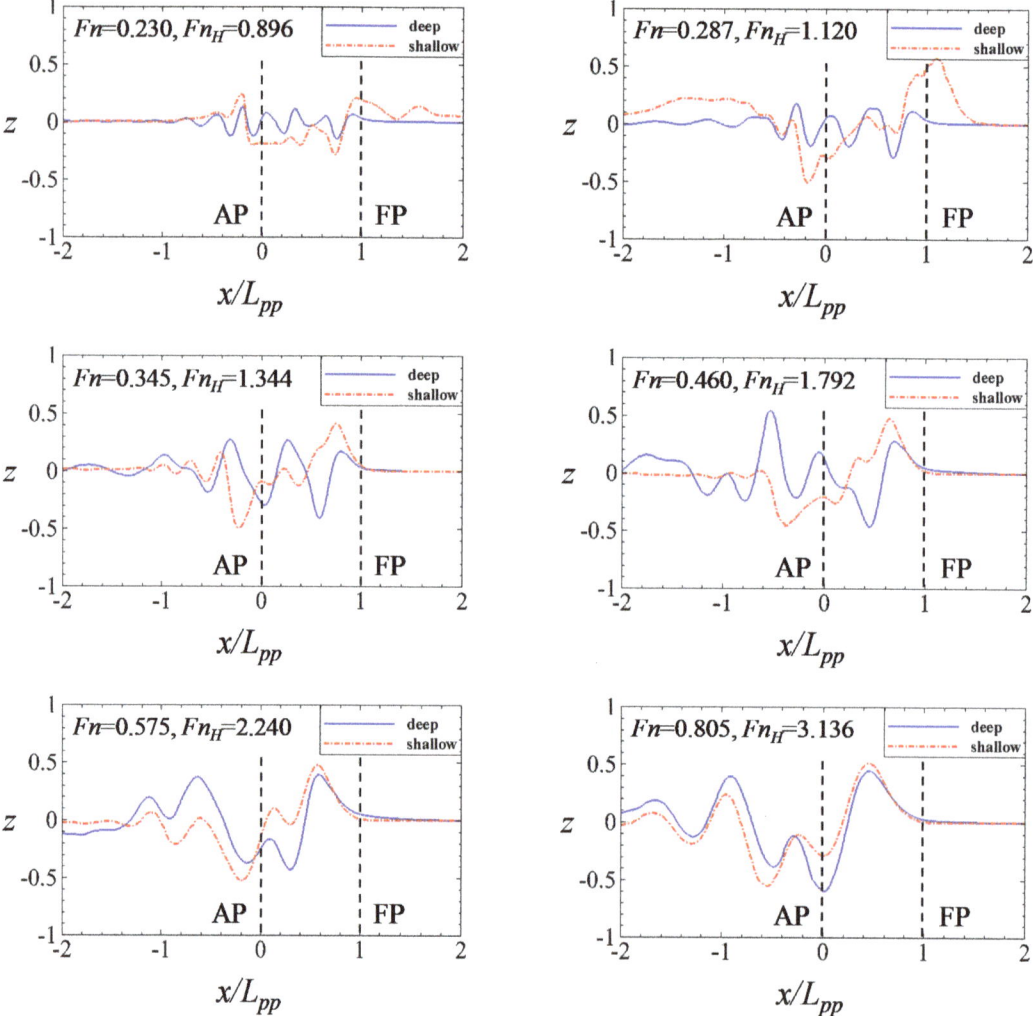

Figure 16. Comparison of the longitudinal wave cuts at the catamaran symmetry plane in deep and shallow water.

As observed from previous wave patterns in Figures 14 and 15, the catamaran generates a remarkable trough right behind the stern of the demihull. The magnitude of this trough can be more clearly demonstrated by the longitudinal wave cuts at the mid-plane of the demihull, as shown in Figure 17. In deep water, the magnitude of the trough reaches its maximum value at $Fn = 0.575$, where the water level difference between FP and AP is also maximised. In shallow water, the trough's magnitudes at trans-critical speeds are significantly larger than those in deep water. The maximum amplitude is achieved at $Fn = 0.287$, where the critical wave is also created in front of the bow, resulting in a remarkably large difference between the water levels at the FP and AF of the catamaran. It is worth emphasising that $Fn = 0.287$ and 0.575 correspond to the speeds where the maximum pressure resistance is produced in shallow and deep water, respectively, as seen from Figure 9a. At supercritical speeds, the trough's amplitude in shallow water becomes smaller than that in deep water, which can be attributed to smaller sinkage and trim created in shallow water.

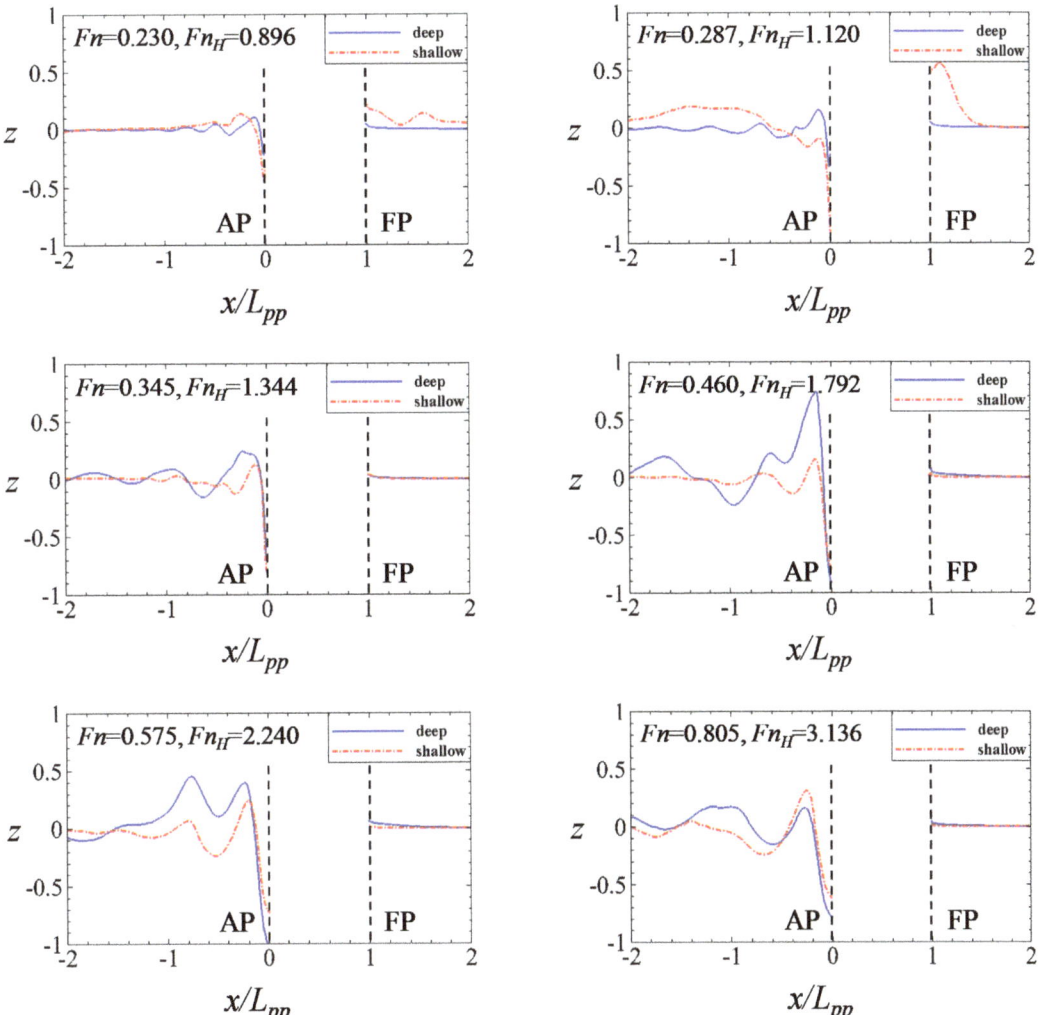

Figure 17. Comparison of the longitudinal wave cuts at the mid-plane of the demihull in deep and shallow water.

4.4. Crossflow Fields

With the wave interference between the demihulls, the flow field created by the demihull becomes non-symmetrical against its mid-plane, which will cause a transverse pressure gradient. This can further lead to a crossflow under the keel of the demihull, which is believed to be one of the main causes of the increase in total resistance [31]. The crossflow fields of the London Demonstrator are plotted in Figure 18, where the positive and negative velocities indicate that the flow moves to the outer and inner regions, respectively. In deep water, the location and strength of the crossflow are closely associated with the wave interaction between the demihulls. At lower Froude numbers, multiple changes of the crossflow direction under the keel can be observed, which corresponds to the existence of multiple waves between the demihulls (see Figures 14 and 16). With the increase of the Froude number, the strength and extension of the crossflow are significantly enhanced, and the locations where the crossflow occurs is also moved towards the stern. This phenomenon was also observed by Zaghi et al. [30] and Farkas et al. [34]. Besides, the number of changes in the crossflow direction is also reduced with increased speed. At higher Froude numbers, significant crossflows are also generated behind the stern. For shallow water scenarios, similar to the deep water cases, the strength of the crossflow is considerably enhanced and the location where the maximum crossflow occurs is also moved towards the stern with increased speed. However, the crossflows created in shallow water are remarkably stronger than the corresponding cases in deep water. Moreover, the phenomenon of multiple changes in crossflow direction observed at lower Froude numbers no longer exists, and for all speeds in shallow water, the crossflow moves from the inner side of the demihull to the outer region.

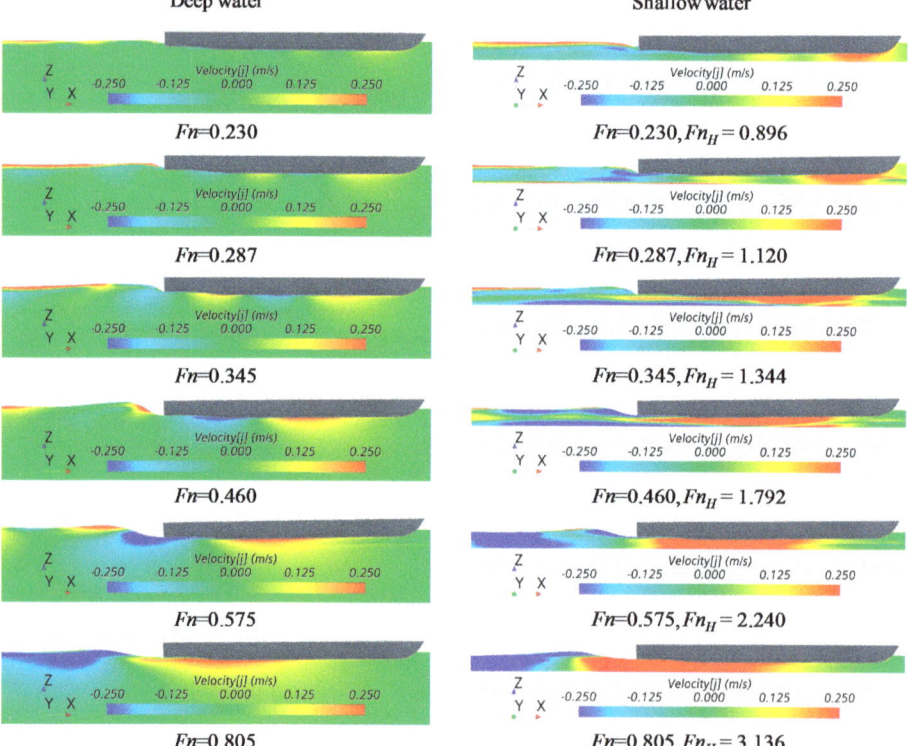

Figure 18. Crossflow fields at the mid-plane of the demihull for deep (**left**) and shallow (**right**) water. Positive and negative velocity values mean the flow moves towards the outer and inner sides of the demihull, respectively.

5. Conclusions

In the present work, the hydrodynamics of a full scale, zero-emission, high-speed catamaran (London demonstrator) in both deep and shallow water was numerically investigated. The numerical methods used in the current study were validated against experimental data of the NPL 4a02 model [9] and the Stavanger demonstrator [2]. For numerical simulations on the London Demonstrator, a blind validation was also carried out in collaboration with HSVA and good agreement was accomplished.

The resistance, sinkage and trim of the London Demonstrator as functions of Froude number (ranged from 0.2 to 0.8) in deep and shallow water were firstly analysed. The total resistance in deep water increased continuously, while in shallow water, a hump was experienced at $Fn = 0.287$ ($Fn_H = 1.12$). Besides, the total resistance in shallow water was higher when $Fn < 0.45$ and became smaller at larger speeds. As the frictional resistance was almost the same in deep and shallow water, i.e., the difference in total resistance was mainly caused by the pressure component. The variations of the pressure resistance were closely related to the behaviours of trim and sinkage. In particular, the maximum trim was accomplished at the Froude number where the pressure resistance was maximised ($Fn = 0.287$ and 0.575 for shallow and deep water, respectively). The largest sinkage in shallow water occurred at the lowest speed, whereas in deep water the sinkage reaches its maxima at a Froude number ($Fn = 0.517$) slightly lower than the one where the maximum trim occurred. Furthermore, the total resistance coefficient curve in deep water showed multiple humps, while only one significant peak near the critical speed was produced in shallow water.

The computed wave patterns, longitudinal wave cuts and crossflow fields were also analysed and correlated with the behaviours of the resistance and motion of the catamaran. In general, for both deep and shallow water scenarios, the crests and troughs generated within the inner region were strengthened and moved astern with the increase of Froude number. In deep water, the maximum pressure resistance was related to creating a secondary trough near the stern of the demihull. In contrast, the mechanism involved in shallow water was due to the generation of a critical wave in front of the catamaran and normal to the moving direction. Moreover, the creation of maximum pressure resistance was also correlated with the largest water level difference between the forward and aft perpendiculars. Crossflows occurred in deep and shallow water scenarios, due to the asymmetrical flow fields between the inner and outer regions. Compared with deep water cases, the crossflows created in shallow water were much stronger. Moreover, the crossflow in shallow water moved towards the outer region for all speeds considered here, whereas, in deep water, changes in crossflow directions were observed.

Author Contributions: Conceptualization, E.B.; methodology, G.S. (Guangyu Shi), A.P. and Y.X.-K.; software, G.S. (Guangyu Shi), A.P. and Y.X.-K.; validation, G.S. (Guangyu Shi), Y.X.-K.; formal analysis, G.S. (Guangyu Shi); investigation, G.S. (Guangyu Shi), A.P., E.B., Y.X.-K. and A.D.P.; resources, E.B.; data curation, G.S. (Guangyu Shi); writing-original draft preparation, G.S. (Guangyu Shi); writing-review and editing, E.B., A.P., H.W. and A.D.P.; visualization, G.S. (Guangyu Shi); supervision, E.B.; project administration, E.B., G.S. (Geoff Symonds); funding acquisition, E.B.. All authors have read and agreed to the published version of the manuscript.

Funding: This research was funded by European Commission (EC), grant number 769303.

Data Availability Statement: Not applicable.

Acknowledgments: This work was funded by the H2020 European Union project "TrAM—Transport: Advanced and Modular" (contract 769303). The authors affiliated with the MSRC greatly acknowledge the funding from DNV and Royal Caribbean Group for the MSRC's establishment and operation. The opinions expressed herein are those of the authors and do not reflect the views of DNV and Royal Caribbean Group. Results of CFD simulations run by the University of Strathclyde were obtained using ARCHIE-WeSt High Performance Computer (www.archie-west.ac.uk (accessed on 23 April 2021)).

Conflicts of Interest: The authors declare no conflict of interest.

Nomenclature

b	Breadth of the demihull
B	Breadth of the catamaran
C_T	Total resistance coefficient
C_F	Frictional resistance coefficient
$C_{F,ITTC}$	Frictional resistance coefficient calculated according to ITTC 1957 correlation line formula
C_P	Pressure resistance coefficient
Fn	Froude number
Fn_H	Depth Froude number
g	Gravity acceleration
H	Water depth
L_{pp}	Length between perpendiculars
Re	Reynolds number
R_T	Total resistance
R_F	Frictional resistance
R_P	Pressure resistance
s	Separation distance between the demihulls
S_{sw}	Static wetted surface area
S_{dw}	Dynamic wetted surface area
T	Draught
U	Ship speed relative to the incoming flow
σ	Sinkage
θ	Trim
AP	Aft Perpendicular
CFD	Computational Fluid Dynamics
FP	Forward Perpendicular
HSVA	Hamburg Ship Model Basin
ITTC	International Towing Tank Committee
LCG	Longitudinal centre of gravity
MSRC	Maritime Safety Research Centre
VCG	Vertical centre of gravity

References

1. Papanikolaou, A.D. Review of the Design and Technology Challenges of Zero-Emission, Battery-Driven Fast Marine Vehicles. *J. Mar. Sci. Eng.* **2020**, *8*, 941. [CrossRef]
2. Papanikolaou, A.; Xing-Kaeding, Y.; Strobel, J.; Kanellopoulou, A.; Zaraphonitis, G.; Tolo, E. Numerical and Experimental Optimization Study on a Fast, Zero Emission Catamaran. *J. Mar. Sci. Eng.* **2020**, *8*, 657. [CrossRef]
3. Mittendorf, M.; Papanikolaou, A.D. Hydrodynamic Hull Form Optimization of Fast Catamarans Using Surrogate Models. *Ship Technol. Res.* **2021**, *68*, 14–26. [CrossRef]
4. Wang, H.; Boulougouris, E.; Theotokatos, G.; Priftis, A.; Shi, G.; Dahle, M.; Tolo, E. Risk Assessment of a Battery-Powered High-Speed Ferry Using Formal Safety Assessment. *Safety* **2020**, *6*, 39. [CrossRef]
5. Molland, A.F.; Wellicome, J.F.; Couser, P.R. Theoretical prediction of the wave resistance of slender hull forms in catamaran configurations. In *Ship Science Report 72*; University of Southampton: Southampton, UK, 1994.
6. Shahjada Tarafder, M.; Suzuki, K. Computation of Wave-Making Resistance of a Catamaran in Deep Water Using a Potential-Based Panel Method. *Ocean Eng.* **2007**, *34*, 1892–1900. [CrossRef]
7. Bari, G.S.; Matveev, K.I. Hydrodynamic Modeling of Planing Catamarans with Symmetric Hulls. *Ocean Eng.* **2016**, *115*, 60–66. [CrossRef]
8. Insel, M.; Molland, A.F. An Investigation into the Resistance Components of High Speed Displacement Catamarans. *R. Inst. Nav. Archit.* **1992**, *134*, 1–20.
9. Molland, A.F.; Wellicome, J.F.; Couser, P.R. Resistance experiments on a systematic series of high speed displacement catamaran forms: Variation of length-displacement ratio and breadth-draught ratio. In *Ship Science Report 71*; University of Southampton: Southampton, UK, 1994.
10. Zaraphonitis, G.; Spanos, D.; Papanikolaou, A. Numerical and Experimental Study on the Wave Resistance of Fast Displacement Asymmetric Catamarans. In Proceedings of the International Conference on High-Performance Marine Vehicles, Hamburg, Germany, 2–5 May 2001.

11. Van't Veer, R. Experimental results of motions hydrodynamic coefficients and wave loads on the 372 catamaran model. In *Report 1129*; Delft University of Technology: Delft, The Netherlands, 1998.
12. Broglia, R.; Bouscasse, B.; Jacob, B.; Olivieri, A.; Zaghi, S.; Stern, F. Calm Water and Seakeeping Investigation for a Fast Catamaran. In Proceedings of the 11th International Conference on Fast Sea Transportation, FAST 2011, Honolulu, HI, USA, 26–29 September 2011; pp. 336–344.
13. Broglia, R.; Jacob, B.; Zaghi, S.; Stern, F.; Olivieri, A. Experimental Investigation of Interference Effects for High-Speed Catamarans. *Ocean Eng.* **2014**, *76*, 75–85. [CrossRef]
14. Bouscasse, B.; Broglia, R.; Stern, F. Experimental Investigation of a Fast Catamaran in Head Waves. *Ocean Eng.* **2013**, *72*, 318–330. [CrossRef]
15. Durante, D.; Broglia, R.; Diez, M.; Olivieri, A.; Campana, E.F.; Stern, F. Accurate Experimental Benchmark Study of a Catamaran in Regular and Irregular Head Waves Including Uncertainty Quantification. *Ocean Eng.* **2019**, *195*, 106685. [CrossRef]
16. Falchi, M.; Felli, M.; Grizzi, S.; Aloisio, G.; Broglia, R.; Stern, F. SPIV Measurements around the DELFT 372 Catamaran in Steady Drift. *Exp. Fluids* **2014**, *55*. [CrossRef]
17. Zaraphonitis, G.; Papanikolaou, A.; Mourkogiannis, D. Hull form optimization of high speed vessels with respect to wash and powering. In Proceedings of the 8th International Conference on Marine Design (IMDC), Athens, Greece, 5–8 May 2003.
18. Souto-Iglesias, A.; Zamora-Rodríguez, R.; Fernández-Gutiérrez, D.; Pérez-Rojas, L. Analysis of the Wave System of a Catamaran for CFD Validation. *Exp. Fluids* **2007**, *42*, 321–332. [CrossRef]
19. Souto-Iglesias, A.; Fernández-Gutiérrez, D.; Pérez-Rojas, L. Experimental Assessment of Interference Resistance for a Series 60 Catamaran in Free and Fixed Trim-Sinkage Conditions. *Ocean Eng.* **2012**, *53*, 38–47. [CrossRef]
20. Danişman, D.B. Reduction of Demi-Hull Wave Interference Resistance in Fast Displacement Catamarans Utilizing an Optimized Centerbulb Concept. *Ocean Eng.* **2014**, *91*, 227–234. [CrossRef]
21. Broglia, R.; Zaghi, S.; Campana, E.F.; Dogan, T.; Sadat-Hosseini, H.; Stern, F.; Queutey, P.; Visonneau, M.; Milanov, E. Assessment of Computational Fluid Dynamics Capabilities for the Prediction of Three-Dimensional Separated Flows: The Delft 372 Catamaran in Static Drift Conditions. *J. Fluids Eng.* **2019**, *141*, 1–28. [CrossRef]
22. Choi, J.E.; Min, K.S.; Kim, J.H.; Lee, S.B.; Seo, H.W. Resistance and Propulsion Characteristics of Various Commercial Ships Based on CFD Results. *Ocean Eng.* **2010**, *37*, 549–566. [CrossRef]
23. Liefvendahl, M.; Fureby, C. Grid Requirements for LES of Ship Hydrodynamics in Model and Full Scale. *Ocean Eng.* **2017**, *143*, 259–268. [CrossRef]
24. Romanowski, A.; Tezdogan, T.; Turan, O. Development of a CFD Methodology for the Numerical Simulation of Irregular Sea-States. *Ocean Eng.* **2019**, *192*, 106530. [CrossRef]
25. Castiglione, T.; Stern, F.; Bova, S.; Kandasamy, M. Numerical Investigation of the Seakeeping Behavior of a Catamaran Advancing in Regular Head Waves. *Ocean Eng.* **2011**, *38*, 1806–1822. [CrossRef]
26. He, W.; Diez, M.; Zou, Z.; Campana, E.F.; Stern, F. URANS Study of Delft Catamaran Total/Added Resistance, Motions and Slamming Loads in Head Sea Including Irregular Wave and Uncertainty Quantification for Variable Regular Wave and Geometry. *Ocean Eng.* **2013**, *74*, 189–217. [CrossRef]
27. Jamaluddin, A.; Utama, I.K.A.P.; Widodo, B.; Molland, A.F. Experimental and Numerical Study of the Resistance Component Interactions of Catamarans. *J. Eng. Marit. Environ.* **2013**, *227*, 51–60. [CrossRef]
28. Yengejeh, M.A.; Amiri, M.M.; Mehdigholi, H.; Seif, M.S.; Yaakob, O. Numerical Study on Interference Effects and Wetted Area Pattern of Asymmetric Planing Catamarans. *J. Eng. Marit. Environ.* **2016**, *230*, 417–433. [CrossRef]
29. Zha, R.S.; Ye, H.X.; Shen, Z.R.; Wan, D.C. Numerical Computations of Resistance of High Speed Catamaran in Calm Water. *J. Hydrodyn.* **2015**, *26*, 930–938. [CrossRef]
30. Zaghi, S.; Broglia, R.; Di Mascio, A. Analysis of the Interference Effects for High-Speed Catamarans by Model Tests and Numerical Simulations. *Ocean Eng.* **2011**, *38*, 2110–2122. [CrossRef]
31. Broglia, R.; Zaghi, S.; Di Mascio, A. Numerical Simulation of Interference Effects for a High-Speed Catamaran. *J. Mar. Sci. Technol.* **2011**, *16*, 254–269. [CrossRef]
32. He, W.; Castiglione, T.; Kandasamy, M.; Stern, F. Numerical Analysis of the Interference Effects on Resistance, Sinkage and Trim of a Fast Catamaran. *J. Mar. Sci. Technol.* **2015**, *20*, 292–308. [CrossRef]
33. Haase, M.; Zurcher, K.; Davidson, G.; Binns, J.R.; Thomas, G.; Bose, N. Novel CFD-Based Full-Scale Resistance Prediction for Large Medium-Speed Catamarans. *Ocean Eng.* **2016**, *111*, 198–208. [CrossRef]
34. Farkas, A.; Degiuli, N.; Martić, I. Numerical Investigation into the Interaction of Resistance Components for a Series 60 Catamaran. *Ocean Eng.* **2017**, *146*, 151–169. [CrossRef]
35. Ferreiro, L.D. The Effects of Confined Water Operations on Ship Performance: A Guide for the Perplexed. *Nav. Eng. J.* **1992**, *104*, 69–83. [CrossRef]
36. Jachowski, J. Assessment of Ship Squat in Shallow Water Using CFD. *Arch. Civ. Mech. Eng.* **2008**, *8*, 27–36. [CrossRef]
37. Tezdogan, T.; Incecik, A.; Turan, O. A Numerical Investigation of the Squat and Resistance of Ships Advancing through a Canal Using CFD. *J. Mar. Sci. Technol.* **2016**, *21*, 86–101. [CrossRef]
38. Terziev, M.; Tezdogan, T.; Oguz, E.; Gourlay, T.; Demirel, Y.K.; Incecik, A. Numerical Investigation of the Behaviour and Performance of Ships Advancing through Restricted Shallow Waters. *J. Fluids Struct.* **2018**, *76*, 185–215. [CrossRef]

39. Feng, D.; Ye, B.; Zhang, Z.; Wang, X. Numerical Simulation of the Ship Resistance of KCS in Different Water Depths for Model-Scale and Full-Scale. *J. Mar. Sci. Eng.* **2020**, *8*, 745. [CrossRef]
40. Chen, X.N.; Sharma, S.D.; Stuntz, N. Wave Reduction by S-Catamaran at Supercritical Speeds. *J. Ship Res.* **2003**, *47*, 145–154. [CrossRef]
41. Castiglione, T.; He, W.; Stern, F.; Bova, S. Effects of Shallow Water on Catamaran Interference. In Proceedings of the 11th International Conference on Fast Sea Transportation, FAST 2011, Honolulu, HI, USA, 26–29 September 2011; pp. 371–376.
42. Moraes, H.B.; Vasconcellos, J.M.; Latorre, R.G. Wave Resistance for High-Speed Catamarans. *Ocean Eng.* **2004**, *31*, 2253–2282. [CrossRef]
43. Zhu, Y.; Ma, C.; Wu, H.; He, J.; Zhang, C.; Li, W.; Noblesse, F. Farfield Waves Created by a Catamaran in Shallow Water. *Eur. J. Mech. B/Fluids* **2016**, *59*, 197–204. [CrossRef]
44. Molland, A.F.; Wilson, P.A.; Taunton, D.J. A systematic series of experimental wash wave measurements for high speed displacement monohull and catamaran forms in shallow water. In *Ship Science Report 122*; University of Southampton: Southampton, UK, 2001. [CrossRef]
45. Molland, A.F.; Wilson, P.A.; Taunton, D.J. Resistance experiments on a systematic series of high speed displacement monohull and catamaran forms in shallow water. In *Ship Science Report 127*; University of Southampton: Southampton, UK, 2003.
46. Gourlay, T. Sinkage and Trim of a Fast Displacement Catamaran in Shallow Water. *J. Ship Res.* **2008**, *52*, 175–183. [CrossRef]
47. Lee, S.H.; Lee, Y.G.; Kim, S.H. On the Development of a Small Catamaran Boat. *Ocean Eng.* **2007**, *34*, 2061–2073. [CrossRef]
48. Castiglione, T.; He, W.; Stern, F.; Bova, S. URANS Simulations of Catamaran Interference in Shallow Water. *J. Mar. Sci. Technol.* **2014**, *19*, 33–51. [CrossRef]
49. Ferziger, J.H.; Peric, M.; Street, R.L. *Computational Methods for Fluid Dynamics*, 4th ed.; Springer Nature Switzerland: Cham, Switzerland, 2020.
50. Stern, S.; Wilson, R.; Shao, J. Quantitative V&V of CFD simulations and certification of CFD codes. *Int. J. Numer. Methods Fluids* **2006**, *50*, 1335–1355.
51. Gatchell, S.; Hafermann, D.; Streckwall, H. Open Water Test Propeller Performance and Cavitation Behaviour Using PPB and FreSCo+. In Proceedings of the Second International Symposium on Marine Propulsors, Hamburg, Germany, 15–17 June 2011.

Article

The Effect of Hull Form Parameters on the Hydrodynamic Performance of a Bulk Carrier

Rui Deng [1,2], Shigang Wang [1,2], Yuxiao Hu [1,2], Yuquan Wang [1,2] and Tiecheng Wu [1,2,*]

[1] School of Marine Engineering and Technology, Sun Yat-sen University, Zhuhai 519000, China; dengr23@mail.sysu.edu.cn (R.D.); wangshg5@mail2.sysu.edu.cn (S.W.); huyx73@mail2.sysu.edu.cn (Y.H.); wangyq297@mail2.sysu.edu.cn (Y.W.)
[2] Southern Marine Science and Engineering Guangdong Laboratory (Zhuhai), Zhuhai 519000, China
* Correspondence: wutch7@mail.sysu.edu.cn; Tel.: +86-187-4514-7797

Citation: Deng, R.; Wang, S.; Hu, Y.; Wang, Y.; Wu, T. The Effect of Hull Form Parameters on the Hydrodynamic Performance of a Bulk Carrier. *J. Mar. Sci. Eng.* **2021**, *9*, 373. https://doi.org/10.3390/jmse9040373

Academic Editor: Kostas A. Belibassakis

Received: 13 March 2021
Accepted: 30 March 2021
Published: 1 April 2021

Publisher's Note: MDPI stays neutral with regard to jurisdictional claims in published maps and institutional affiliations.

Copyright: © 2021 by the authors. Licensee MDPI, Basel, Switzerland. This article is an open access article distributed under the terms and conditions of the Creative Commons Attribution (CC BY) license (https://creativecommons.org/licenses/by/4.0/).

Abstract: In this study, the effect of joint optimization of the principal dimensions and hull form on the hydrodynamic performance of a bulk carrier was studied. In the first part of the joint optimization process, fast principal-dimension optimization of the origin parent ship considering the integrated performance of ship resistance, seakeeping, and maneuverability, as well as their relationships with the principal dimensions were analyzed in detail based on the ship resistance, seakeeping qualities, and maneuverability empirical methods of Holtrop and Mennen, Bales, and K and T indices, respectively. A new parent ship was chosen from 496 sets of hulls after comprehensive consideration. In the remaining part, a further hull form optimization was performed on the new parent ship according to the minimum wave-making resistance. The obtained results demonstrate that: (a) For the case in which the principal dimension of the original parent-type ship is different from that of the owner's target ship, within the bounds of the relevant constraints from the owner, an excellent parent ship can be obtained by principal-dimension optimization; (b) the joint optimization method considering the principal dimension and hull form optimization can further explore the optimization space and provide a better hull.

Keywords: principal-dimension optimization; ship resistance; seakeeping; maneuverability; Holtrop and Mennen's empirical methods; towing tank test

1. Introduction

To reduce maritime greenhouse gas (GHG) emissions to reach the International Maritime Organization (IMO) 2050 target, new energy-efficient ships are urgently needed. Although these energy-efficient designs have a higher newbuild cost, the savings on fuel consumption and, in turn, the cost, tend to be considerably larger than the additional newbuild cost [1–4].

Among the various aspects of ship performance, hull form optimization has long focused on minimizing ship resistance. For certain hull forms, hull form optimization can reduce resistance. Sariöz [5] presented an optimization approach to be used in the preliminary design stage to create a high-quality ship hull form geometry. Hong et al. [6] developed a self-blending method to modify and optimize a bulbous bow. The shape of the bulbous bow of a fishing vessel was optimized, and the resistance was reduced by 2%. Rotteveel et al. [7] analyzed the optimization of propulsion power for various water depths using a parametric inland ship stern shape. Cerka et al. [8] presented a numerical simulation of hull form optimization of a multi-purpose catamaran-type research vessel based on the method of successive approximations. Deng et al. [9] used nonlinear programming and genetic algorithms to optimize the hull form and achieved promising results. Cheng et al. [10] used a new hull surface automatic modification method based on Delaunay triangulation to perform hull form optimization, which can significantly improve the optimization efficiency. Hou [11] presented the hull form optimization design method for minimum Energy Efficiency Operation Index (EEOI), and four case studies were conducted to verify the feasibility and superiority of the novel approach. Zheng et al. [12] took numerical functions and the surface combatant model DTMB 5415 as the research objectives for knowledge extraction by combining the partial correlation analysis and self-organizing map (SOM) based on optimization data. Kim et al. [13] studied an efficient and effective hull surface modification technique for the Computational Fluid Dynamics (CFD)-based hull form optimization. Numerical results obtained in this study have shown that the present hull surface modification technique can produce smooth hull forms with reduced drag effectively and efficiently in the CFD-based hull form optimization. Feng et al. [14] performed an experimental and numerical study of multidisciplinary design optimization to improve the resistance performance and wake field quality of a vessel. Lin et al. [15] set up an automatic design optimization of a small waterplane area twin-hull (SWATH) that provides accurate flow prediction and is integrated into the optimization module. They obtained lower resistance than the original hull, which shows the effectiveness of the optimization. Seok et al. [16] applied the design of experiments and CFD to improve the bow shape of a tanker hull. The results show that the added resistance of the improved hull form is reduced by 52%. Priftis et al. [17] applied a holistic optimization design approach to study the parametric design and multi-objective optimization of ships under uncertainty. Papanikolaou et al. [18] performed a numerical and experimental optimization study on a fast, zero-emission catamaran. Jeong et al. [19] proposed two methods for comparing the mesh deformation method for hull form optimization. Various bow shapes of the Japan Bulk Carrier were applied to validate the applicability of the methods. The proposed mesh deformation method was efficient and effective for CFD-based hull form optimization.

However, in general, hull form optimization does not result in significant changes to the original hull form, and the corresponding effect on the resistance reduction becomes increasingly limited with the improvement in the hull form. Compared with hull form optimization, the principal dimensions of the hull can have a more significant impact on the hydrodynamic performance of the ship; however, they are usually determined by the usage requirements, parent ship dimensions, and other constraints in the initial stage of ship design, following which the modification of the principal dimensions is seldom considered. Therefore, few researchers have conducted studies on resistance reduction based on principal-dimension optimization. Zhang et al. [20] used regression analysis to study the sensitivity of the resistance to the principal dimension of the hull form; the principal dimension parameters with the most significant effects on the total resistance were identified, and the ship resistance was significantly reduced by changing the principal dimensions. Pechenyuk [21] proposed a wave-based optimization method for hull form design, which changes the displacement volume distribution by varying the principal dimensions and thus optimizes the transverse and scattered waves induced. The optimized design of the hull provided the best displacement volume distribution, and the resistance was reduced by 8.9% compared with that of the parent ship. Lindstad et al. [1,4,22–24] studied how hull forms can be made more energy efficient for realistic sea conditions by modifying the main ratios among beam, draught, and length to reduce the block coefficients while keeping the cargo-carrying capacity unchanged. In addition to resistance optimization, Ouahsine et al. [25,26] proposed a numerical method based on c the combination of a mathematical model of nonlinear transient ship maneuvering motion in the horizontal plane and mathematical programming techniques; this method was validated by the turning circle and zigzag maneuvers based on experimental data of sea trials of the 190,000 dwt oil tanker. Subsequently, they developed a numerical model to predict ship maneuvering in a confined waterway using a nonlinear model with optimization techniques to identify the hydrodynamic coefficients accurately.

Some studies have been performed for hull form and principal-dimension optimization of ships. Few scholars have conducted relevant research on the joint optimization of principal dimensions and hull form of ships considering the integrated performance of ship resistance, seakeeping, and maneuverability. Thus, there are still some important aspects that need to be investigated further regarding this topic, such as the accuracy and applicability of empirical methods for the rapid prediction of ship resistance, seakeeping, and maneuverability; accuracy correction of empirical methods for given ship types; and relationships of resistance, seakeeping, and maneuverability performance with the principal dimensions.

In this study, the effect of joint optimization of the principal dimensions and hull form on the hydrodynamic performance of a bulk carrier (origin parent ship) is studied based on empirical methods and towing tank tests, considering the integrated performance of ship resistance, seakeeping, and maneuverability. First, empirical methods of ship resistance, seakeeping, and maneuverability are introduced, and then the accuracy correction of the resistance empirical method based on CFD for the given ship is studied. Second, the resistance, seakeeping, and maneuverability of 496 sets of hulls with different principal dimensions are calculated using the modified empirical methods, and the relationships of resistance, seakeeping, and maneuverability of the hull with the principal dimensions are analyzed in detail. Thereafter, a new parent ship with L = 136.0 m and B = 18.38 m is chosen through the systematic analysis of principal-dimension optimization. Finally, further hull form optimization and verification based on the new parent ship by the towing tank test are presented. The remainder of this paper is organized as follows. Section 1 discusses the literature review of the form and principal-dimension optimization of ships. The geometric model and offset point information of the parent ship are described in Section 2. In Section 3, the ship resistance, seakeeping qualities, and maneuverability empirical methods of Holtrop and Mennen, Bales, and K and T indices are described, respectively. The accuracy correction of Holtrop and Mennen's empirical method based on CFD for the given ship type is studied in detail. Section 4 presents the relationships between resistance, seakeeping, and maneuverability performance with the principal dimensions. Section 5 describes the optimization procedure. Further hull foam optimization and verification based on the selected new parent ship are discussed in Section 6. Section 7 provides a summary of this study.

2. Geometric Model and Information of the Parent Ship

In this study, a bulk carrier was treated as the origin parent ship, with a length of 132 +m, width of 18.2 m, and draft of 5.9 m, block coefficient of 0.6025, displacement of 8806.6 t, and designed speed of 19 kn. The 3-dimensional geometric model and the offset points used for calculation are shown in Figure 1.

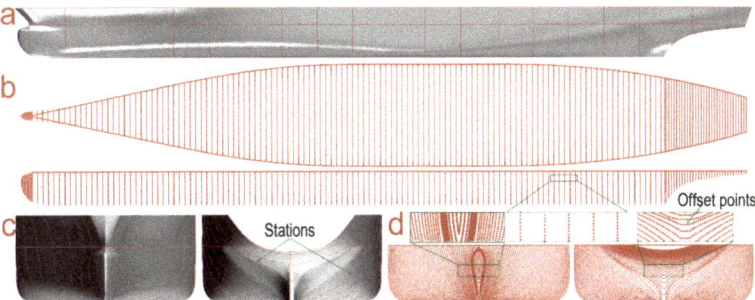

Figure 1. Geometric model and offset points of the parent ship: (**a**) Side view of the geometric model of the parent ship; (**b**) top and side views of the offset points of the parent ship; (**c**) front and stern views of the geometric model of the parent ship; (**d**) front and stern views of the offset points of the parent ship.

The required offsets were extracted and calculated by the software GAMBIT which is a registered trademark of Fluent, Inc (now owned by ANSYS Inc, Canonsburg, PA, USA) (Figure 1). The stations were set every 0.2 m for the bow, every 1.0 m for the hull, and every

0.5 m for the stern, such that the underwater part of the hull was divided into 161 stations from the bow apex to the stern. A total of 70 offset points were obtained for each station line, and the maximum distance between the offset points was approximately 0.14 m. A total of 13,651 offset points were obtained from the waterline, and the maximum distance between the offset points was approximately 0.44 m. This yielded a total of 24,921 offset points to ensure that the hull geometric information was accurately captured.

3. Methodology

Owing to the large Reynolds numbers of full-scale ships, the numerical calculation of their viscous wake fields based on the CFD method requires many cells and specific turbulence models, 2-phase flow models, and degree of freedom motion models, leading to a high threshold of numerical skills and long computation time. Therefore, this method is not applicable for the comparison of multiple schemes in the preliminary design stage. In contrast, existing empirical methods based on regression analysis of model tests and trial data of many ships have good usability and are less time-consuming. Although the calculation accuracy for a particular hull form is limited, it can accurately reflect the changes in the hydrodynamic performance of the ship as the principal dimension changes. Therefore, in this study, the empirical methods of Holtrop and Mennen, Bales, and K and T exponents were used in principal-dimension optimization to calculate the resistance, seakeeping, and maneuverability of a series of hull forms.

3.1. Holtrop and Mennen's Empirical Methods of Ship Resistance

At present, there are many empirical formula methods for resistance, such as Ayre's method, Lap–Keller's method, and Holtrop and Mennen's method. Ayre and Lap–Keller's methods are based on the statistical data of ship types of the 1940s and 1950s. Thus, obvious errors arise from new types of ships in the estimation after the late 1980s. In the early 1980s, Holtrop and Mennen developed a resistance prediction method based on regression analysis of model tests and trial data of Marine Research Institute Netherlands (MARIN), the model basin in Wageningen, The Netherlands [27–31]. Holtrop and Mennen's method was arguably the most popular method for estimating the resistance and horsepower of displacement-type ships. It was based on the regression analysis of a vast range of model tests and trial data, which provided wide applicability [32]. Holtrop and Mennen's method defines the total resistance as:

$$R_T = R_F + R_P + R_W, \qquad (1)$$

where R_T is the total resistance, and R_F, R_P, and R_W represent the frictional, pressure, and wave resistances, respectively. The friction resistance is corrected by introducing the form factor k, which affects the estimation of the residuary resistance, and the pressure resistance is included in the friction resistance. The frictional resistance R_F is computed on the basis of the international towing tank conference (ITTC) 1957 model–ship correlation line coefficient C_F as the resistance of a flat plate with wetted surface S:

$$R_F(1+k) = R_F + R_P, \qquad (2)$$

$$R_F = 1/2 C_F \rho V^2 S, \quad C_F = 0.075/(\log Re - 2)^2, \qquad (3)$$

To estimate the wave resistance R_W, Holtrop defines R_W as the range of Froude numbers into 3 sections:

$$R_W = \begin{cases} c_1 c_2 c_5 \rho g \nabla \cdot e^{m_1 Fr^d + m_4 \cos(\lambda Fr^{-2})} & \text{if } Fr \leq 0.4 \\ R_W(0.4) + \frac{(20Fr-8)}{3}[R_W(0.55) - R_W(0.4)] & \text{if } 0.4 < Fr \leq 0.55 \\ c_{17} c_2 c_5 \rho g \nabla \cdot e^{m_3 Fr^d + m_4 \cos(\lambda Fr^{-2})} & \text{if } Fr > 0.55 \end{cases} \qquad (4)$$

In Equations (2)–(4), ρ is the density of sea (fresh) water, V is the velocity of the ship, ∇ is the volumetric displacement, and Re and Fr are the Reynolds and Froude numbers, respectively. Furthermore, c_1, c_2, c_5, c_{17}, d, λ, m_1, m_3, and m_4 are coefficients for the wave resistance computation in Equation (4), and the detailed description, definition, and calculation equations of the above coefficients can be found in references [27–32].

For the empirical formula methods proposed in the ship resistance evaluation method section, the methods should be first compared to determine the one to be applied. The resistance of 3 types of ships [33] (25,000 t tanker, 82,000 t bulk carrier, and 900 TEU container ship) was predicted by the empirical formula methods and compared with the experimental data, as shown in Figure 2.

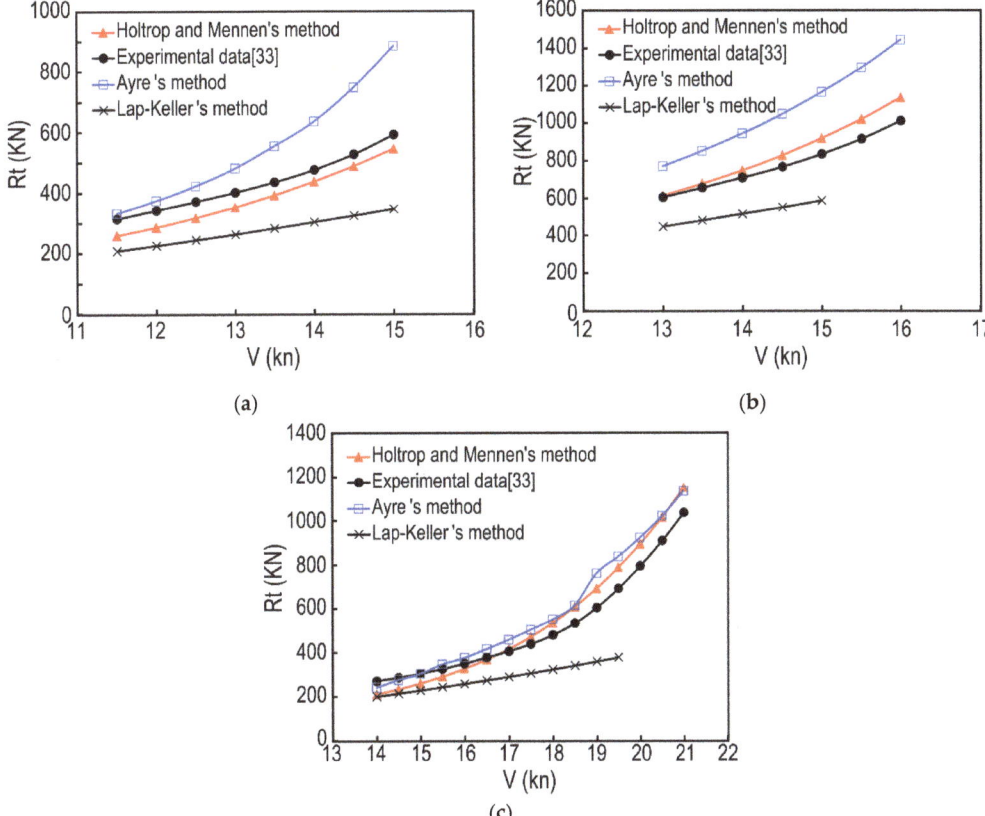

Figure 2. Comparison between the results of the Ayre's, Lap–Keller's, Holtrop and Mennen's methods and experimental data. (**a**) 25,000 t tanker; (**b**) 82,000 t bulk carrier; (**c**) 900 TEU container ship.

From the comparison of the results in Figure 2, it can be observed that the calculation results of various empirical formula methods can basically maintain the tendency as the experimental results and can reflect the resistance characteristics of the ship. Owing to the different applicability of each method, the errors for different ship types are also different; Ayre and Lap–Keller's methods have better accuracy at low speeds and gradually become misaligned as the speed increases. These methods are based on the statistical data of ship types in the 1940s and the 1950s, and the resistance estimation errors for emerging ship types are relatively large. Although the Holtrop and Mennen's method has certain errors, it is generally better than the other two methods. In chronological order, this method was also the latest resistance empirical formula method, which has certain credibility for the estimation of modern ship types. Therefore, in terms of resistance prediction, the

Holtrop and Mennen's method can be recommended as a credible method for estimating the resistance of the target ship.

3.2. Brief Description of Bales's Empirical Method for Ship Seakeeping Performance

Bales calculated the seakeeping properties of 20 destroyers, used 6 geometric characteristics C_{WF}, C_{WA}, T_d/L, C/L, C_{VPF}, and C_{VPA} as variables for regression analysis, and established the relationship between the seakeeping rank factor R and geometric characteristics. The seakeeping prediction model proposed by N.K. Bales [34,35] was adopted by the ship design department of the US Navy and was later promoted and can be used in a variety of ship types. The rank factor R is defined as follows:

$$\hat{R} = 8.422 + 45.104 C_{WF} + 10.078 C_{WA} - 378.465 \frac{T_d}{L} + 1.273 \frac{C}{L} - 23.501 C_{VPF} - 15.875 C_{VPA} \quad (5)$$

where $\hat{R}$ is the estimated value of R; C is the distance from Station 0 to the cut-up point; T_d and L are the draft and length between the perpendiculars of the ship; C_{WF} and C_{WA} represent the water-plane coefficients forward and aft of amidships, respectively; and C_{VPF} and C_{VPA} are the vertical prismatic coefficients forward and aft of amidships, respectively. The rank factor R indicates the degree of seakeeping performance: A larger value indicates better performance. The detailed description, definition, and calculation equations of the above coefficients can be found in references [34,35].

3.3. Brief Description of K and T Indices Empirical Methods for Ship Maneuverability

Nomoto [36,37] studied the problem of ship maneuverability from the viewpoint of control engineering based on the linear equation of ship maneuverability motion and regarded the various maneuvering motions caused by changing the rudder angle as the response of the output maneuvering motion to the input rudder angle. In addition, a second-order maneuvering motion equation was derived, which was also called K. Nomoto's model. The exponents K and T of K. Nomoto's model can define the maneuverability of the ship, which has a clear physical meaning. The K index reflects the turning ability of the ship and is called the turning ability index; the T index represents the ship's rapid response to the rudder and navigation stability and is called the turning lag index. K and T are collectively referred to as the ship's maneuverability index. Ships with good maneuverability should have a large positive K value and a small positive T value. Zhang et al. [38,39], based on the research of Hong [40] and Yao [41] by increasing the number and types of statistical ships, and considering the influence of nonlinear factors between the data volumes, used the parameters of 59 ships as samples, established the quaternion second-order polynomial regression mathematical model, and obtained a statistical regression formula. The results of this formula were compared with the Z-shaped experimental results of the ship, which were not in the statistical samples, to verify the validity of the equation. The estimation formulae for K and T are defined as follows:

$$\hat{K} = 47.875 - 2.64 \frac{L}{B} + 0.004 \frac{LT_d}{A_R} + 66.589 C_b^2 - 112.702 C_b + 3.826 C_b \frac{L}{B} - 0.393 C_b \frac{B}{T_d}, \quad (6)$$

$$\hat{T} = 26.464 + 0.408 C_b \frac{LT_d}{A_R} - 0.033 \frac{L}{B} \frac{LT_d}{A_R} - 79.114 C_b + 0.757 \frac{L}{B} + 46.129 C_b^2. \quad (7)$$

where $\hat{K}$ and $\hat{T}$ are the estimated values of K and T, respectively; T_d, B, and L are the draft, breadth, and length between the perpendiculars of the ship, and C_b and A_R represent the block coefficient and rudder area, respectively. The maneuverability index P is defined as $P = K/T$, in which a larger P-value indicates better ship maneuverability.

3.4. Accuracy Correction of Holtrop and Mennen's Empirical Method Based on CFD for the Given Ship Type

In the principal-dimension optimization part, the resistance performance was the most important aspect of the hydrodynamic performance, followed by seakeeping and maneuverability. Accuracy correction was only performed for the empirical method of resistance. Because the ship used in the establishment of Holtrop and Mennen's empirical method was somewhat different from the one used in this study, and as described in Section 3.1, directly using this method to calculate the resistance of a ship will result in certain potential errors. Therefore, it was necessary to improve the accuracy of Holtrop and Mennen's empirical method according to the ship used in this study. In the process of correction for the empirical method of resistance, the toolbox commercial CFD software STAR CCM+ was used to calculate the total resistance of the parent ship at speeds of 9, 11, 13, 15, 17, 19, 21, and 23 kn, and the detailed experience and description of the numerical calculation strategy can be found in our previous research [42,43]. The total resistance was composed of friction and residual resistances. Because the frictional resistance in the Holtrop and Mennen's method was calculated using the ITTC-1957 formula, it can be assumed that the frictional resistance was correct after considerable experience, and the error of the method only comes from the residual resistance. Subsequently, the residual resistance was separated from the numerical results, and the residual resistance calculated by Holtrop was compared. The ratio of the two parts was used to establish a correction coefficient related to the Fr number, and then the residual resistance term of Holtrop and Mennen's empirical method was corrected. Finally, principal-dimension optimization was performed based on the modified method with acceptable accuracy.

A comparison between the prediction results of the original and modified Holtrop and Mennen's method and the CFD results of the full-scale ship is presented in Figure 3.

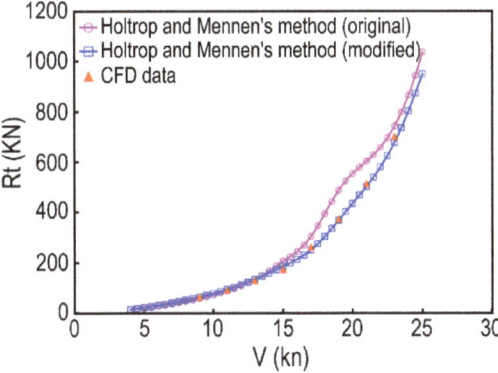

Figure 3. Comparison between the results of the original and modified Holtrop and Mennen's method and the CFD data.

As indicated by Figure 3, the modified Holtrop and Mennen's method demonstrates good agreement with the CFD results, better pertinence, and accuracy. Thus, it can be adopted as a reliable approach for subsequent research. Although the correction coefficient for Holtrop and Mennen's empirical formula in this study is only suitable for the parent ship and is not applicable to all the ships, the correction strategy employed can be implemented for specific ship types and has universal applicability.

4. Relationships of Resistance, Seakeeping, and Maneuverability Performance with the Principal Dimensions

The variation of the principal dimensions in the study was constrained to within ±15% and +15% of the original ship length, beam: The length varied between 112.2 m ≤ L ≤ 151.8 m, with one length selected every 1% of the baseline length (132 m), for a total of 31 lengths;

the ship beam varied within the range of 18.20 m ≤ B ≤ 20.93 m, with one ship beam selected every 1% of the baseline ship beam (18.20 m), for a total of 15 ship beams. The ship's length and beam were considered as the main variables, and the draft was considered as a secondary variable. The draft was determined after selecting different ship lengths and beams, while keeping the displacement constant at 8600 t, yielding a total of 496 sets of hulls with different principal dimensions. Among them, the change in principal dimensions met the regulations and owner requirements, and the general arrangement of the ship. The relationships of the total hull resistance, seakeeping, and maneuverability with the length and beam at the design speed (19 kn) are shown in Figure 4.

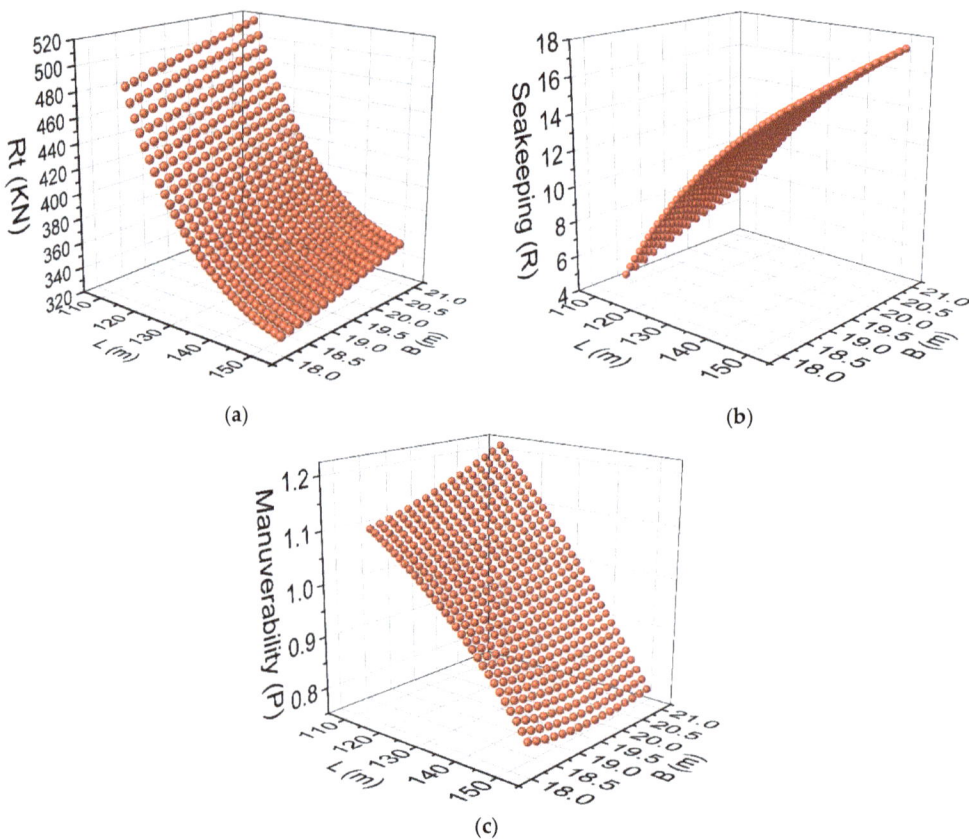

Figure 4. Relationships of various hydrodynamic performances with the principal dimensions: (**a**) ship resistance; (**b**) rank factor R of seakeeping performance; (**c**) maneuverability index P ($P = K/T$).

As shown in Figure 4, while the displacement was kept constant, the resistance monotonically increased with the hull beam and monotonically decreased with increasing ship length. The seakeeping index rapidly increased with ship length and slightly increased with increasing hull beam, which may be resulted from a decrease in the draft. The maneuverability index exhibited opposite change tendencies: as the ship length increased, the maneuverability index first increased and then decreased, whereas as the ship beam increased, the maneuverability index first decreased and then increased.

5. Optimization Procedure

5.1. Rough Selection from Large Amounts of Data

A rough selection from large amounts of data set of performance is discussed in this subsection. The schematic diagram of the optimization process is shown in Figure 5. First of all, we established 496 sets of hulls with different principal dimensions based on the original parent ship, with the variation interval of the principal dimension as the constraint condition. The variation interval of the principal dimension was described in detail in Section 4. Secondly, the hull resistance, seakeeping, and maneuverability indices were calculated for 496 selected sets of principal dimensions using resistance, seakeeping qualities, and maneuverability empirical methods of Holtrop and Mennen, Bales, and K and T indices, respectively. In addition, we got a data set of ship performance. Thirdly, a rough selection from large amounts of data was carried out based on constraints and selection conditions. The constraints and selection conditions were as follows: (a) The resistance of the hulls should be smaller than that of the parent ship at the design speed (19 kn); (b) the change ranges of seakeeping and maneuverability are within 21% and 9%, respectively. Finally, we got five sets of hulls through the rough selection.

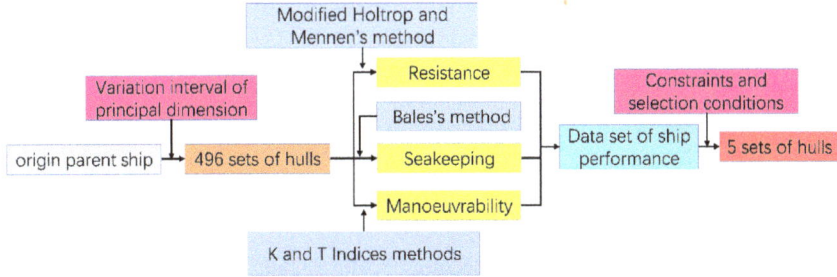

Figure 5. Schematic diagram of rough selection from large amounts of data in the optimization process.

The performance parameters for the five sets of principal dimensions and the parent ship at a speed of 19 kn are presented in Table 1.

Table 1. Performance parameters for the different principal dimensions and parent ship at 19 kn.

NO.	L (m)	B (m)	T (m)	C_B	C_P	C_{WP}	S (m^2)	RT (KN)	Seakeeping (R)	Maneuverability (P)
1	141.2	18.38	5.555	0.6027	0.6158	0.7089	2838	345.1	13.90	0.9051
2	139.9	18.38	5.608	0.6025	0.6514	0.7114	2822	347.9	13.62	0.9173
3	138.6	18.38	5.661	0.6025	0.6507	0.7596	2806	350.9	13.32	0.9291
4	137.3	18.38	5.715	0.6025	0.6502	0.7161	2790	354.2	13.03	0.9406
5	136.0	18.38	5.771	0.6023	0.6496	0.7814	2774	357.7	12.72	0.9518
Original parent ship	132.0	18.20	6.003	0.6025	0.6256	0.7724	2723	368.1	11.57	0.9851

As indicated in Table 1, the hulls of the five selected principal-dimension combinations had a lower resistance in the 16–20 kn speed range and also larger seakeeping and maneuverability indices. This indicates that the vessels of these five principal-dimension combinations had better seakeeping and maneuverability.

5.2. Effect Analysis of Principal-Dimension for the Selected Five Sets of Hulls

To analyze the effects of principal-dimension optimization on the resistance, seakeeping, and maneuverability, the resistance of the five hull forms and the parent ship within the speed range of 4–25 kn were analyzed, and the results are shown in Figure 6.

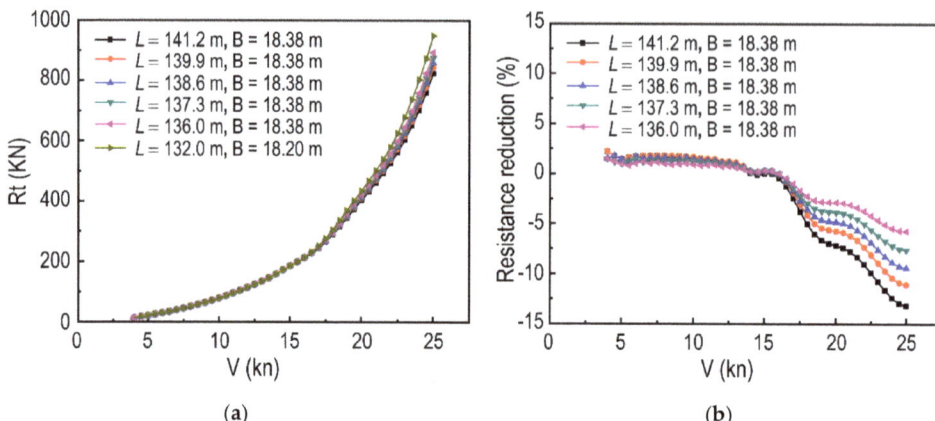

Figure 6. Comparison of the total resistance and resistance reduction rates before and after principal-dimension optimization: (**a**) total resistance; (**b**) resistance reduction rates.

As shown in Figure 6, when the speed was lower than 17.5 kn, the effect of principal-dimension optimization was insignificant, and the resistance of the optimized hull form was not significantly lower than that of the parent ship. When the speed was higher than 17.5 kn, the effect of the principal-dimension optimization was more significant, and the resistance of the hull forms after the principal dimensions were changed was significantly lower than that of the parent ship. Furthermore, the resistance is decreased when the ship length is increased, mainly because wave-making resistance can be reduced by increasing the ship length at high speed. Additionally, the ship beams were identical for the five hull forms, with good resistance performance. This implies that for the hull form adopted in this study, if the displacement remains adopted and the principal dimensions are altered to reduce the resistance at high speed, an optimal value can be obtained for the ship beam. In addition, it is helpful to increase the ship length and reduce the draft for resistance reduction at high speed. For the selection of one hull form from the five sets of principal-dimension combinations, if the key consideration is the resistance at the design speed (19 kn), the principal dimensions of the hull form with the minimum resistance are L = 141.24 m and B = 18.38 m, indicating a 6.7% reduction in resistance, 18.4% improvement in seakeeping, and 8% reduction in maneuverability compared with the parent ship.

Based on the comparison of the resistance performance before and after the principal-dimension optimization, the seakeeping and maneuverability of the five selected hull forms within the range of 16–20 kn were further compared, and the results are shown in Figure 7. When the ship beam was fixed, the seakeeping gradually increased with the increasing ship length, and the seakeeping index was maximized when ship length was L = 141.2 m, indicating that increasing the ship length and reducing the draft can improve the seakeeping under the same displacement (see Figure 7a). As shown in Figure 7b, the maneuverability exhibited the opposite tendency when increasing ship length within the speed range of interest. When the speed was less than 17 kn, the maneuverability index was increased with the ship length, indicating better maneuverability; when the speed was higher than 17 kn, the maneuverability index was decreased with ship length. Hence, the maneuverability did not vary monotonically with ship length over a wider speed range, and increasing the ship length at high speed did not benefit the maneuverability of the examined hull form. The variation of the maneuverability with respect to the ship beam was more significant when the ship beam was smaller than 17 m. For the hull with L = 141.2 m and B = 18.38 m, the maneuverability was good at a low speed but poor at high speed. The hull with L = 136.0 m and B = 18.38 m yielded the best maneuverability at high speed, and the maneuverability was improved by 5.16% when the speed was 19 kn compared with the case of the hull with L = 141.2 m and B = 18.38 m.

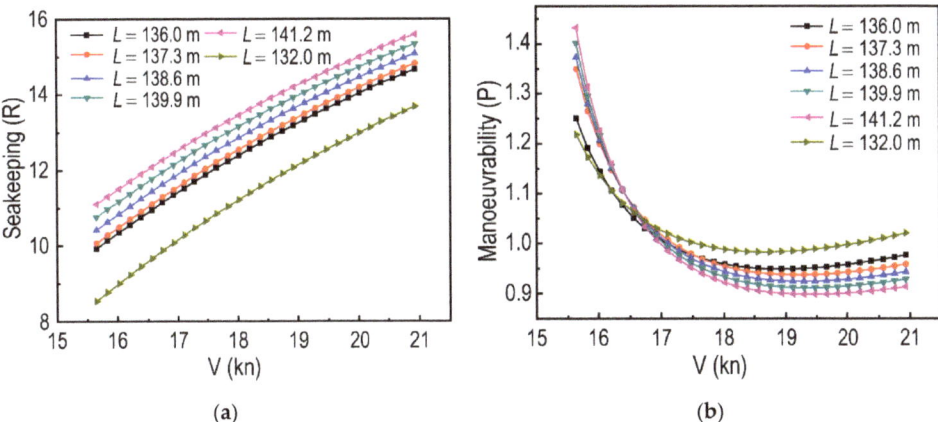

Figure 7. Variation of seakeeping and maneuverability with the principal dimensions: (**a**) Seakeeping; (**b**) maneuverability.

5.3. Further Selection from the Five Sets of Principal-Dimension Combinations

A further selection from the five sets of principal-dimension combinations is discussed in this subsection. According to the law of actual ship manufacturing cost, the increase of ship length under the same conditions will directly lead to the increase in shipbuilding cost. In the further selection process, in addition to considering the ship's resistance, seakeeping, and maneuverability performance, it also comprehensively considers the additional construction costs caused by the increase in the length of the ship.

By comprehensively considering the ship's length, resistance, seakeeping, and maneuverability, we have established a comprehensive optimization index, and the comprehensive optimization index Z is defined as follows:

$$Z = -2.5\frac{L_i - L_0}{L_0} + \frac{|Rt_i - Rt_0|}{Rt_0} + 0.8\frac{R_i - R_0}{R_0} + 0.8\frac{P_i - P_0}{P_0} \tag{8}$$

Note: Compared with the parent ship, the five ship types have negative effects on the relative comprehensive optimization indexes of ship length and maneuverability, and "−" should be added, and the weights of length, resistance seakeeping, and maneuverability are 2.5, 1, 0.8, and 0.8, respectively.

Where L_i, Rt_i, R_i, and P_i are the length, resistance, rank factor of seakeeping performance and maneuverability index of the selected five sets of hulls, and L_0, Rt_0, R_0, and P_0 are the length, resistance, rank factor of seakeeping performances and maneuverability index of the parent ship, respectively. The calculation results of the comprehensive index of the selected five sets of hulls are shown in Table 2.

Table 2. Comprehensive optimization calculation results of the selected five sets of hulls.

NO.	L (m)	B (m)	$\frac{L_i - L_0}{L_0}$	$\frac{|Rt_i - Rt_0|}{Rt_0}$	$\frac{R_i - R_0}{R_0}$	$\frac{P_i - P_0}{P_0}$	Z
1	141.2	18.38	0.0697	0.0625	0.2014	−0.0812	−0.01559
2	139.9	18.38	0.0598	0.0549	0.1772	−0.0688	−0.00788
3	138.6	18.38	0.0500	0.0467	0.1513	−0.0568	−0.0027
4	137.3	18.38	0.0402	0.0378	0.1263	−0.0452	0.00218
5	136.0	18.38	0.0303	0.0283	0.0994	−0.0338	0.00503

According to the comparative analysis of the five optimized hull forms, the hull form with L = 141.2 m and B = 183.38 m exhibited a good ship resistance but also a loss of maneuverability due to the increase in the ship length, which increased the production costs. By comprehensively considering the foregoing factors, it was found that while the displacement was kept constant and only the principal dimensions were changed, five hull forms outperformed the parent ship with regard to the resistance performance, seakeeping, and maneuverability. The ship beam of these five hull forms was 18.38 m, which was higher than that of the parent ship. This is acceptable because an increase in the beam is helpful for improving stability. Both the resistance performance and seakeeping were improved with an increase in the length, but the maneuverability exhibited an initial improvement, followed by deterioration with increasing length. In this study, only the displacement was kept constant. However, in practice, an excessive ship length increased the lightship weight and reduced its effective loading capacity. Furthermore, a long and slender hull requires further strengthening of the hull structure. Therefore, although the main objective of this study was to improve the resistance performance, it was also necessary to consider the improvement of other aspects of the performance and the practical value of the optimized hull form. According to the calculation results in Table 2, it can be seen that the No.5 ship hull has the largest comprehensive optimization index Z. Therefore, after comprehensive consideration, the optimized hull form (new parent ship) with L = 136.0 m and B = 18.38 m was chosen.

6. Further Hull Foam Optimization and Verification Based on the Selected New Parent Ship

After the principal dimensions were optimized, further hull form optimization was performed on the new parent ship (L = 136.0 m and B = 18.38 m) according to the minimum wave-making resistance. Several previous studies reported results in this area, and thus will not be described here. After hull form optimization, no significant changes were made to the hull form. The optimized hull form was compared with the new parent ship, as shown in Figure 8, in which the red dashed line indicates the body plan lines of the optimized hull form, and the solid black line indicates the body plan line of the new parent ship. On this basis, the ship model of the new parent ship and the optimized hull form were created at a scale ratio of $\lambda = 22$, and a towing tank test was performed. The main parameters of the optimized hull ship in the model and the full scale are presented in Table 3.

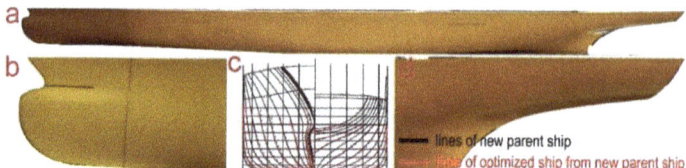

Figure 8. Ship model for towing tank experiment: (**a**) side view of the ship model of the optimized hull from new parent ship; (**b**) sketch of the bow; (**c**) front and stern views of the ship plan line of hulls; (**d**) sketch of the stern.

Table 3. Main parameters of the optimized hull from the new parent ship.

Parameters	Symbols	Model Scale	Ship Scale
Length overall (m)	Loa	6.5818	147.80
Length on waterline (m)	Lwl	6.1791	136.00
Beam (m)	B	0.8354	18.38
Depth (m)	D	0.4095	9.00

Figure 9 shows the resistance of the hull before and after optimization, including the experimental data, empirical formula data, and resistance reduction. The red line is the experimental data of the new parent ship, the green line is the experimental data of the optimized hull of the new parent ship, and the purple diamond points are the Holtrop and Mennen's empirical formula (modified) data. Because the new parent ship was formed by the principal-dimension optimization of the original parent ship, the new parent ship had undergone a round of resistance optimization. Under the condition that the hull form type did not change significantly, the maximum resistance reduction of the hull optimized from the new parent ship was 1.5%. In addition, a comparison between the experimental data and Holtrop and Mennen's empirical formula (modified) data of the new parent ship shows that the accuracy correction of the empirical method based on CFD for the given ship type is reliable.

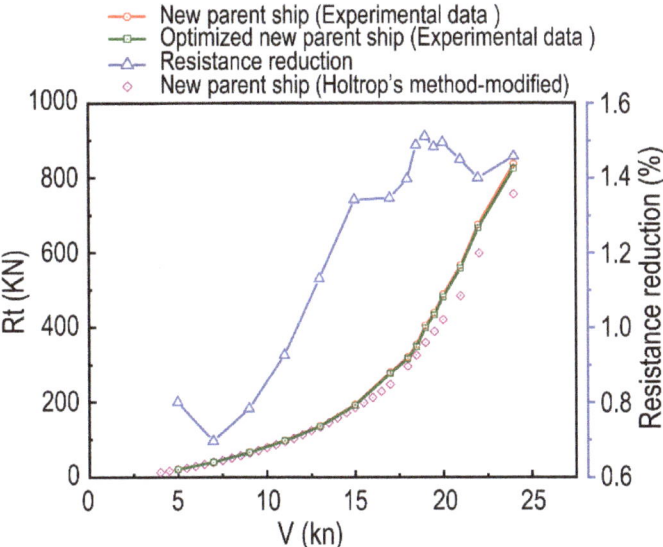

Figure 9. Resistance of the hull before and after optimization, including experimental data, empirical formula data, and resistance reduction.

Figure 10 shows the waveform of the hull in the towing tank before and after optimization, including the bow wave, shoulder wave, and wave scars. From the waveform information of the same viewing angle in Figure 9, it can be seen that the optimized ship has a smaller wave-making range. Because the length of the waterline remains unchanged before and after the optimization, the frictional resistance (1957 ITTC empirical formula) of the ship remains unchanged, and the reduction in the resistance of the ship is mainly reflected in the residual resistance component, including the wave-making resistance. The above waveform information also provides this evidence.

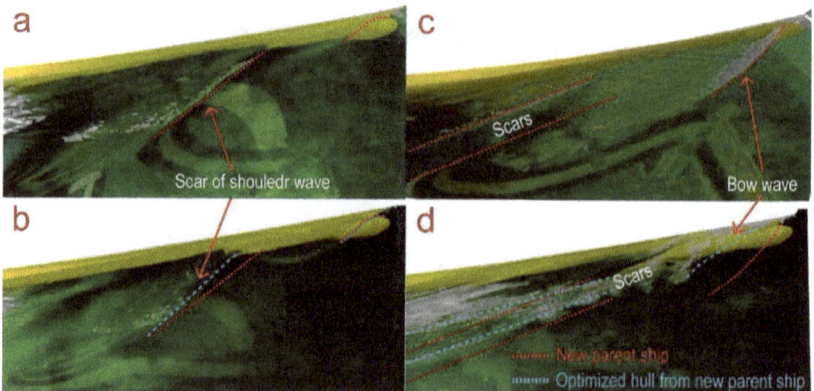

Figure 10. Wave form of the ships under different speeds: (**a**) Wave form of the new parent ship at the speed of 13 kn; (**b**) wave form of the optimized hull from new parent ship at the speed of 13 kn; (**c**) wave form of the new parent ship at the speed of 19 kn; (**d**) wave form of the optimized hull from new parent ship at the speed of 19 kn.

7. Conclusions

Joint optimization of the principal dimensions and hull form on the hydrodynamic performance of a bulk carrier (origin parent ship) considering the integrated performance of ship resistance, seakeeping, and maneuverability was studied based on empirical methods and towing tank tests. The following conclusions were drawn:

1. Holtrop and Mennen's method is arguably the most popular method for estimating the resistance of displacement-type ships. The results obtained by the modified Holtrop method based on CFD for the given ship in the present study exhibited good agreement with the CFD and experimental data of the towing tank, good pertinence accuracy.
2. Variations of the principal dimensions affected ship resistance, seakeeping, and maneuverability. Within the requirements of regulations, owner's requirements, and general arrangement of the ship, principal-dimension optimization can improve the performance of the original parent ship and provide a new parent ship for further hull form optimization.
3. The joint optimization method considering the principal dimension and hull form optimization can further explore the optimization space and provide a better hull.

Some research limitations in this paper: All the optimization in this paper were performed conditions of calm water, without considering the influence of wind and waves; Optimization targets only focus on resistance, seakeeping, and maneuverability. There is no verification and attempt to adopt more updated empirical formula methods.

Author Contributions: Conceptualization, R.D. and T.W.; methodology, R.D.; software, R.D.; validation, Y.H. and Y.W.; formal analysis, R.D.; investigation, R.D.; resources, T.W.; data curation, T.W.; writing—original draft preparation, S.W., R.D. and T.W.; writing—review and editing, R.D. and T.W.; visualization, S.W.; supervision, R.D.; project administration, R.D.; funding acquisition, R.D. and T.W. All authors have read and agreed to the published version of the manuscript.

Funding: This research was funded by the Key-Area Research and Development Program of Guangdong Province (Grant No. 2020B1111010002); National Natural Science Foundation of China (Grant No. 51679053; 12002404); Opening Fund for State Key Laboratory of Shipping Technology and Security; Guangdong Basic and Applied Basic Research Foundation (2019A1515110721), the China Postdoctoral Science Foundation (No. 2019M663243); the Fundamental Research Funds for the Central Universities (No.20lgpy52).

Data Availability Statement: Data is contained within the present article.

Conflicts of Interest: The authors declare no conflict of interest. The funders had no role in the design of the study; in the collection, analyses, or interpretation of data; in the writing of the manuscript, or in the decision to publish the results.

References

1. Lindstad, H.; Jullumstro, E.; Sandaas, I. Reductions in cost and greenhouse gas emissions with new bulk ship designs enabled by the Panama Canal expansion. *Energy Policy* **2013**, *59*, 341–349. [CrossRef]
2. Lindstad, H. Assessment of Bulk Designs Enabled by the Panama Canal Expansion. *Trans. Soc. Nav. Arch. Mar. Eng.* **2015**, *121*, 590–610.
3. Lindstad, H.E.; Rehn, C.F.; Eskeland, G.S. Sulphur Abatement Globally in Maritime Shipping. *Transp. Res. Part D* **2017**, *57*, 303–313. [CrossRef]
4. Lindstad, E.; Bø, T.I. Potential power setups, fuels and hull designs capable of satisfying future EEDI requirements. *Transp. Res. Part D* **2018**, *63*, 276–290. [CrossRef]
5. Sariöz, E. An optimization approach for fairing of ship hull forms. *Ocean Eng.* **2006**, *33*, 2105–2118. [CrossRef]
6. Hong, Z.C.; Zong, Z.; Li, H.T.; Hefazi, H.; Sahoo, P.K. Self-blending method for hull form modification and optimization. *Ocean Eng.* **2017**, *146*, 59–69. [CrossRef]
7. Rotteveel, E.; Hekkenberg, R.; Auke, V.D.P. Inland ship stern optimization in shallow water. *Ocean Eng.* **2017**, *141*, 555–569. [CrossRef]
8. Cerka, J.; Mickeviciene, R.; Ašmontas, Ž.; Norkevičius, L.; Žapnickas, T.; Djačkov, V.; Zhou, P. Optimization of the research vessel hull form by using numerical simulation. *Ocean Eng.* **2017**, *139*, 33–38. [CrossRef]
9. Deng, R.; Huang, D.B.; Yu, L.; Cheng, X.K.; Liang, H.G. Research on factors of a flow field affecting catamaran resistance calculation. *J. Harbin Eng. Univ.* **2011**, *32*, 141–147.
10. Cheng, X.D.; Feng, B.W.; Liu, Z.Y.; Chang, H.C. Hull surface modification for ship resistance performance optimization based on Delaunay triangulation. *Ocean Eng.* **2018**, *153*, 333–344. [CrossRef]
11. Hou, Y.H. Hull form uncertainty optimization design for minimum EEOI with influence of different speed perturbation types. *Ocean Eng.* **2017**, *140*, 66–72. [CrossRef]
12. Zheng, Q.; Chang, H.C.; Feng, B.W.; Liu, Z.Y.; Zhan, C.S. Research on knowledge-extraction technology in optimization of ship-resistance performance. *Ocean Eng.* **2019**, *179*, 325–336.
13. Kim, H.; Chi, Y. A new surface modification approach for CFD-based hull form optimization. *J. Hydrodyn. Ser. B* **2010**, *22*, 520–525. [CrossRef]
14. Feng, Y.K.; Chen, Z.G.; Dai, Y.; Wang, F.; Cai, J.Q.; Shen, Z.X. Multidisciplinary optimization of an offshore aquaculture vessel hull form based on the support vector regression surrogate model. *Ocean Eng.* **2018**, *166*, 145–158. [CrossRef]
15. Lin, Y.; Yang, Q.; Guan, G. Automatic design optimization of SWATH applying CFD and RSM model. *Ocean Eng.* **2019**, *172*, 146–154. [CrossRef]
16. Seok, W.; Kim, G.H.; Seo, J.; Rhee, S.H. Application of the design of experiments and computational fluid dynamics to bow design improvement. *J. Mar. Sci. Eng.* **2019**, *7*, 226. [CrossRef]
17. Priftis, A.; Boulougouris, E.; Turan, O.; Atzampos, A. Multi-objective robust early stage ship design optimization under uncertainty utilising surrogate models. *Ocean Eng.* **2020**, *197*, 106850. [CrossRef]
18. Papanikolaou, A.; Xing-Kaeding, Y.; Strobel, J.; Kanellopoulou, A.; Zaraphonitis, G.; Tolo, E. Numerical and Experimental Optimization Study on a Fast, Zero Emission Catamaran. *J. Mar. Sci. Eng.* **2020**, *8*, 657. [CrossRef]
19. Jeong, K.L.; Jeong, S.M. A Mesh Deformation Method for CFD-Based Hull form Optimization. *J. Mar. Sci. Eng.* **2020**, *8*, 473. [CrossRef]
20. Zhang, H.; Liu, Z.Y.; Zhan, C.S.; Feng, B.W. A sensitivity analysis of a hull's local characteristic parameters on ship resistance performance. *J. Mar. Sci. Technol.* **2016**, *21*, 592–600. [CrossRef]
21. Pechenyuk, A.V. Optimization of a hull form for decrease ship resistance to movement. *Comput. Res. Model.* **2017**, *9*, 57–65. [CrossRef]
22. Lindstad, H.; Sandaas, I.; Steen, S. Assessment of profit, cost, and emissions for slender bulk vessel designs. *Transp. Res. Part D Transp. Environ.* **2014**, *29*, 32–39. [CrossRef]
23. Lindstad, H.E.; Sandaas, I. Emission and Fuel Reduction for Offshore Support Vessels through Hybrid Technology. *J. Ship Prod. Des.* **2016**, *32*, 195–205. [CrossRef]
24. Lindstad, E.; Alterskjær, S.A.; Sandaas, I.; Solheim, A.; Vigsnes, J.T. Open Hatch Carriers—Future Vessel Designs & Operations. In Proceedings of the Conference proceedings SNAME 2017, Houston, TX, USA, 24 October 2017.
25. Du, P.; Ouahsine, A.; Toan, K.T.; Sergent, P. Simulation of ship maneuvering in a confined waterway using a nonlinear model based on optimization techniques. *Ocean Eng.* **2017**, *142*, 194–203. [CrossRef]
26. Khanh, T.T.; Ouahsine, A.; Naceur, H.; Wassifi, K.E. Assessment of ship maneuverability by using a coupling between a nonlinear transient maneuvering model and mathematical programming techniques. *J. Hydrodyn.* **2013**, *25*, 788–804. [CrossRef]
27. Holtrop, J. A statistical analysis of performance test results. *Int. Shipbuild. Prog.* **1977**, *24*, 23–28. [CrossRef]
28. Holtrop, J. A statistical re-analysis of resistance and propulsion data. *Int. Shipbuild. Prog.* **1984**, *31*, 272–276.

29. Holtrop, J. A statistical resistance prediction method with a speed dependent form factor. In Proceedings of the Scientific and Methodological Seminar on Ship Hydrodynamics (SMSSH '88), Varna, Bulgaria, 1 October 1988.
30. Holtrop, J.; Mennen, G. A statistical power prediction method. *Int. Shipbuild. Prog.* **1978**, *25*, 253–256. [CrossRef]
31. Holtrop, J.; Mennen, G. An approximate power prediction method. *Int. Shipbuild. Prog.* **1982**, *29*, 166–170. [CrossRef]
32. Birk, L. *Fundamentals of Ship Hydrodynamics: Fluid Mechanics, Ship Resistance and Propulsion*; John Wiley & Sons: Hoboken, NJ, USA, 2019; pp. 611–627.
33. Deng, R. Optimization of ship principal dimensions based on empirical methods. In *House Report*; Harbin Engineering University: Harbin, China, 2017; pp. 29–30.
34. Bales, N.K.; Cumins, W.E. The Influence of Hull Form on seakeeping. *Trans. SNAME* **1970**, *78*, 00007480.
35. Bales, N.K. Optimizing the seakeeping performance of destroyer type hulls. In Proceedings of the 13th Symposium on Naval Hydrodynamics, Tokyo, Japan, 6–10 October 1980.
36. Nomoto, K.; Taguchi, K.; Honda, K.; Hirano, S. On the Steering Qualities of Ships. *J. Zosen Kiokai* **1956**, *99*, 75–82. [CrossRef]
37. Nomoto, K.; Taguchi, K. On Steering Qualities of Ships (2). *J. Zosen Kiokai* **1957**, *101*, 57–66. [CrossRef]
38. Zhang, X.K.; Li, Y.K. Prediction of Ship Maneuverability Indices. *Navig. China* **2009**, *32*, 96–101.
39. Li, Z.B.; Zhang, X.K.; Zhang, Y. Prediction of maneuver ability indecis for ships using SPSS. *Mar. Technol.* **2007**, *32*, 2–5.
40. Hong, B.G.; Yu, Y. Ship's, K and T indices statistics analysis. *J. Dalian Marit. Univ.* **2000**, *26*, 29–33.
41. Yao, J.; Ren, Y.Q.; Li, X. Statistical Analysis of Fishing Vessel's Maneuverability Indexes K and T. *Navig. China* **2003**, *54*, 31–33.
42. Guo, C.Y.; Wu, T.C.; Zhang, Q.; Lou, W.Z. Numerical simulation and experimental studies on aft hull local parameterized non-geosim deformation for correcting scale effects of nominal wake field. *Brodogradnja* **2017**, *68*, 77–96. [CrossRef]
43. Deng, R.; Huang, D.B.; Li, J.; Cheng, X.K.; Lei, Y. Discussion on flow field CFD factors influencing catamaran resistance calculation. *J. Mar. Sci. Appl.* **2010**, *9*, 187–191. [CrossRef]

Journal of
Marine Science and Engineering

Article

Mixed-Fidelity Design Optimization of Hull Form Using CFD and Potential Flow Solvers

Gregory J. Grigoropoulos *, Christos Bakirtzoglou, George Papadakis and Dimitrios Ntouras

Laboratory of Ship and Marine Hydrodynamics, School of Naval Architecture and Marine Engineering, National Technical University of Athens (N.T.U.A.), 15780 Athens, Greece; chrisbak@naval.ntua.gr (C.B.); papis@fluid.mech.ntua.gr (G.P.); ntourasd@fluid.mech.ntua.gr (D.N.)
* Correspondence: gregory@central.ntua.gr

Abstract: The present paper proposes a new mixed-fidelity method to optimize the shape of ships using genetic algorithms (GA) and potential flow codes to evaluate the hydrodynamics of variant hull forms, enhanced by a surrogate model based on an Artificial Neural Network (ANN) to account for viscous effects. The performance of the variant hull forms generated by the GA is evaluated for calm water resistance using potential flow methods which are quite fast when they run on modern computers. However, these methods do not take into account the viscous effects which are dominant in the stern region of the ship. Solvers of the Reynolds-Averaged Navier-Stokes Equations (RANS) should be used in this respect, which, however, are too time-consuming to be used for the evaluation of some hundreds of variants within the GA search. In this study, a RANS solver is used prior to the execution of the GA to train an ANN in modeling the effect of stern design geometrical parameters only. Potential flow results, accounting for the geometrical design parameters of the rest of the hull, are combined with the aforementioned trained meta-model for the final hull form evaluation. This work concentrates on the provision of a more reliable framework for the evaluation of hull form performance in calm water without a significant increase of the computing time.

Keywords: optimization; genetic algorithms; artificial neural networks; meta-models; multilevel optimization; potential flow; viscous flow

1. Introduction

Hydrodynamic hull form optimization is a very demanding task in terms of computer and time resources. In general, it is a multi-disciplinary process to take into account resistance, propulsion, seakeeping, and maneuvering characteristics of a vessel related to different sea states and wind directions as stated in Grigoropoulos et al. [1,2]. However, even in the single objective case, where only calm water resistance is handled, the use of fine grids for CFD evaluation is too time-consuming to be used in all steps of optimization via a Genetic Algorithm (GA). In this respect, metamodels or surrogate models have been widely used in several engineering contexts, such as structural optimization, aeronautics, aerospace and ground or waterborne vehicles, including stochastic applications and uncertainty quantification.

Thus, the necessity to reduce the computational effort in the optimization process without sacrificing the accuracy of the outcome has led to the extensive use of metamodels. The latter are based on the number of high-fidelity evaluations required, since the computational cost of the resulting algorithms is highly reduced [3]. Typical surrogate models are polynomial regression, kriging method, artificial neural networks (ANN), support vector machines (SVM), or radial basis functions (RBF) [4–6].

Once built, they are very fast (split seconds vs. hours of simulation). Of course, it is imperative to be ensured that a chosen surrogate model approximates the simulation sufficiently well to replace them for the design task at hand, at least at the level needed for engineering purposes at early stages when trends and dependencies need to be understood [7].

There are various ways to incorporate metamodels within an EA, so there are various metamodel-assisted evolutionary algorithms (MAEAs). Many relevant papers are based on the use of offline trained metamodels, i.e., metamodels which are trained separately from the evolution. On the other hand, in the variant of metamodel-assisted EAs (MAEAs) with online trained metamodels, these are trained on the fly separately for each new population member [8].

This paper proposes a mixed methodology to optimize the hull form for resistance in calm water, using both potential and viscous flow codes, both with grids of suitable density utilizing a limited number of viscous evaluations carried a priori (offline) to formal optimization. It is well known that potential flow codes implementing Boundary Element Method (BEM) are quite efficient and reliable in modeling the effect of various geometrical design parameters on the hydrodynamic performance of ships in calm water for the major part of the hull form except for the stern region. In the latter area, the viscous phenomena dominate and the potential flow modeling is poor. The 3D, time domain, and Rankine source potential flow code SWAN2 2002 is used for the potential flow calculation of the wave making resistance R_w. The potential flow results for the bow and the middle part of the hull form are combined with a surrogate model based on an ANN trained by viscous flow results to account for the effect of the stern local design parameters on its hydrodynamic performance. The in-house (U)RANS solver MaPFlow, described in detail in Papadakis et al. [9,10], a cell centered CFD Solver that uses both structured and unstructured grids, is suitable for the viscous flow calculations.

On the basis of the aforementioned discussion, an Artificial Neural Network (ANN) is established to account for the effect of the stern only geometrical parameters, assuming the parent form for the rest of the hull. This methodology assumes that the effect of the stern design parameters on the performance of the whole hull is not altered significantly when the bow and the middle design variables are modified. In other words, this assumption is expressed in terms of the (at least partial) independence of the effects of the stern design variables from the rest ones. This assumption is verified by the comparison of the performance of the parent and the optimized hull using viscous flow calculations. A further check is also included in the paper to verify the reliability of the ANN.

It is expected that a small number of variants of the stern hull form geometry as evaluated by MaPFlow provide sufficient and reliable training to the ANN. The number of the required evaluations depends, of course, on the hull form, magnitude of variances and the complexity of vessel shape in the stern region. The latter is responsible for the number of the geometrical parameters that should contribute to the hull form optimization.

To be more specific, the mixed fidelity optimization procedure, presented in this paper, is performed in two separate steps, firstly, for the stern region and following for the middle and bow region. Once, the ANN is trained sufficiently to constitute a reliable metamodel, the first optimization step consists of ANN function's minimization, in order to derive the combination of stern parameters that minimize hull viscous resistance R_{TOTAL}, as predicted by the ANN. On the second step, the bow and middle ship geometrical variables are optimized while maintaining the parent form for the stern of the hull in order to minimize wave resistance R_w. Finally, all the geometrical variables, those in the bow and the middle section of the hull, as well as those in the stern region, as tuned by the two optimization cycles are combined in one fair hull. This final hull is evaluated by direct comparison with CFD modeling employing a grid of the same density. The KRISO Container Ship (KCS) has been used as a test case.

The major contribution of the work is that limited viscous flow calculations are carried out a priori (offline) to train the ANN in order to model efficiently stern viscous effects which dominate in relatively low Froude numbers, such as in the case examined. In this way, a reliable optimization framework is achieved with minimum resources to estimate stern viscous pressure effects, encompassing at the same time the advantages of potential flow codes to guide the optimization of the rest of the hull form for wave resistance.

2. Hydrodynamic Hull Form Optimization

The problem of hydrodynamic hull form optimization, focusing on its performance in calm and rough water, has been studied since the early 1980s by many authors. The use of parametric models with genetic algorithms to carry out a dual objective optimization for calm water resistance and seakeeping is described in detail by Grigoropoulos et al. [11].

The methodology is ruled out by the reliability of the computer codes used for the evaluation of the hydrodynamic performance of the variant hull forms and resources needed. They actually drive the optimization algorithm to a realistic optimum solution. Although viscous flow calculations provide more reliable estimates of calm water performance, they are too time-consuming to be used in each of the hundreds of steps of genetic algorithms. Thus, a way to tackle this difficulty is to use potential flow solvers in the optimization procedure and to verify the result by a viscous flow code. However, potential flow codes are not reliable in the stern region of the ship, where viscous flow phenomena prevail. In order to remedy this situation, it is proposed to use an Artificial Neural Network as a meta-model to account the effect of variations of the stern shape. This ANN is trained using a few runs of the viscous flow code.

3. The Optimization Procedure

3.1. Optimization Method

The problem we have to solve is an optimization problem where the displacement is allowed to vary up to 1% of its initial value. The space of the design variables' variance has been specified by trials during a preparatory stage, to ensure realistic hull form variants. The mathematical formulation of the optimization problem is as follows:

$$min\ R(H_i, Fn, SS)$$

$$w.r.t. l \leq i \leq u\ and\ D\Delta \leq const$$

where

R : Hull's resistance as evaluated by potential solver or ANN.
i: Design variable vector; l and u are the lower and upper bounds, respectively.
H_i : Hull shape as affected by the changes in i parameters.
Fn : Froude number of the vessel.
SS : Sea state, calm water for cases evaluated.
$D\Delta$: Variation in the volume of displacement.

For the optimization process, the NSGA II (Non-Dominated Sorting Genetic Algorithm II), Figure 1, as described by Deb et al. [12], and which is provided inside CAESES software [13], is selected. The architecture of the artificial neural network used for the training is discussed in the next paragraph. Details of the genetic algorithm used in the optimization process are briefly described below:

- A number of variant geometries is generated.
- An equal number of off-springs is formed.
- The total number of parents and offspring is then sorted to levels according to non-domination.
- The geometries of each level are ranked with respect to their crowded distance of each solution in the population.
- A new generation is being produced with a population number equal to the initial one.
- Steps 2 to 5 are repeated.

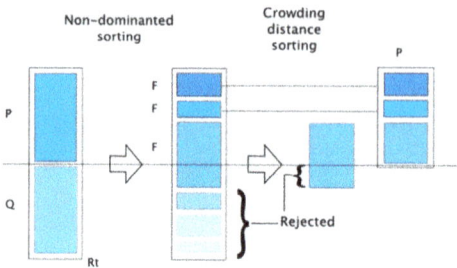

Figure 1. NSGA-II procedure.

For the two optimization cycles performed, an adequate number of generations and off-spring was selected in order to ensure convergence for all parameters. A total of 448 variants were evaluated concerning the three stern design variables parameters in the first step, comprised of 32 generations with 14 offspring each. In the second step, where the five bow and middle-part hull design variables were investigated, 37 generations with 24 offspring each produced in total 888 alternative hulls. The mutation and crossover probability rates were set to 0.04 and 0.92, respectively, in both cases. In both steps, the modification of the design variables is controlled by the GA, while CAESES automatically creates fair ship lines for each combination of them. The theoretical background of CEASES is described by Harries and Abt [14].

3.2. Design Modification Procedure

CAESES software has been selected for the parametric design and variation of the parent hull.

The initial geometry is represented by a set of basic curves providing topological information in the longitudinal direction (design waterline, centerline, deck-line) and a set of 32 section curves. All of them are either F-splines or B-splines. F-splines are used to describe areas or characteristic lines subjected to variation, which directly affect the geometrical hull form parameters to be optimized.

The geometry is split into three regions: the main hull, the stern region, and the bow bulb, assigning specific design variables for each of them in order to ease the optimization process. Hull form is described by different kinds of surfaces which reflect the changes at the parameters under investigation. Surfaces are generated either by interpolating the parametric-modeled section curves or by using the so-called engine curves. The approximation of the initial surfaces is very satisfactory and allows for the establishment of the eight design variables in total. Five of them refer to the bow bulb and the main hull, while the remaining three at the stem region. Figure 2 depicts the geometry delivered by CAESES software, while Figure 3 presents the control lines and the surface of the bow bulb.

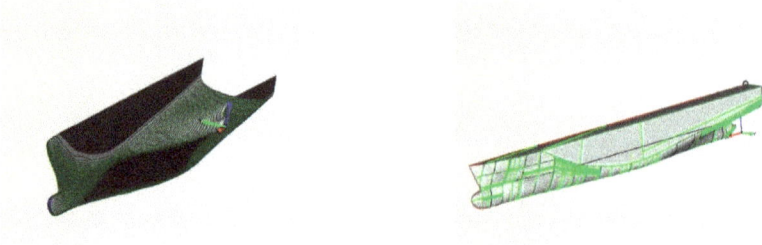

Figure 2. The geometry of KRISO containership as derived by CAESES software.

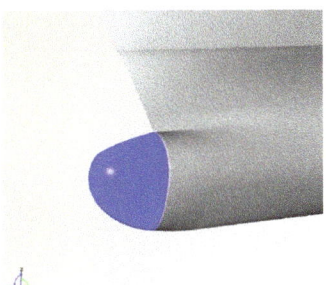 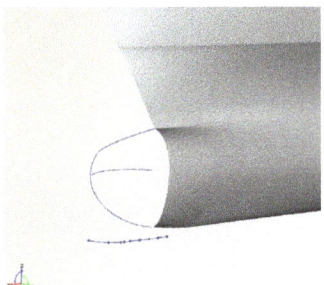

Figure 3. The control lines and the surface of the bow bulb.

Table 1. The initial values and limits of variation of design variables.

Design Variable	Lower Limit	Initial Value	Upper Limit
dX8%_from FP (m)	−1.6	0.0	1.5
FOS_dZ (m)	−1.2	0.0	1.0
Bulb_dL (m)	−0.8	0.0	1.3
Angle_WL (°)	150	170	180
Angle_Prof (°)	65	81	90
TransomLow_zPos (m)	10.75	11.003	11.3
Curve_xPos (m)	7.6	8.9	10.2
TubeEnd_xPos (m)	4.65	5.113	5.65

The design variables of Table 1 are described in the following:

dX8% aft FP: The longitudinal shift of the frame located 8% of ship length L aft of FP.
FOS_dZ: Vertical variation of the lower point of the Flat-Of-Side (FOS) area.
Bulb_dL: Change of bulbous bow total length.
Angle_WL: Angle of waterline at design draft T.
Angle_Prof: Angle of rise of bulbous bow profile curve.
TransomLow_zPos: Vertical position of transom lowest point (Figure 4).
Curve_xPos: Profile stern curve (Figure 4).
TubeEnd_xPos: Variation of stern tube axis length (Figure 4).

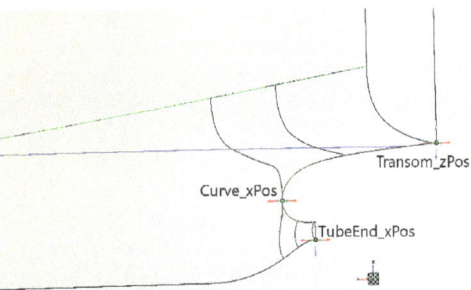

Figure 4. The three parameters set at the stern region.

The range of the design variables and their values for the parent hull form are presented in Table 1. The constraints of the geometrical variables have been specified by a trial and error method to reduce the number of non-realistic or, more generally, invalid variant hull forms. However, this does not mean that all the variants generated during the optimization process are realistic hulls.

3.3. The KRISO Containership

The characteristics and the floating conditions of the containership under investigation have been proposed by the Korean Research Organization KRISO (former MOERI), and they are given in Table 2. The service speed of the vessel has been specified to be 24 kn corresponding to Froude number 0.26.

Table 2. Main particulars and floating conditions of the KRISO containership.

The KRISO Container Ship KCS and Its Tested Model *			
		Ship	Model
Scale	-	1:1	1:37.89
LPP	(m)	230.0	6.0702
LWL	(m)	232.5	6.1357
BWL	(m)	32.2	0.8498
TM	(m)	10.8	0.2850
WS	(m^2)	9501	6.618
∇	(m^3)	52,062	0.957
CB	-	0.651	0.651
LCG from AP	(m)	111.6	2.945
VCG from BL	(m)	14.324	0.378

* The detailed geometry database of the ship is provided at the site of Tokyo 2015 CFD workshop website [15].

However, this vessel has not been constructed in physical scale. Only models of this hull form have been tested in towing tank facilities. In the current study, the full-scale vessel has been modeled.

3.4. Potential Flow Calculation

The 3D panel, Rankine source, time domain, linear code SWAN2 2002 [16,17] has been used to carry out the potential calculations. The mesh generation of the free-surface and the body surface of the hull is an internal routine of SWAN2 2002. The spline sheet of the body surface is defined by 45 nodes in a direction parallel to the x-axis, corresponding to a number of 44 panels and by 13 nodes on the y-axis' perpendicular. The domain of the free-surface accounted in the calculations extends 0.5 L_{BP} upstream, 1.5 L_{BP} downstream, and 1.0 L_{BP} in the transverse distance (athwartships). Figure 5 presents the spline sheet of the free-surface (a) and the body (b).

3.5. Viscous Flow Calculation

Regarding the CFD solver, an in-house (U)RANS solver, MaPFlow, was employed. MaPFlow is a cell centered CFD solver that can use both structured and unstructured grids, capable of solving compressible flows, as well as fully incompressible flows using the artificial compressibility method. For the reconstruction of the flow field, a 2nd order piecewise linear interpolation scheme is used. The limiter of Venkatakrishnan [18] is utilized when needed. The viscous fluxes are discretized using a central 2nd order scheme.

Turbulence closures implemented on MaPFlow include the one-equation turbulence model of Spalart (SA) [19] as well as the two-equation turbulence model of Menter (k-ω SST) [20]. Regarding laminar to turbulent transition modeling, the correlation γ-Reθ model of Langtry and Menter [21] has been implemented.

MaPFlow can handle both steady and unsteady flows. Time integration is achieved in an implicit manner permitting large CFL numbers. The unsteady calculations use a 2nd order time accurate scheme combined with the dual time-stepping technique to facilitate convergence. MaPFlow is able to handle moving/deforming geometries through the arbitrary Eulerian Lagrangian formulation.

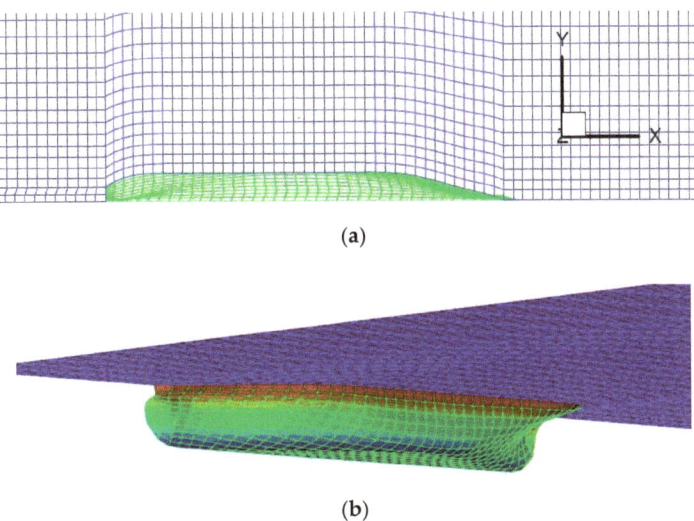

Figure 5. Spline sheet on the computational greed (**a**) free surface; (**b**) body surface.

Regarding the free surface treatment, the Volume of Fluid (VOF) method is employed, and two phase flows are described by two immiscible fluids with their interface being defined implicitly as a discontinuity in the density field. The system of equations is solved in a non-segregated manner, utilizing the Kunz Preconditioner, as discussed by Yue and Wu [22], to remove density dependencies from the system's eigenvalues.

For the CFD simulations, in order to reduce the computational domain, half of the hull is resolved with symmetry conditions applied on the side. The high Reynolds number of full scale simulations poses a significant challenge for CFD simulations since fully resolved simulations are computationally prohibitive. Following a grid-independence study, a grid consisting of approximately 5 million cells is employed. In the wall region, wall functions are employed; nevertheless, a structured-like region composed of 25 layers is used around the solid boundary. Lastly, the hull was resolved using approximately 200,000 elements. A snapshot of the computational grid employed can be seen in Figure 6.

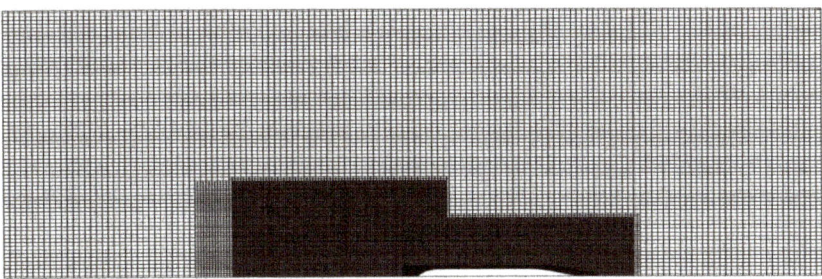

Figure 6. Top view of the computational grid used in the free surface region. The grid is refined to accurately capture the resulting wave system.

The span of the domain was 5 L_{BP} in the streamwise direction, 3 L_{BP} in the side direction, and 6 L_{BP} in the vertical direction. On the ship hull, a no-slip condition was applied, symmetry (zero gradient in the normal direction) conditions were applied on the symmetry plane, while a freestream condition was imposed on the rest of the domain. Additionally, a damping zone was adopted to avoid reflections from the generated wave system.

Regarding the y+ values, the average y+ was 150, while a maximum of 300, and due to that, wall functions were adopted. Unfortunately, this was a mandatory compromise cost-wise in order to make full scale simulations feasible.

For all the CFD simulations, a time step of 0.1 s using a second-order implicit scheme is used which yields a convective CFL around 3. Nevertheless, it was adopted to save computational time since the flows considered here converge to a steady state. It is evident from Figure 7 that both the time step and the grid spacing selected are tuned in order to properly capture the resulting wave system.

Figure 7. Contour of the resulting wave system for the original KCS hull case—Contour of Elevation of the water level. Half of the model was resolved to save computational resources.

3.6. Artificial Neural Network (ANN)

During the last decade, the application of Artificial Neural Networks has stepped up in every scientific field. From plain vanilla networks to unsupervised deep convolutional networks, ANN is able to model or detect complex nonlinear relationships within systems without using the physics of the system. Furthermore, they are a valuable tool, since they can bridge fragmented data to efficiently identify system characteristics or make up for a lack of analytical relations within complex systems.

Artificial neural network theory is based on the analysis of biological nervous systems consisting of neurons and their connections. A mathematical model of a neural network is created, based on this structure and signal transmission. ANNs are composed of internal parameters to be specified through the process of training. Such parameters are the weights by which the inputs of each neuron are multiplied so that the corresponding output emerges. Explicitly, the output of the neuron is calculated by the sum of all the inputs, weighted by the weights of the connections from the inputs to the neuron. Additionally, a bias term is introduced to this sum. This weighted sum is often called the activation, which is, then, passed through a (usually nonlinear) activation function to produce the output. Ultimately, the output of the last neuron, subsequently the output of the model in general, is compared against actual values and the difference between predicted and real values of the same parameter is estimated through a metric function. This part of the optimization algorithm is critical. The training aims at the minimization of the mean difference, or loss as it is called in the ANN field, by updating the weights in each iteration.

In the field of naval architecture, artificial neural networks (ANN) have gained popularity. In recent years, applications of ANNs for modelling and predicting vessel hull form [23], calm water resistance [24,25], added resistance in waves [26], speed and fuel consumption [27], maneuverability qualities [28], and seakeeping characteristics [29] are presented in several studies. In most cases, the use of artificial neural networks offers satisfactory results.

Compared to the above-mentioned publications, this paper's distinctive feature lies in the relatively low number of input data available for the training of an ANN. As discussed in the Introduction, the ANN is trained with only 27 examples as input, which correspond to the 27 (=3^3) combinations of three values per variable, the two selected limiting ones, and the value of the parent hull for the three chosen stern geometrical variables affecting the optimization scheme. Usually, the stern design variables are two to four and three values per variable are sufficient to train a reliable ANN, taking into account that the variation of the variables is limited. The use of limited CFD calculations is a major advantage of the proposed methodology. The derived ANN is handled by a GA, which, after 440 evaluations, reaches the optimum combination of the stern variables.

Selection of a suitable artificial neural network structure is probably the hardest part of the problem and critical to obtaining accurate predictions. Since dealing with a regression problem, the multilayer perceptron (MLP) concept was applied consisting of an input, hidden, and output layer, as well as utilizing a backpropagation learning algorithm. The development of the ANN was performed in Python assisted by the Tensorflow/Keras neural network library.

The usual search process for the optimal neural network goes through the following steps: data normalization, division of data set, selection of ANN model architecture, and finally the assessment of ANN model results. In this study, the small number of available training data significantly hindered this process. Input data were normalized using a custom Min/Max scalar function centered around parent hull resistance value with a 20% reserve. This reserve was used in order to ease the first step of optimization process, ANN's function minimization search, by allowing us to extrapolate values beyond observed ranges. Moreover, during ANN model's training, no validation set was used. It was decided to use every data point available for the more efficient training of the network and take the risk to validate the model's prediction at the final stage of optimization procedure via CFD calculations:

$$\text{Norm. Function} := \frac{\left(\left(R_{parent\_hull} - R_{variant}\right) + \left(R_{parent\_hull} - R_{min}\right) \times 1.2\right)}{(R_{max} - R_{min}) \times 1.2} \quad (1)$$

In order to identify the ANN architecture that is better suited to the problem, many trials were conducted with different configurations. The number of input neurons was set to three, representing the three variables set at the vessel stern region, and the output node was set to one referring to the target value of CFD calculation. The rest of the ANN configuration as determined by the number of hidden layers, the number and type of neurons that comprise each one of them, the training algorithm, learning rate, and the backpropagation optimizer method went through exhaustive numerical experiments, probably an inevitable stage when developing an ANN. An overview of the performance of the best ANN models is presented in Table 3. The number of neurons at each layer and their activation functions can also be seen.

Table 3. Overview of the best ANN performance.

Code				Training Set	
	Input Layer	Hidden Layer (s) *	Output Layer	MSE	MAE
N-1	3	6(S)–4(S)	1(S)	2.1×10^{-3}	3.3×10^{-2}
N-2	3	6(S)	1(S)	4.3×10^{-2}	1.7×10^{-1}
N-3	3	6(S)–3(R)	1(S)	4.4×10^{-2}	1.7×10^{-1}
N-4	3	6(S)–3(R)–3(R)	1(S)	4.0×10^{-2}	1.6×10^{-1}
N-5	3	12(S)	1(S)	4.6×10^{-2}	1.9×10^{-1}

* Activation Functions: R—ReLU (Rectified Linear Unit), S—Sigmoid.

In this work, the Mean Squared Error (MSE) function was used as the loss or cost function under minimization during models' training. Notice should be kept, though, on the Mean Absolute Error (MAE) as well. The progression of MSE and MAE values during training is presented in Figure 8.

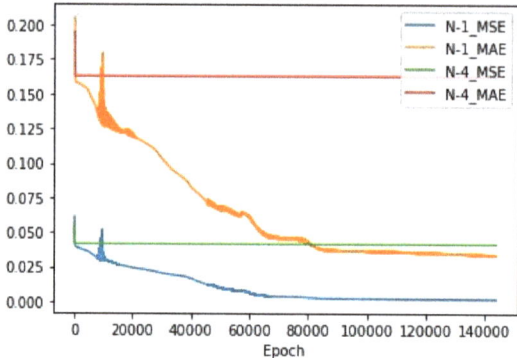

Figure 8. MSE and MAE during networks' training.

The network (N-1) that showed the lowest MSE and MAE error was selected in order to evaluate hull meta-models. It consists of the three stern parameters as inputs, two hidden layers comprised of six and four nodes, respectively, and the output layer. In our effort to overcome the problem of vanishing gradients and saddle points, the sigmoid function was used as an activation function at input, first and output layer of the network, while tanh was utilized at the second layer. Stochastic gradient descent with momentum was selected against ADAM as the backpropagation optimizer method, learning rate, and momentum were set to 0.16 and 0.7, respectively. Selected ANN model's architecture and performance are presented at Figures 9 and 10. The Pearson coefficient was calculated at 0.975.

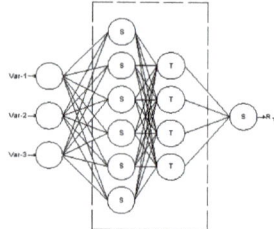

Figure 9. Architecture of the selected ANN model.

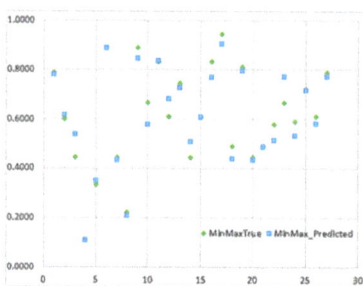

Figure 10. ANN input vs. predicted values.

The small size of the training data led to an increase of epochs performed in order to get satisfactory results. For the best configuration found, the number of epochs was set at the scale of 150,000. Despite the large number of epochs, the training procedure required less than 5 min to complete at an i7 9700K.

4. Results

The results obtained from the implementation of the above-described methodology are presented in this chapter.

In the below figures, the outcome of the first step of optimization procedure is presented. After the training of the ANN model, an NSGA-II optimization algorithm was executed for the minimization of the ANN function modeling vessel total viscous resistance R_{TOTAL}. In Figure 11a–c, the convergence of the three input parameters representing the variables at the hull stern is observed. Moreover, in Figure 11d, it is evident that the ANN function was minimized. This minimization translates to a 9% reduction at vessel's R_{TOTAL} compared to the parent hull as evaluated by the ANN model.

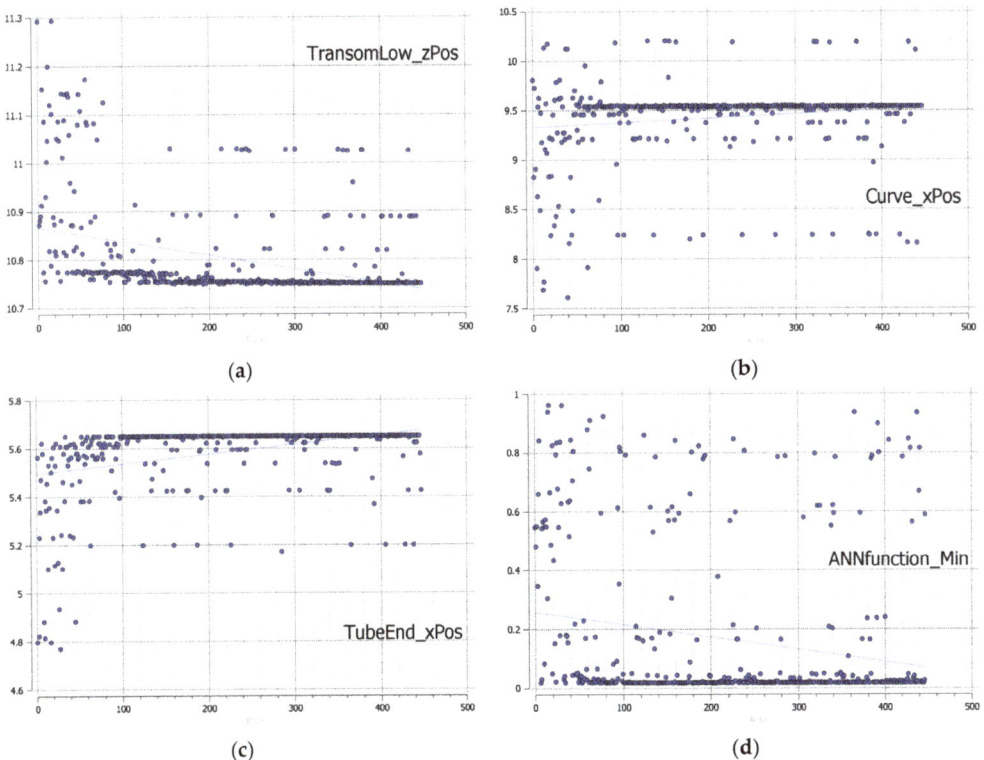

Figure 11. (**a**–**d**) Convergence of the three stern parameters (**a**–**c**) and the minimization of ANN's function (**d**).

To verify this result, a direct CFD evaluation of the hull form with the optimum stern variables was performed. Dynamic sinkage and trim as calculated by the potential flow solver were used in the viscous evaluation.

As presented in Table 4, geometrical parameters, assessed by the ANN and optimized via NSGA-II at the vessel stern region, contributed to the reduction of R_{TOTAL} by 15% after validation by MaPFlow, using the same discretization scheme and turbulence model for the evaluation of the parent and the optimum hull form. However, as stated in Section 3.5, the y+ of the simulations was relatively large; thus, it must be validated using successively denser grids in the hull region. Apart from that, results obtained by the trained neural network are quite satisfactory despite the small size of the training dataset. The small discrepancy in the R_{TOTAL} reduction between network prediction and high-fidelity viscous evaluation verifies the reliability of the ANN meta-model.

Table 4. Results of the first step of optimization.

Model	TransomLow_zPos (m)	Curve_xPos (m)	TubeEnd_xPos (m)	R_{TOTAL} (kN)
Parent Hull	11.003	8.900	5.113	1780
Opt_Hull_Step-1	10.750	9.539	5.651	1611(ANN)/1510(CFD)

In the 2nd step of the optimization procedure, the five design variables at the bow and the middle region of the vessel went through another NSGA-II optimization cycle. Figures 12 and 13. This time, hull variants were evaluated via the potential solver SWAN2 as described in Section 3.4. It should be noted that the vessel stern region was kept at parent shape while only the five parameters were varied. Reduction of wave pressure resistance has been relatively small. Results are presented at following Table 5.

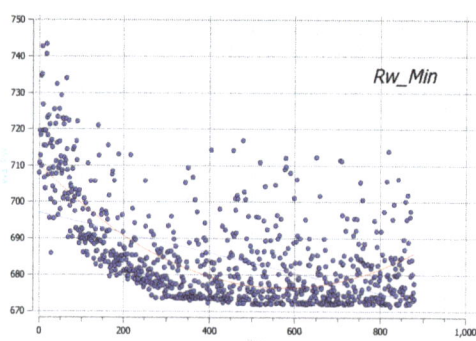

Figure 12. Convergence of the 2nd step of optimization.

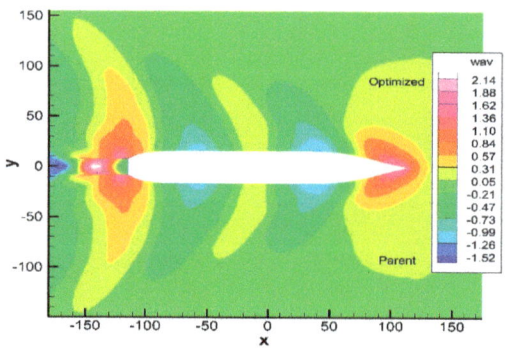

Figure 13. Comparison of wave elevation distribution by SWAN2.

Table 5. Results of the second step of optimization.

Model	dX8%fromFP (m)	FOS_dZ (m)	Bulb_dL (m)	Angle_WL (°)	Angle_Prof (°)	R_w (kN)
Parent Hull	0.0	0.0	0.0	170	81	703
Opt_Hull_Step-2	−0.054	0.986	−0.684	171.493	89.877	676

As a final step, the optimum design variables as those that emerged from both optimization steps were combined to derive the overall optimized hull form, presented in Figure 14. Its performance was assessed via viscous calculation. Results were quite encouraging and are presented in Table 6.

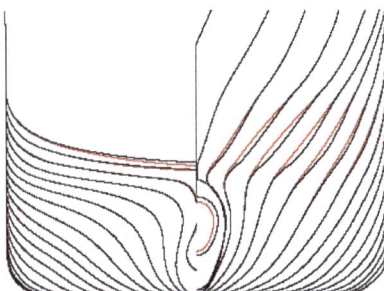

Figure 14. Comparison of optimized (red) and original hull stations.

Table 6. Comparison of the parent and final optimized hull.

Quantity	Parent Hull	Final Hull
R_{TOTAL}	1780	1460
Cb	0.6507	0.6500
Cp	0.9029	0.9019
Cm	0.7206	0.7206
LCB	111.92	111.84
WS (at speed)	9764	9739

5. Conclusions

On the basis of the results presented in the previous sections, it seems that the proposed mixed fidelity method is quite efficient in optimizing the hull form for calm water resistance of ships using genetic algorithms and ANN. The method combines the capabilities of the potential flow codes to evaluate the hydrodynamic effects of the geometrical design parameters of the fore and middle body of a ship with the strength of the viscous flow codes to estimate the respective effect of the design variables in the aft body. ANN was able to produce results of sufficient accuracy to be useful for the preliminary prediction of vessel resistance despite a small number of training inputs. It is worth underlining that the development of a case specific neural network model is needed. Besides traditional programming languages, though, online platforms and libraries for development of artificial neural networks are widely available nowadays. Depending on the number of inputs/outputs, the magnitude of their variances and size of training data an ANN can be effortlessly integrated at the optimization process

Employing automated artificial neural network software can greatly accelerate the process of designing and training an ANN capable of modeling hull resistance problems. The agreement between the optimum value of the objective function and the results of the RANS solver is satisfactory. This is considerably easier and quicker than traditional statistical methods. Whilst it is important to choose a reasonable artificial neural network architecture, the exact number of neurons in the hidden layer is not too critical. Quality and quantity of input data are the key factors.

The final outcome of the optimization procedure is validated by evaluating the optimum hull form via the RANS solver. As mentioned above, dynamic sinkage and trim were utilized for the viscous evaluation. For a more precise evaluation, though, the actual hull wetted surface should have been taken into account. Full scale simulations and subsequent large Reynolds number still pose a significant challenge for numerical experiments since fully resolved simulations are computationally prohibitive, especially if combined into an optimization scheme. Therefore, the urge to enhance the fidelity of optimization schemes with accurate and efficient methods for metamodels to integrate viscous effects into potential flows results with different tools such as ANNs is a challenging task.

It should be noted that the potential of the present approach can be further explored by increasing the number of training data to improve the fidelity of the ANN. Similar ANNs could be used to account for other aspects of ship performance, such as seakeeping and maneuverability characteristics, which, however, are currently estimated satisfactorily by fast codes using strip theory or 3D potential flow methods. Finally, to ensure the significance of the optimization simulations, results shall be validated by towing tank experiments. The construction of scaled models of the parent and the optimum hull forms and their testing in the towing tank would provide more confidence about the efficiency of the optimization scheme.

Author Contributions: Conceptualization, G.J.G.; Supervision, G.J.G.; Methodology, G.J.G., C.B.; Viscous flow calculations, G.P., D.N.; Potential flow calculations, C.B.; ANN development, C.B.; Writing G.J.G., G.P. and C.B. All authors have read and agreed to the published version of the manuscript.

Funding: This research received no external funding.

Institutional Review Board Statement: Not applicable.

Informed Consent Statement: Not applicable.

Data Availability Statement: The data of KCS containership presented in this study are openly available at "Tokyo 2015 A Workshop on CFD in Ship Hydrodynamics": https://t2015.nmri.go.jp/ (accessed on 2 September 2021).

Conflicts of Interest: The authors declare no conflict of interest.

References

1. Grigoropoulos, G.J. Hull Form optimization for hydrodynamic performance. *Mar. Technol.* **2004**, *41*, 167–182.
2. Grigoropoulos, G.J.; Chalkias, D. Hull form Optimization in Calm and Rough Water. *Comput.-Aided Des. J.* **2010**, *42*, 977–984. [CrossRef]
3. Volpi, S.; Gaul, N.; Diez, M.; Song, H.; Iema, U.; Campana, E.; Choi, K.; Stern, F. Development and validation of a dynamic metamodel based on stochastic radial basis functions and uncertainty quantification. *Struct. Multidiscip. Optim.* **2015**, *51*, 347–368. [CrossRef]
4. Sclavounos, P.; Yu, M. Artificial Intelligence machine Learning in marine Hydrodynamics. In Proceedings of the International Conference on Ocean, Offshore and Arctic Engineering, Madrid, Spain, 17–22 June 2018.
5. Chen, X.; Diez, M.; Kandashamy, M.; Zhang, Z.; Campana, E.; Stern, F. High-fielity global optimization of shape design by dimensionality reduction, metamodels and deterministic particle swarm. In *Engineering Optimization*; Taylor & Francis: Abingdon, UK, 2014; pp. 473–494.
6. Harries, S.; Abt, C. CAESES—The HOLISHIP platform for process integration and design optimization. In *A Holistic Approach to Ship Design*; Springer: Berlin/Heidelberg, Germany, 2019; pp. 276–291.
7. Papanikolaou, A.; Flikkema, M.; Harries, S.; Marzi, J.; Le Nena, R.; Torben, S.; Yrjanainen, A. Tools and applications for the holistic ship design. In Proceedings of the 8th Transport Research Arena, Helsinki, Finalnd, 27–30 April 2020.
8. Giannakoglou, K. Design of optimal aerodynamic shapes using stochastic optimization methods and computational intelligence. *Prog. Aerosp. Sci.* **2002**, *38*, 43–76. [CrossRef]
9. Papadakis, G. Development of a Hybrid Compressible Vortex Particle Method and Application to External Problems Including Helicopter Flows, NTUA. 2016. Available online: https://dspace.lib.ntua.gr/xmlui/handle/123456789/40024?locale-attribute=en (accessed on 22 July 2021).
10. Papadakis, G.; Filippas, E.; Ntouras, D.; Belebassakis, K. Effects of viscocity and non-linearity on 3D flapping foil thruster for marine applications. In Proceedings of the OCEANS 2019, Marseille, France, 17–20 June 2019.
11. Grigoropoulos, G.J.; Perdikari, T.; Asouti, V.; Giannakoglou, K. MDO of Hull Forms Using Low-Cost Evolutionary Algorithms. In Proceedings of the NATO-RTO, Advanced Vehicle Technology Panel AVT-173, Sofia, Bulgaria, 19 May 2011.
12. Deb, K.; Pratap, A.; Agarwal, S.; Meyarivan, T. A Fast and Elitist Multiobjective Genetic Algorithm: NSGA-II. *IEEE Trans. Evol. Comput.* **2002**, *6*, 182–198. [CrossRef]
13. CAESES. *CAESES Software Manual*; Friendship Systems GmBH: Potsdam, Germany, 2020.
14. Harries, S.; Abt, C. Parametric Curve Design applying fairness criteria. In Proceedings of the International Workshop on Creating Fair and Shape-Preserving Curves and Surfaces, Potsdam/Berlin, Germany, 14–17 September 1997.
15. Tokyo. A Workshop on CFD in Ship Hydrodynamics. 2015. Available online: https://t2015.nmri.go.jp/ (accessed on 2 September 2021).
16. Sklavounos, P. Computation of Wave Ship Interactions. *Adv. Mar. Hydrodyn. Comput. Mech.* **1995**, 2618. Available online: http://resolver.tudelft.nl/uuid:da01c58e-1285-414b-83da-e31b812b11dc (accessed on 25 October 2021).

17. SWAN2. *User Manual: Ship Flow Simulation in Calm Water and in Waves*; Boston Marine Consulting Inc.: Boston, MA, USA, 2002.
18. Venkatakrishnan, V. On the accuracy of limiters and convergence to steady state solutions. In Proceedings of the 31st Aerospace Sciences Meeting, Reno, NV, USA, 14–17 June 1993.
19. Spalart, P.; Allmaras, S.; Reno, J. A One-Equatlon Turbulence Model for Aerodynamic Flows Boeing Commercial Airplane Group. In Proceedings of the 30th Aerospace Sciences Meeting and Exhibit, Reno, NV, USA, 6–9 January 1992; AIAA: Reston, VA, USA, 1992.
20. Menter, F. Two-equation eddy-viscosity turbulence models for engineering applications. *AIAA J.* **1994**, *32*, 1598–1605. [CrossRef]
21. Langtry, R.; Menter, F. Correlation-Based Transition Modeling for Unstructured Parallelized Computational Fluid Dynamics Codes. *AIAA J.* **2009**, *47*, 2894–2906. [CrossRef]
22. Yue, D.; Wu, S. An improvement to the Kunz preconditioner and numerical investigation of hydrofoil interactions in tandem. *Int. J. Comput. Fluid Dyn.* **2018**, *32*, 167–185. [CrossRef]
23. Taniguchi, T.; Ichinose, Y. Hull form design support tool based on machine learning. In Proceedings of the 19th Conference on Computer and IT Applications in the Maritime Industry (COMPIT), Pontignano, Italy, 17–19 August 2020.
24. Grabowska, K.; Szczuko, P. Ship resistance prediction with Artificial Neural Networks. In Proceedings of the Signal Processing: Algorithms, Architectures, Arrangements, and Applications (SPA), Poznan, Poland, 23–25 September 2015.
25. Margari, V.; Kanellopoulou, A.; Zaraphonitis, G. On the use of Artificial Neural Networks for the calm water resistance prediction of MARAD Systematic Series' hullforms. *Ocean. Eng.* **2018**, *165*, 528–537. [CrossRef]
26. Cepowski, T. The prediction of ship added resistance at the preliminary design stage by the use of an artificial neural network. *Ocean. Eng.* **2020**, *195*, 106657. [CrossRef]
27. Tarelko, W.; Rudzki, K. Applying artificial neural networks for modelling ship speed and fuel consumption. *Neural Comput. Appl.* **2020**, *32*, 17379–17395. [CrossRef]
28. Abramowski, T. Application of artificial neural networks to assessment of ship manoeuvrability qualities. *Pol. Marit. Res.* **2008**, *15*, 15–21. [CrossRef]
29. Martins, P.; Lobo, V. Estimating Maneuvering and Seakeeping Characteristics with Neural Networks. In Proceedings of the OCEANS 2007—Europe, Aberdeen, UK, 29 September–4 October 2007.

Article

Multi-Objective Optimization of Jet Pump Based on RBF Neural Network Model

Kai Xu [1], Gang Wang [2,*], Luyao Zhang [3], Liquan Wang [1], Feihong Yun [1], Wenhao Sun [1], Xiangyu Wang [1] and Xi Chen [4,5]

1. College of Mechanical and Electrical Engineering, Harbin Engineering University, Harbin 150001, China; xukai0705@163.com (K.X.); wangliquan@hrbeu.edu.cn (L.W.); yunfeihong@hrbeu.edu.cn (F.Y.); swh1053749521@163.com (W.S.); wangxiangyu325@126.com (X.W.)
2. College of Shipbuilding Engineering, Harbin Engineering University, Harbin 150001, China
3. Shanghai Marine Diesel Engine Research Institute, Shanghai 200000, China; freedom__me@163.com
4. College of Information and Communication Engineering, Harbin Engineering University, Harbin 150001, China; chenxi_1113652@hrbeu.edu.cn
5. College of Mechanical and Electrical Engineering, Heilongjiang Institute of Technology, Harbin 150050, China
* Correspondence: wanggang@hrbeu.edu.cn

Citation: Xu, K.; Wang, G.; Zhang, L.; Wang, L.; Yun, F.; Sun, W.; Wang, X.; Chen, X. Multi-Objective Optimization of Jet Pump Based on RBF Neural Network Model. *J. Mar. Sci. Eng.* **2021**, *9*, 236. https://doi.org/10.3390/jmse9020236

Academic Editor: Md Jahir Rizvi

Received: 21 December 2020
Accepted: 18 February 2021
Published: 23 February 2021

Publisher's Note: MDPI stays neutral with regard to jurisdictional claims in published maps and institutional affiliations.

Copyright: © 2021 by the authors. Licensee MDPI, Basel, Switzerland. This article is an open access article distributed under the terms and conditions of the Creative Commons Attribution (CC BY) license (https://creativecommons.org/licenses/by/4.0/).

Abstract: In this study, an annular jet pump optimization method is proposed based on an RBF (Radial Basis Function) neural network model and NSGA-II (Non-Dominated Sorting Genetic Algorithm) optimization algorithm to improve the hydraulic performance of the annular jet pump applied in submarine trenching and dredging. Suction angle, diffusion angle, area ratio and flow ratio were selected as design variables. The computational fluid dynamics (CFD) model was used for numerical simulation to obtain the corresponding performance, and an accurate RBF neural network approximate model was established. Finally, the NSGA-II algorithm was selected to carry out multi-objective optimization and obtain the optimal design variable combination. The results show that the determination coefficient R^2 of the two objective functions (jet pump efficiency and head ratio) of the approximate model of the RBF neural network were greater than 0.97. Compared with the original model, the optimized model's suction angle increased, and the diffusion angle, flow ratio and area ratio decreased. In terms of performance, the head ratio increased by 30.46% after the optimization of the jet pump, and efficiency increased slightly. The proposed jet pump performance optimization method provides a reference for improving the performance of other pumps.

Keywords: optimization; annular jet pump; RBF neural network; NSGA-II optimization algorithm

1. Introduction

The annular jet pump utilizes a high-speed working fluid to entrain the low-speed fluid and realize the pressurization process. This has the advantage of no moving parts in the suction channel and a simple, reliable, and easily accessible structure that is especially applicable for pumping fluids containing a great quantity of solid particles (mineral, live fish, gravel, etc.). In addition, with the rapid development of offshore oil and natural gas resource exploration, underwater trenchers have become a research focus in the ocean engineering. The jet pump is the key technology of underwater trencher design [1–3]. With the development of engineering, higher requirements are being put forward for the performance of jet pumps in various aspects. However, the jet pump's complex flow field and relatively poor performance restrict its development. Therefore, research into the internal mechanisms, optimization methods and performance improvements of jet pumps are important.

Shimizu [4] carried out experimental research on different annular jet pumps and the cavitation performance of annular jet pumps. Kwon [5] researched the suction angle influence on the flow field characteristic via 2D model simulations and found the efficiency calculated by the RNG k-ε model had a smaller error according to the experimental data. Deng [6] constructed a solid–liquid two-phase flow equation using the RNG k-ε model and concluded that changing the structure of a diffuser to reduce the reverse velocity vector of solid–liquid two-phase flow is a useful method for improving a jet pump's comprehensive performance. Yang [7] researched the influence of the jet pump's nozzle shape on critical back pressure and the entrainment rate, concluding that a non-circular nozzle could improve the performance of jet pump. Lyu [8] analyzed a two-factor reciprocal action and single-factor effect on performance along with internal field characteristics of the annular jet pump using to the design of experiments (DOE) method. Deng [9] designed the improved annular jet pump, which had a performance increase of about 10%. Wang [10] proposed a streamlined jet pump to increase pumping efficiency. The results show that the streamlined annular jet pump efficiency could be increased by up to 1.2%. Gao [11] investigated the suction half-angle influence on flow-rate ratio, pressure ratio and gas pumping performance, and the simulated results show that the suction half-angle had a major impact. Xu [12] studied a jet pump's internal flow field using a large eddy simulation, and systematically analyzed the field with instantaneous and time average aspects. It could be concluded that the potential core increased linearly with increases in the flow ratio; however, the instantaneous velocity distribution was more complicated and unordered. Zou [13] researched the installation mode's influence on performance through simulation technology. They used two models of vertical installation and horizontal installation based on three turbulence models and simulated jet pump flow field characteristics. The results showed that the jet pump was more efficiently installed vertically. Elger [14] studied the effect of the annular jet pump area ratio on the reflux region and introduced the dimensionless parameter, momentum ratio, to study the generation and disappearance of reflux in an annular jet pump.

These studies mostly used computational fluid dynamics (CFD) techniques and focused on improving the efficiency, ignoring how to improve the efficiency and head ratio of a jet pump at the same time. Multi objective optimization methods were classified into two types. The first type converted the multi-objective function to a single-objective function by using targets, utilities, preferences, or weights. These methods demanded a priori selection of targets, utilities, or weights for each optimization objective [15,16]. Due to the lack of a rigorous weight selection method in practical problems, it was very hard to decide which weight factors to use. Further, the weighted sum method could not get Pareto points in nonconvex regions [17]. The second type could determine some discrete points as an approximate Pareto optimal frontier for designers who did not presuppose preferences. In this type, an evolutionary algorithm was the most widely and successfully used method [18–20]. These methods generated lots of Pareto points for designers. Compared with the jet pump, there has been multi-objective research into centrifugal pumps.

In the wake of developments in CFD technology and optimization algorithms, many kinds of research into centrifugal pump multi-objective optimization have determined its optimal performance by constructing an approximate model. The optimization method based on an approximate model and intelligent algorithm has a low calculation cost and can research properties of the response function more comprehensively [21]. Barthelemy [22] reviewed the application of approximate models in terms of structural optimization. Approximate models have been applied in multi-objective optimization, such as the response surface method [23,24], artificial neural networks [25,26] and radial basis function [27,28]. Zhang [29] put forward a centrifugal pump optimization method based on the Kriging model, the optimization result of which was in good agreement with the experimental data. Safikhani [30] put forward a multi-objective optimization method of a centrifugal pump using a genetic algorithm combined with an approximate model to get the Pareto optimal solution of centrifugal pump efficiency and the head ratio. Wang [31] carried out a numerical simulation to obtain performance data and used a neural network to construct a prediction model using the structural parameters of head and efficiency performance. It was used as the fitness evaluation model from a particle swarm optimization algorithm.

The optimal value was determined in the sample space to obtain the Pareto solution. Zhao [32] used the NSGA-II algorithm based on the back-propagating neural network model to obtain the pareto optimal front of two conflicting objectives of low specific speed centrifugal pump efficiency and cavitation safety margin.

In this paper, an annular jet pump multi-objective optimization based on CFD simulation, an RBF neural network model and the NSGA-II optimization algorithm improves the efficiency and the head ratio of jet pump, realizing the maximum optimization of jet pump performance.

2. Working Principle and Structure of Jet Pump

The schematic drawing of the annular jet pump is shown in Figure 1. The working principle of the jet pump is the Venturi effect. The primary flow is injected into the throat at high speed from the power source along the pressure pipe. The air is taken away, forming a vacuum near the nozzle due to the viscous interaction. Under the action of external atmospheric pressure, the secondary flow is sucked up through the suction chamber and blended with the high-speed primary flow. When the primary flow transmits a lot of energy to the secondary flow, the primary flow decelerates and the secondary flow speeds up. The mixing process between the two flows is mostly completed as they achieve a uniform velocity in the end of the throat. In the diffuser, the velocity gradually slows down, with the pressure close to the ambient pressure.

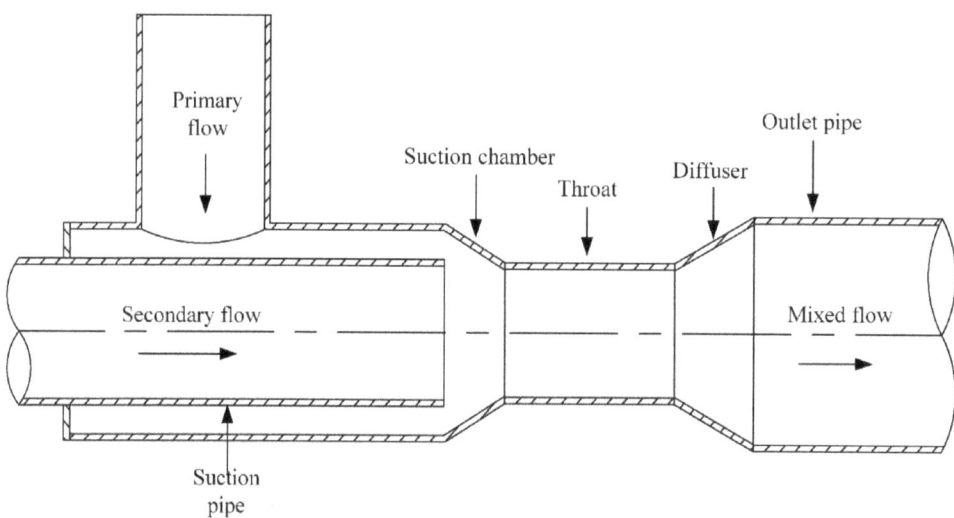

Figure 1. Structural schematic drawing of annular jet pump.

The efficiency and head are decided by the turbulent effective mixing of the two flows. Because the fluid is limited by space, the structure parameters of the jet pump have a great influence on the turbulent mixing effect. Compared to other jet pumps, the annular jet pump efficiency is improved because the energy and momentum exchange of fluid in the short throat is relatively faster and the friction loss in the throat is minimized.

Dimensionless parameters, pressure ratio h and efficiency η are introduced as the objectives. The equations [4] are as below:

$$h = \frac{\Delta p_o}{\Delta p_w} = \frac{\left(p_o + \gamma_o \frac{V_o^2}{2g} + \gamma_o z_o\right) - \left(p_s + \gamma_s \frac{V_s^2}{2g} + \gamma_s z_s\right)}{\left(p_w + \gamma_w \frac{V_w^2}{2g} + \gamma_w z_w\right) - \left(p_o + \gamma_o \frac{V_o^2}{2g} + \gamma_o z_o\right)} \tag{1}$$

$$\eta = \frac{Q_s}{Q_w} \cdot \frac{\Delta p_o}{\Delta p_w - \Delta p_o} = q\frac{h}{1-h} \qquad (2)$$

where

$$q = \frac{Q_s}{Q_w} \qquad (3)$$

In above equations, q is the flow ratio, p is the static pressure, Q is the volume flow rate, γ is the unit weight, g is the gravitational acceleration, z is the positional water head, V is the sectional average velocity and footnotes w, s and o represent the primary flow at the inlet, the secondary flow at the inlet and the mixture at the outlet, respectively.

In addition, the area ratio m is also introduced:

$$m = \frac{A_s}{A_w} \qquad (4)$$

where A is the nozzle outlet area.

3. Modeling and Numerical Simulation

3.1. Modeling

The main components of annular jet pump include a suction pipe, suction chamber, throat and diffuser, so these parts are selected as the calculation domain. The original model parameters [4] are $w = 4$ mm, $m = 2.27$, $L_t = 179$ mm, $r = 21.5$ mm, $r_0 = 27.5$ mm, $r_t = 38$ mm, $\alpha = 18°$, $\beta = 5.8°$. This computational domain is shown in Figure 2. A 2D axisymmetric model and a 3D model have been applied to the jet pump simulation and the 2D simulation and 3D results showed a reasonable agreement in terms of efficiency and head ratio relative to the experimental results [33]. Suction angle α, diffusion angle β and area ratio m, plus the flow ratio q, are among the structural parameters chosen as the design parameters (Figure 2).

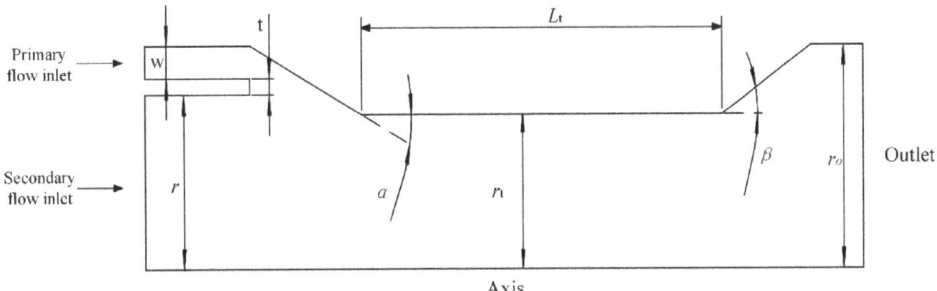

Figure 2. The computational domain.

3.2. CFD modeling and Verification

3.2.1. CFD Model

The mixing process of two fluids in the jet pump is very complex. In order to facilitate the study, the assumptions are as below:
(1) the fluid medium is steady and incompressible;
(2) there is no heat transfer between fluid and the environment;
(3) the influence of the jet pump's wall roughness is neglected;
(4) the buoyancy influence is neglected.

The continuity and momentum equations are listed below:

$$\frac{\partial(\rho\gamma)}{\partial x_i} = 0 \qquad (5)$$

$$\frac{\partial(\rho u_j u_i)}{\partial x_j} = \frac{\partial}{\partial x_j}\left[\mu\frac{\partial u_i}{\partial x_j} - \rho\overline{u_i u_j}\right] - \frac{\partial p}{\partial x_i} \qquad (6)$$

where Reynolds stresses are

$$\frac{\partial(\rho u_j u_i)}{\partial x_j} = \frac{\partial}{\partial x_j}\left[\mu\frac{\partial u_i}{\partial x_j} - \rho\overline{u_i u_j}\right] - \frac{\partial p}{\partial x_i} \qquad (7)$$

In these equations, u_i is the velocity component, x_i is the space coordinate, δ is boundary layer thickness, μ is dynamic viscosity, μ_t is turbulent viscosity, and k is turbulent dynamic energy.

The shear layer between the primary flow and the secondary flow leads to a turbulent flow in the jet pump. Thus, choosing the proper CFD simulation scheme to guarantee the precision of the simulation results is very necessary. By comparing several CFD simulation schemes and experimental data, the standard wall function and realizable k-ε model can accurately simulate the characteristics and calculate the efficiency [33].

The realizable k-ε model put forward by Shih [34] is different from the standard k-ε model in two important aspect: (1) the realizable k-ε model adds an additional turbulent viscosity calculation formula; and (2) a modified transport equation for the dissipation rate, ε, has been deduced. Comparing the realizable k-ε model with the standard k-ε model, the modified transport equation achieves great progress in simulating flow field characteristics such as rotation, vortex and strong streamline curvature. The equation is as follows:

$$\frac{\partial(\rho k)}{\partial t} + \frac{\partial(\rho k u_j)}{\partial x_j} = \frac{\partial}{\partial x_j}\left[\left(\mu + \frac{\partial u_t}{\sigma_k}\right)\frac{\partial(k)}{\partial x_j}\right] + G_k + G_b - \rho\varepsilon - Y_M + S_k \qquad (8)$$

$$\frac{\partial(\rho\varepsilon)}{\partial t} + \frac{\partial(\rho\varepsilon u_j)}{\partial x_j} = \frac{\partial}{\partial x_j}\left[\left(\mu + \frac{\partial u_t}{\sigma_\varepsilon}\right)\frac{\partial \varepsilon}{\partial x_j}\right] + \rho C_1 S\varepsilon - \rho C_2 \frac{\varepsilon^2}{k+\sqrt{v\varepsilon}} + C_{1\varepsilon}\frac{\varepsilon}{k}C_{3\varepsilon}G_b + S_\varepsilon \qquad (9)$$

where

$$C_1 = \max\left[0.43, \frac{\eta}{\eta+5}\right], \eta = S\frac{k}{\varepsilon}, S = \sqrt{2S_{ij}S_{ij}} \qquad (10)$$

where S represents strain rate magnitude; G_k and G_b are the generation of turbulent kinetic energy due to the mean velocity gradients and buoyancy, respectively; Y_m is the contribution of the fluctuating dilatation to the overall dissipation rate in compressible turbulence; C_2 and C_ε represent constants; σ_ε and σ_k represent the turbulent Prandtl numbers of ε and k, respectively; and S_k and S_ε represent customized source terms.

The eddy viscosity, μ_t, is

$$\mu_t = \rho C_\mu \frac{k^2}{\varepsilon} \qquad (11)$$

C_μ is a variable in the realizable k-ε model as follows:

$$C_\mu = \frac{1}{A_0 + A_S \frac{kU^*}{\varepsilon}} \qquad (12)$$

And

$$\begin{aligned}
U^* &\equiv \sqrt{S_{ij}S_{ij} + \widetilde{\Omega}_{ij}\widetilde{\Omega}_{ij}} \\
\widetilde{\Omega}_{ij} &= \Omega_{ij} - 2\varepsilon_{ijk}\omega_k \\
\Omega_{ij} &= \overline{\Omega}_{ij} - \varepsilon_{ijk}\omega_k \\
A_0 &= 4.04, A_s = \sqrt{6}\cos\varphi \\
\varphi &= \tfrac{1}{3}\cos^{-1}\left(\sqrt{6}W\right), W = \frac{S_{ij}S_{jk}S_{ki}}{\widetilde{S}^3}, \widetilde{S} = \sqrt{S_{ij}S_{ij}}, S_{ij} = \tfrac{1}{2}\left(\frac{\partial u_j}{\partial x_i} + \frac{\partial u_i}{\partial x_j}\right)
\end{aligned} \qquad (13)$$

where $\overline{\Omega}_{ij}$ is the mean rate-of-rotation tensor, and C_μ is a function of the mean strain and rotation rates, the angular velocity of the system rotation and the turbulence fields.

The fluid medium used in this simulation is water. The unstructured grids generated by MESH are two-dimensional mixed elements. The element number is about 65,000. The inlets of the two flows are set as velocity inlet boundary conditions, with outflow used for the outlet. Simulations of the CFD program are carried out by the ANSYS Fluent R18.0 using the finite volume method. The convection terms spatial discretization method is set as the second order upwind scheme. The pressure-velocity coupling scheme is SIMPLE.

The number of complete convergence iterations is 3000. In order to guarantee the reliability of the simulation results, the working conditions in the CFD process are regulated to ensure the calculation results have strict consistency. The residuals for momentum equations drop to 10^{-6}, and the continuity equation drops to 10^{-5}. Boundary layer grids are quad and the y+ is below 300, with an average around 200.

3.2.2. Verification of CFD simulation

The feasibility of the selected CFD model and algorithm applied in further research of the annular jet pump can be provided through a comparison of the calculation data with the experimental results [4]. To verify the mesh independence based on experimental case 1 ($q = 0.3$), the mesh is set to coarse (38,174), medium (60,823) and fine (78,551). The calculation results are shown in Table 1. The maximum error is 1.28% and the mesh independence is verified. In order to give consideration to the calculation speed and accuracy, the mesh is set to medium in the subsequent simulation.

Table 1. The results with different numbers of elements.

Number of Elements	h	e
coarse	0.5793	0.0924
medium	0.5798	0.0926
fine	0.5826	0.0936

The comparison is as shown in Figure 3. It can be concluded from Figure 3 that the error between the numerical data and the experimental results is very small. The maximum error of efficiency is 0.039, and the maximum error of the head ratio is 0.017. The CFD scheme selected in this paper demonstrates accuracy and reliability in calculating the performance of an annular jet pump, and can be used in subsequent simulations.

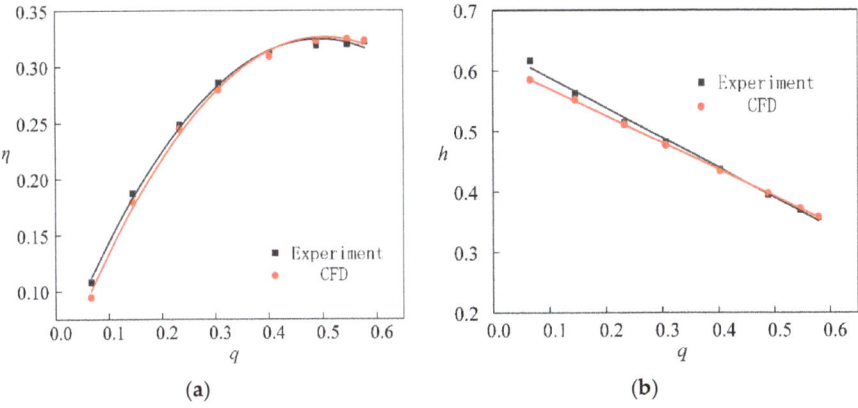

Figure 3. Experiment and simulation data. (**a**) Efficiency; (**b**) head ratio.

4. Hybrid Algorithm and Optimization Process

4.1. Optimization Algorithm Design

The annular jet pump optimization method proposed in this paper consists of several algorithms. A flow chart is shown in Figure 4. This method uses the optimum space filling (OSF) [35] experimental design method to get the global optimal point in the nonlinear optimization space, and adopts a non-dominated sorting genetic algorithm (NSGA-II) to optimize the local detail based on the RBF neural network model. The basic steps are as follows:

1. Given the space of four design variables, 80 uniformly distributed sample points are generated by the OSF method.
2. According to the sampling point, the CFD software Fluent is used to simulate the annular jet pump with different structural parameters. Through numerical simulation, the efficiency and head ratio of the annular jet pump are calculated.
3. The neural network model is constructed via the RBF function. The structural parameters of the annular jet pump obtained at the sampling point in step 1 are the input variables, and the efficiency η and head ratio h of the jet pump obtained in step 2 are the output variables.
4. The RBF neural network model constructed in step 3 is verified, then the predicted values and simulated values are compared. If the error between the two sets of data is very small, move on to step 5; otherwise, return to step 3 and continue to update the RBF neural network model.
5. Based on the RBF neural network model, the NSGA-II optimization algorithm is used to get the optimal solution of the structural parameters of the annular jet pump.
6. According to the design parameters of the optimal solution, an optimization model is generated. The CFD simulation and optimization results of the optimization model are verified with each other.

4.2. DOE Method

The premise of determining optimization method reliability in this paper is to guarantee the reliability of the approximate model. DOE plays a decisive role in the quality of the approximate model.

Compared with traditional DOE methods, OSF is used in this paper to generate sample points in the design space. OSF is an improvement of the Latin hypercube sampling algorithm (LHS) that develops an efficient global optimal search algorithm, the enhanced stochastic evolutionary algorithm (ESE). The criteria for evaluating optimality is based on an entropy criterion, CL_2 criterion, and a Φ_p criterion. The proposed algorithm is much more efficient compared with existing techniques in terms of the computation time, number of exchanges needed for generating new designs, and the achieved optimality criteria [35]. OSF improves the randomness of LHS and makes all sample points evenly distributed in the design space as much as possible to ensure good space-filling and uniformity. Therefore, OSF can better reflect the mapping relationship between factors and responses, which makes the fitting of factors and responses more accurate with the minimum sample points.

Figure 5a shows that the sampling points generated by LHS are randomly generated, and Figure 5b shows that the sampling points generated by OSF are evenly distributed. It can be seen that LHS lacks sample points in the upper right corner of the design space, which do not reflect the relationship between the factor and the response in this region. The use of OSF makes the sample points in the whole design space more uniform.

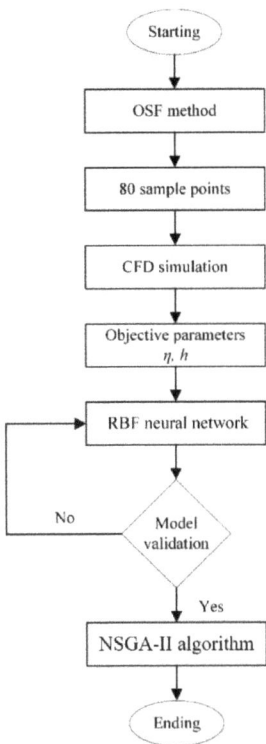

Figure 4. The optimization flow.

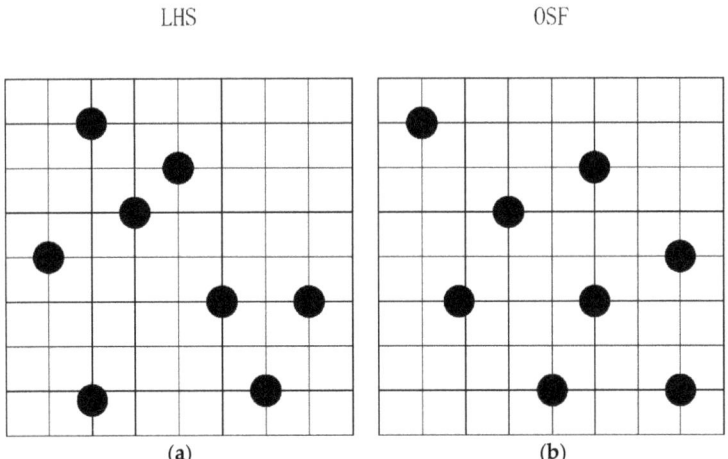

Figure 5. Design of experiments (DOE) method. (**a**) Latin hypercube sampling algorithm (LHS); (**b**) optimum space filling algorithm (OSF).

4.3. Approximate Model

The RBF model has a strong ability to approximate complex nonlinear functions. Because of its fast learning speed without a mathematical hypothesis or black box characteristics, the model has been extensively applied in function approximation, pattern recognition, financial systems, signal processing, power systems, expert systems, military systems, image processing and computer visions, medical control, and optimization [36,37]. In this paper, an RBF neural network is used to map the approximate function between the parameters of the jet pump and the objectives.

As illustrated in Figure 6, there are usually three layers in this neural network. On the left is the input layer with four neurons, which represent four design parameters. In the middle is the hidden layer with nine neurons; this contains a non-linear radial basis activation function. On the right is an output layer with two neurons that represent two objective parameters.

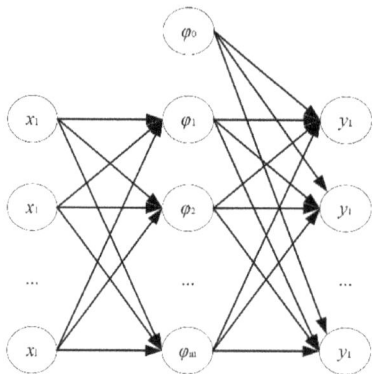

Figure 6. RBF model structure.

The radial basis function commonly applied in an RBF neural network is the Gaussian function, so the activation function of the RBF neural network can be described as

$$R(x_p - c_i) = \exp\left(-\frac{1}{2\sigma^2}\|x_p - c_i\|^2\right) \quad (14)$$

where $\|x_p - c_i\|$ is the Euclidean norm, σ is the variance of the Gaussian function and c_i is the center of the Gaussian function.

According to the RBF neural network structure, the output of the RBF neural network is

$$y_i = \sum_{i=1}^{h} \omega_{ij} \exp\left(-\frac{1}{2\sigma^2}\|x_p - c_i\|^2\right) \quad j = 1, \cdots, n \quad (15)$$

where $x = (x_1^p, x_2^p, \ldots, x_n^p)$ is a p-th input sample, $p = 1, 2, \ldots, P$; P is the total number of samples; ω_{ij} is the connection weight from hidden layer to output layer; $i = 1, 2, \ldots, h$; h is the number of hidden layer nodes; and y_i is the actual output of the j-th output node of the network corresponding to the input sample.

4.4. Establishment of Sample Database

Table A1 shows the design parameters of some sample points generated by OSF, as well as the corresponding jet pump efficiency and head ratio. The number of training samples should be at least 10 times the number of input variables [29]. Together with the number of test samples, a total of 80 samples are used. Figure 7 shows the three-dimensional distribution of 80 samples. It can be seen that the design parameters of sample points are evenly distributed in the parameter space. The best sample point of efficiency is No. 52 (q = 0.4354, m = 1.84, α = 27.74°, β = 4.76°), and the corresponding jet pump efficiency is 0.332. The best sample point for head ratio is No. 27 (q = 0.3728, m = 5.292, α = 34.98° and β = 5.292°), with a corresponding jet pump head ratio of 0.4329.

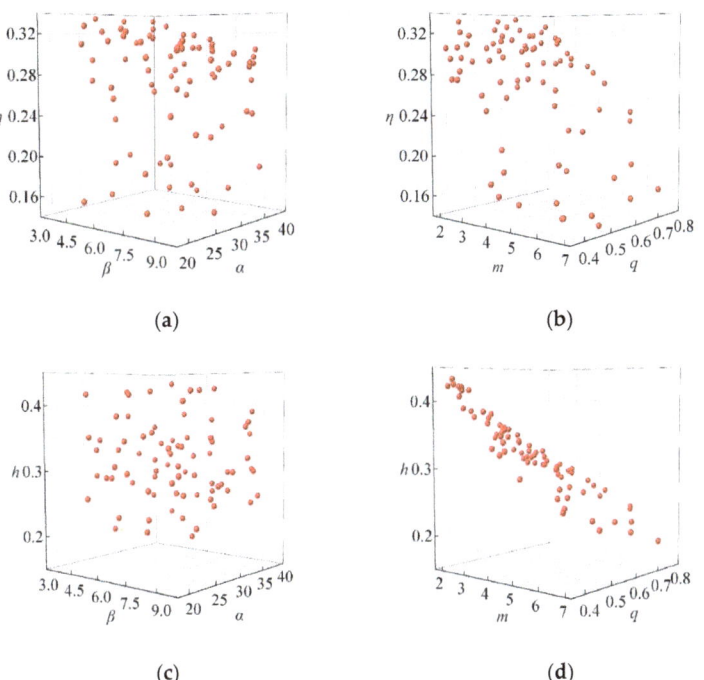

Figure 7. Performance of different sampling points. (**a**) η corresponding to α and β; (**b**) η corresponding to m and q; (**c**) h corresponding to α and β; (**d**) h corresponding to m and q.

4.5. Multi Objective Optimization Algorithm

The jet pump objective functions are expressed as

$$\begin{cases} \text{Maximize} & \eta = f(\alpha, \beta, m, q) \\ \text{Maximize} & h = f(\alpha, \beta, m, q) \end{cases} \quad (16)$$

which are subject to

$$\begin{cases} 1.68 \leq m \leq 7.18 \\ 0.35 \leq q \leq 0.8 \\ 18° \leq \alpha \leq 60° \\ 4° \leq \beta \leq 10° \end{cases} \quad (17)$$

Traditional optimization algorithms contain a common weighting algorithm and constraint method. These methods convert the multi-objective function to a single-objective function by giving weight, then optimizing the single-objective function. This kind of algorithm is too ideal, and the optimization result is not good. Intelligent algorithms include common particle swarm optimizations, annealing algorithms, genetic algorithms and so on. These algorithms can reflect the essence of multi-objective optimization problems, and have been widely applied in recent years.

The NSGA-II [38] algorithm is used to optimize the jet pump in this paper. Compared with the simple GA (Genetic Algorithm), NSGA-II classifies the population according to the dominant relationship between individuals, giving higher-rated individuals a greater opportunity to pass on to the next generation. Compared with NSGA, NSGA-II has the following modifications: (1) A fast non-dominated sorting method to reduce the computational complexity of the algorithm. (2) A crowded comparison operator to replace the fitness sharing strategy that needs to specify the sharing radius. This is used as the criterion when comparing individuals at the same level after quick sorting, so that individuals of the Pareto solution set are extended to the whole Pareto domain and evenly distributed to maintain the diversity of the population. (3) An elite strategy to expand the sampling space. The parent population and its offspring are combined in the selection of the next generation's population. This strategy helps retain the superior individuals from the parent generation in the subsequent generation. The best individuals will therefore not be lost, and the optimization level will be rapidly improved via the hierarchical storage of all the individuals in the population. This principle is shown in Figure 8.

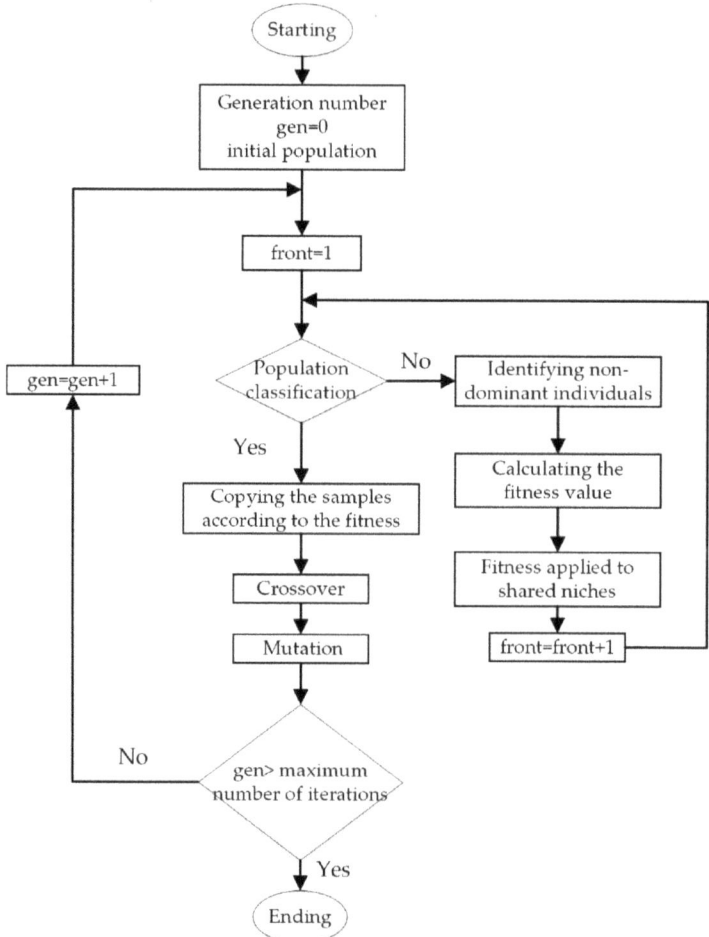

Figure 8. NSGA-II algorithm schematic.

5. Results and Discussion

5.1. Error Analysis of RBF Model

The design variables of 80 sample points generated by the OSF and the corresponding CFD simulation results are used as inputs and outputs to establish an RBF neural network approximate model. Specifically, the parameters α, β, m and q of the jet pump are input variables, and the corresponding efficiency η and head ratio h are the output variables. In this paper, Isight is used to train the RBF neural network model. Type of Basis Function is set as radial. The remaining parameters use default values. We use cross validation, with 65 sample points used as the training set and 15 sample points as the test set. Figure 9 shows the degree of fitting between the predicted value of the objective function and the simulated value of CFD. On the diagonal, the predicted value is equal to the CFD value. It can be seen that all the points distribute around the diagonal.

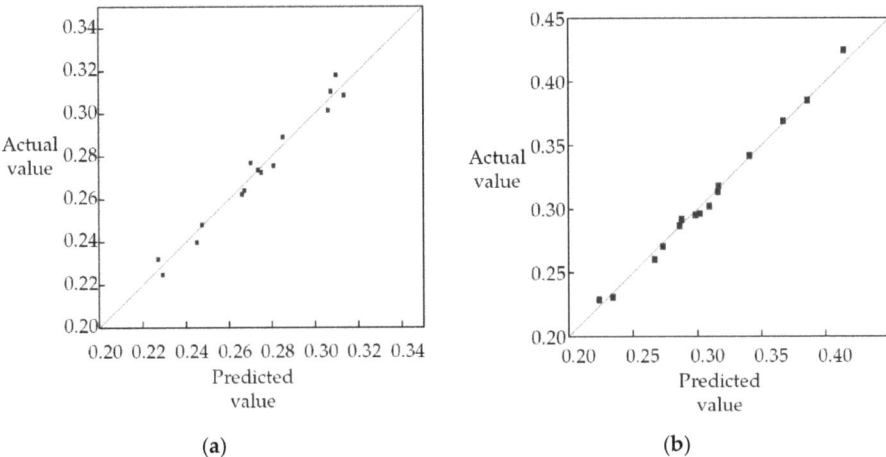

Figure 9. Error analysis. (**a**) Efficiency; (**b**) head ratio.

The average error of efficiency and head ratio is 4.8% and 1.7%, respectively. The maximum error of efficiency and head ratio are 8.6% and 6.1%, respectively. For further verification of the RBF model's accuracy, a statistical measure of objective function, correlation coefficient R^2, is introduced to estimate the approximate degree [39]. An R^2 statistic that close to 1 indicates that a large proportion of the variability in the response is due to the regression. The correlation coefficient is defined as follows:

$$R^2 = 1 - \frac{\sum\limits_{i=1}^{n}(y_i - Y)^2}{\sum\limits_{i=1}^{n}(y_i - \overline{y})^2} \qquad (18)$$

After calculation, the R^2 of efficiency and head ratio are 0.97033 and 0.99286, respectively, which meets the engineering requirement.

5.2. Optimization Results

In this paper, the crossover probability is set as 0.9, the number of generations is 20 and the initial population size is 36. After 721 genetic iterations, global optimization is carried out. Figure 10 shows the changing efficiency and head ratio trends for the annular jet pump, with an increase in iteration numbers during the optimization process. According to the trends observed during the optimization process for the two optimization objectives, it can be seen that the fluctuation range of efficiency has a small iteration number, while the fluctuation range of head ratio has a large iteration number.

Using the NSGA-II algorithm, the non-dominated solutions are obtained and the Pareto optimal frontier is shown after 721 iteration generations (Figure 11). This shows that the Pareto optimal solutions (red dots in the picture) are densely, continuously and smoothly distributed on a convex curve, indicating that NSGA-II has a strong ability to approximate Pareto solutions in the sample space.

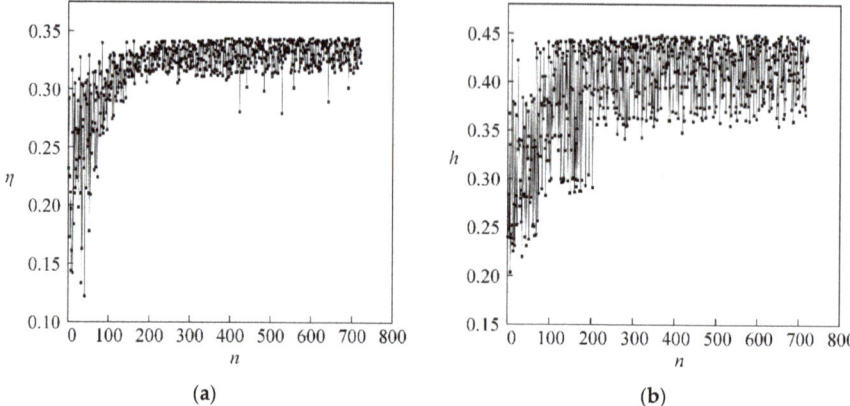

Figure 10. Trends of pump efficiency and head ratio during the searching process. (**a**) Efficiency; (**b**) head ratio.

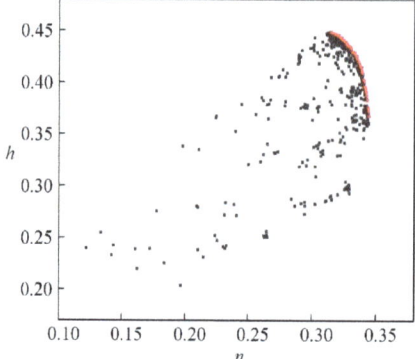

Figure 11. Optimization results of the NSGA-II algorithm.

It can be seen from the red Pareto optimal solutions in Figure 11 that the head ratio has a monotone decrease with increased efficiency in the Pareto optimal frontier, and there are some conflicts between the two objectives in this paper. The two objectives either cannot be compared, or may not be the optimal solutions for all objectives. The mapping primal points of these non-dominated solutions in the decision space, called Pareto solutions, are non-inferior.

Table 2 shows the comparison of structural parameters between the optimization scheme and the original scheme, and Table 3 shows a comparison of performance optimization objectives between the optimization scheme and the original scheme.

Table 2. Comparison of design parameters between original scheme and optimized scheme.

	$\alpha(°)$	$\beta(°)$	m	q
Original scheme	18	5.8	2.27	0.5789
Optimized scheme	25.02	5.1768	1.6823	0.3702

Table 3. Comparison of optimization objectives between the original scheme and optimized scheme.

	η	h
Original scheme	0.3325	0.3648
Optimized scheme	0.3558	0.4758

According to the two tables of parameters, it can be seen that, in terms of structure, the suction angle of the optimization model is increased, the diffusion angle is decreased and the flow ratio and area ratio are decreased, while in terms of performance, the head ratio of the optimization model is increased by 30.46%, and the efficiency is increased by 7%.

5.3. Analysis of Internal Flow Field Optimization of Jet Pump

Figure 12 shows a pressure field comparison between the original model and the optimized model. The pressure distribution in the suction chamber and throat of the optimized model is similar to that of the original model, but the high pressure region of the optimized model is widely distributed in the diffuser and outlet; that is, kinetic energy is converted into pressure energy earlier in the diffusion. In general, the inlet pressure is similar between the two models. The pressure gradient along the axial direction of the jet pump is more obvious and the outlet pressure is higher in the optimized model, which helps to improve the head ratio of the jet pump.

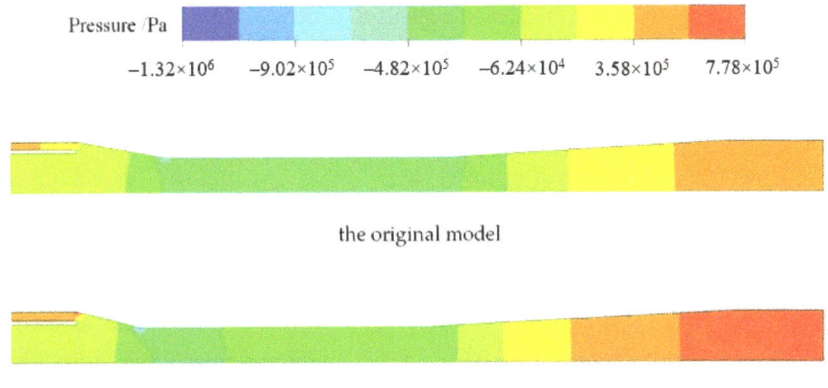

Figure 12. Pressure distribution of annular jet pump.

6. Conclusions

In this study, an annular jet pump optimization method was proposed based on a CFD simulation, RBF neural network model and NSGA-II optimization algorithm. The CFD simulation scheme was created and verified with experiment data, ensuring the reliability of the simulation results. The RBF neural network model was established by the OSF and verified by the CFD simulation results. In the end, the parameters of the annular jet pump were optimized using the NSGA-II algorithm.

According to the results of above, the following conclusions can be presented:

(1) An RBF neural network approximation model was constructed to analyze the efficiency and head ratio of an annular jet pump. The determination coefficients R^2 of the two objectives were greater than 0.97, and the accuracy of the model is reliable.
(2) The NSGA-II algorithm was used to optimize the annular jet pump. In terms of structure, the suction angle increased, the diffusion angle decreased and the flow ratio and area ratio decreased compared with the original model, while in terms of performance, the head ratio increased by 30.46% and efficiency is increased by 7%.
(3) The optimization method based on the RBF neural network model and the NSGA-II optimization algorithm was able obtain the optimal design parameter combination in the global design space of an annular jet pump, which can be applied to other kinds of pumps. However, due to the errors of the CFD and RBF models, this method needs the support of experimental data.

Author Contributions: Formal analysis, K.X. and G.W.; Methodology, K.X.; Writing—original draft preparation, K.X. and G.W.; Writing—review and editing, L.Z., L.W., F.Y., W.S. and X.C.; Validation, L.W. and X.W.; Supervision L.W.; Funding acquisition, G.W. and F.Y. All authors have read and agreed to the published version of the manuscript.

Funding: This research was funded by National Key Research and Development Project (Grant No. 2018YFF01012900), National Natural Science Foundation of China (Grant No. 51779059, 52001116, 52001089, 51779064), National Natural Science Foundation of Heilongjiang Province (Grant No. YQ2020E028, YQ2020E033), China Postdoctoral Science Foundation (Grant No. 2020M670889, 2018M630343), Fundamental Research Funds for the Central Universities (Grant No. 3072020CF0702, 3072020CFT0105, 3072020CFT0704), and the School Land Integration Development Project of Yantai (Grant No. 2019XDRHXMPT29).

Institutional Review Board Statement: Not applicable.

Informed Consent Statement: Not applicable.

Data Availability Statement: Not applicable.

Acknowledgments: The authors gratefully acknowledge the financial support from National Key Research and Development Project (Grant No. 2018YFF01012900), National Natural Science Foundation of China (Grant No. 51779059, 52001116, 52001089, 51779064), National Natural Science Foundation of Heilongjiang Province (Grant No. YQ2020E028, YQ2020E033), China Postdoctoral Science Foundation (Grant No. 2020M670889, 2018M630343), Fundamental Research Funds for the Central Universities (Grant No. 3072020CF0702, 3072020CFT0105, 3072020CFT0704), and the School Land Integration Development Project of Yantai (Grant No. 2019XDRHXMPT29).

Conflicts of Interest: The authors declare no conflict of interest.

Appendix A

Table A1. Design parameters and efficiency of DOE sample points.

q	m	$\alpha(°)$	$\beta(°)$
0.5835	4.256	20.22	9.164
0.7259	1.789	30.54	7.646
0.538	2.454	18.84	8.178
0.4013	4.078	20.5	8.936
0.3899	4.876	36.1	8.026
0.6462	3.019	31.92	4.228
0.4297	1.959	21.34	6.734
0.3842	4.349	25.24	4.988
0.3671	2.649	26.08	8.556
0.3614	2.491	34.44	7.722
0.743	2.185	36.66	4.608
0.7089	3.688	24.12	4.076
0.6861	1.883	37.5	9.012
0.4639	2.824	29.98	4.152
0.669	2.872	40	8.406
0.6291	6.163	35.54	5.14
0.7886	5.383	23.02	5.67
0.6747	1.723	30.82	5.444
0.7316	2.919	20.78	5.974
0.35	4.547	27.18	7.266
0.6918	1.985	19.68	7.114
0.612	3.995	28.58	6.126
0.7544	2.529	29.14	5.898
0.7658	3.835	27.46	7.494
0.7829	3.915	36.94	7.038
0.6348	1.812	38.32	6.582
0.3728	2.28	34.98	5.292
0.5551	2.155	33.04	7.95

Table A1. Cont.

q	m	α(°)	β(°)
0.6804	6.347	21.9	7.494
0.6576	2.012	29.42	9.772
0.6063	6.953	23.3	5.368
0.5038	6.737	31.08	6.886
0.7772	4.447	31.36	5.216
0.4753	3.419	33.6	6.506
0.5949	2.968	33.32	9.62
0.5722	2.313	18.28	5.518
0.7601	2.691	32.76	8.784
0.5095	3.482	26.64	7.798
0.4468	5.992	28.58	9.088
0.4867	3.019	39.16	8.482
0.5494	2.382	28.02	6.05
0.4639	2.067	31.64	10
0.7715	5.524	23.84	9.392
0.612	2.248	24.96	4
0.4127	3.296	32.48	9.696
0.7487	1.933	22.74	5.064
0.407	3.18	19.4	6.658
0.5835	3.76	18.56	6.81
0.5323	5.249	36.38	9.468
0.5209	2.735	24.68	9.848
0.4354	1.835	27.74	4.76
0.481	1.702	33.88	6.278
0.7943	2.185	37.22	6.962
0.7032	7.178	32.2	7.114
0.6234	4.994	28.86	8.86
0.6006	4.652	37.78	7.342
0.5437	3.548	22.18	4.684
0.3557	4.166	34.44	4.836
0.4582	5.829	38.88	5.746
0.7373	3.237	19.12	8.33
0.6975	3.125	25.8	9.924
0.5778	1.681	23.56	5.822
0.3956	1.767	29.7	8.102
0.4241	1.908	21.62	9.316
0.5665	1.745	25.52	8.254
0.5266	3.296	38.06	4.532
0.6519	2.568	26.36	7.874
0.3785	2.608	27.18	6.202
0.4981	6.536	22.46	7.418
0.5152	5.117	30.26	4.304
0.5608	2.04	34.7	4.38
0.7203	4.76	35.82	9.24
0.4411	1.859	38.6	8.708
0.6633	2.096	19.94	9.544
0.8	2.124	24.4	8.632
0.7146	3.616	39.44	4.912
0.4525	5.675	18	5.518
0.6405	2.779	35.26	6.43
0.4924	2.347	39.72	6.354
0.4184	2.417	21.06	4.456

References

1. Yapıcı, R.; Aldas, K. Optimization of water jet pumps using numerical simulation. *Proc. Inst. Mech. Eng. Part A J. Power Energy* **2013**, *227*, 438–449. [CrossRef]
2. Meakhail, T.; Teaima, I. Experimental and numerical studies of the effect of area ratio and driving pressure on the performance of water and slurry jet pumps. *Proc. Inst. Mech. Eng. Part C J. Mech. Eng. Sci.* **2012**, *226*, 2250–2266. [CrossRef]

3. Long, X.; Xu, M.; Wang, J.; Zou, J.; Bin, J. An experimental study of cavitation damage on tissue of Carassius auratus in a jet fish pump. *Ocean Eng.* **2019**, *174*, 43–50. [CrossRef]
4. Shimizu, Y.; Nakamura, S.; Kuzuhara, S. Studies of the Configuration and Performance of Annular Type Jet Pumps. *J. Fluids Eng.* **1987**, *3*, 205–212. [CrossRef]
5. Kwon, O.B.; Kim, M.K.; Kwon, H.C.; Bae, D.S. Two-dimensional Numerical Simulations on the Performance of an Annular Jet Pump. *J. Vis.* **2002**, *5*, 21–28. [CrossRef]
6. Deng, H.; Liu, X.; Ma, W. Flow Field Analysis for the Diffuser Outlet of Jet Pump Used in the Drain Sand of Petroleum Well. *J. Jilin Univ.* **2010**, *40*, 689–693. [CrossRef]
7. Yang, X.; Long, X.; Yao, X. Numerical investigation on the mixing process in a steam ejector with different nozzle structures. *Int. J. Therm. Sci.* **2012**, *56*, 95–106. [CrossRef]
8. Lyu, Q.; Xiao, Z.; Zeng, Q.; Xiao, L.; Long, X. Implementation of design of experiment for structural optimization of annular jet pumps. *J. Mech. Sci. Technol.* **2016**, *30*, 585–592. [CrossRef]
9. Deng, X.; Dong, J.; Wang, Z.; Tu, J. Numerical analysis of an annular water–air jet pump with self-induced oscillation mixing chamber. *J. Comput. Multiph. Flows* **2017**, *9*, 47–53. [CrossRef]
10. Wang, X.; Chen, Y.; Li, M.; Xu, Y.; Wang, B.; Dang, X. Numerical Study on the Working Performance of a Streamlined Annular Jet Pump. *Energies* **2020**, *13*, 4411. [CrossRef]
11. Gao, G.; Xing, Y.; Wang, Y. Effect of Nozzle Throat Geometry on Flow Field in Liquid Gas Jet Pump: A Simulation Study. *Chin. J. Vac. Sci. Technol.* **2020**, *40*, 174–179. [CrossRef]
12. Xu, M.; Yang, X.; Long, X.; Lyu, Q. Large eddy simulation of turbulent flow structure and characteristics in an annular jet pump. *J. Hydrodyn.* **2017**, *2*, 702–715. [CrossRef]
13. Zou, C.H.; Li, H.; Tang, P.; Xu, D.H. Effect of structural forms on the performance of a jet pump for a deep well jet pump. In Proceedings of the Computational Methods and Experimental Measurements XVII, International Conference on Computational Methods and Experimental Measurements 17th, Opatija, Croatia, 5 May 2015. [CrossRef]
14. Elger, D.; Taylor, S.; Liou, C. Recirculation in an Annular-Type Jet Pump. *J. Fluids Eng.* **1994**, *116*, 735–740. [CrossRef]
15. Keeney, R.; Raifa, H. Decisions with Multiple Objectives: Preferences and Value Tradeoffs. *Health Serv. Res.* **1978**, *13*, 1093–1094. [CrossRef]
16. Marler, R.; Arora, J. Survey of multi-objective optimization methods for engineering. *Struct. Multidiscip. Optim.* **2004**, *26*, 369–395. [CrossRef]
17. Chen, W.; Wiecek, M.; Zhang, J. Quality utility: A compromise programming approach to robust design. *ASME J. Mech. Des.* **1999**, *121*, 179–187. [CrossRef]
18. Luh, G.; Chueh, C.; Liu, W. MOIA: Multi-objective immune algorithm. *Eng. Optim.* **2003**, *35*, 143–164. [CrossRef]
19. Deb, K.; Jain, S. Multi-Speed Gearbox Design Using Multi-Objective Evolutionary Algorithms. *ASME J. Mech. Des.* **2003**, *125*, 609–619. [CrossRef]
20. Saitou, K.; Cetin, O. Decomposition-Based Assembly Synthesis for Structural Modularity. *ASME J. Mech. Des.* **2004**, *126*, 234–243. [CrossRef]
21. Ma, X.; Li, Y.; Yan, L. Comparsion review of traditional multi-objective optimization methods and multi-objective genetic algorithm. *Electr. Drive Autom.* **2010**, *3*, 48–50. [CrossRef]
22. Barthelemy, J.; Haftka, R. Approximation concepts for optimum structural design—A review. *Struct. Multidiscip. Optim.* **1993**, *5*, 129–144. [CrossRef]
23. Alexandras, A. Stochastic subset optimization incorporating moving least squares response surface methodologies for stochastic sampling. *Adv. Eng. Softw.* **2012**, *44*, 3–14. [CrossRef]
24. Gholap, A.; Khan, J. Design and multi-objective optimization of heat exchangers for refrigerators. *Appl. Energy* **2007**, *84*, 1226–1239. [CrossRef]
25. Verstraete, T.; Alsalihi, Z.; Braembussche, R. Multidisciplinary Optimization of a Radial Compressor for Microgas Turbine Applications. *J. Turbomach.* **2010**, *132*, 031004. [CrossRef]
26. Naseri, M.; Othman, F. Determination of the length of hydraulic jumps using artificial neural networks. *Adv. Eng. Softw.* **2012**, *48*, 27–31. [CrossRef]
27. Frédéric, M.; Luis, A.; Ichiro, H. Efficient preconditioning for image reconstruction with radial basis functions. *Adv. Eng. Softw.* **2007**, *38*, 320–327. [CrossRef]
28. Sun, H.; Schafer, M. Reduced order model assisted evolutionary algorithms for multi-objective flow design optimization. *Eng. Optim.* **2011**, *43*, 97–114. [CrossRef]
29. Zhang, Y.; Hu, S.; Wu, J. Multi-objective optimization of double suction centrifugal pump using Kriging metamodels. *Adv. Eng. Softw.* **2014**, *74*, 16–26. [CrossRef]
30. Safikhani, H.; Khalkhali, A.; Farajpoor, M. Pareto Based Multi-Objective Optimization of Centrifugal Pumps Using CFD, Neural Networks and Genetic Algorithms. *Eng. Appl. Comp. Fluid. Mech.* **2011**, *5*, 37–48. [CrossRef]
31. Wang, C.; Hu, B.; Feng, Y.; Liu, K. Multi-objective optimization of double vane pump based on radial basis neural network and particle swarm. *Trans. Chin. Soc. Agric. Eng.* **2019**, *35*, 25–32. [CrossRef]
32. Zhao, A.; Lai, Z.; Wu, P. Multi-objective optimization of a low specific speed centrifugal pump using an evolutionary algorithm. *Eng. Optim.* **2016**, *48*, 1251–1274. [CrossRef]

33. Sheha, A.A.A.; Nasr, M.; Hosien, M.A.; Wahba, E.M. Computational and Experimental Study on the Water-Jet Pump Performance. *J. Appl. Fluid Mech.* **2018**, *11*, 1013–1020. [CrossRef]
34. Shih, T.; Liou, W.; Shabbir, A.; Yang, Z.; Zhu, J. A new k-ε eddy viscosity model for high reynolds number turbulent flows. *Comput. Fluids* **1995**, *24*, 227–238. [CrossRef]
35. Jin, R.; Chen, W.; Sudjianto, A. An efficient algorithm for constructing optimal design of computer experiments. *J. Stat. Plan. Infer.* **2005**, *134*, 268–287. [CrossRef]
36. Amini, A.; Taki, M.; Rohani, A. Applied improved RBF neural network model for predicting the broiler output energies. *Appl. Soft Comput.* **2020**, *87*, 106006. [CrossRef]
37. Wang, Y.; Chen, Z.; Zu, H.; Zhang, X. An Optimized RBF Neural Network Based on Beetle Antennae Search Algorithm for Modeling the Static Friction in a Robotic Manipulator Joint. *Math. Probl. Eng.* **2020**, *2020*, 1024–1034. [CrossRef]
38. Deb, K.; Amrit, P.; Sameer, A.; Meyarivan, T. A fast and elitist multiobjective genetic algorithm: NSGA-II. *IEEE Trans. Evol. Comput.* **2002**, *6*, 182–197. [CrossRef]
39. Kajero, O.; Thorpe, R.; Yao, Y.; Hill, W.; David, S.; Chen, T. Meta-model-based calibration and sensitivity studies of computational fluid dynamics simulation of jet pumps. *Chem. Eng. Technol.* **2017**, *40*, 1674–1684. [CrossRef]

Article

Reanalysis of the Sydney Harbor RiverCat Ferry

Lawrence J. Doctors

The University of New South Wales, Sydney 2052, Australia; L.Doctors@UNSW.edu.au

Abstract: In this paper, we revisit the hydrodynamics supporting the design and development of the RiverCat class of catamaran ferries operating in Sydney Harbor since 1991. More advanced software is used here. This software accounts for the hydrodynamics of the transom demisterns that experience partial or full ventilation, depending on the vessel speed. This ventilation gives rise to the hydrostatic drag, which adds to the total drag of the vessel. The presence of the transom also creates a hollow in the water. This hollow causes an effective hydrodynamic lengthening of the vessel, which leads to a reduction in the wave resistance. Hence, a detailed analysis is required in order to optimize the size of the transom. It is demonstrated that the drag of the vessel and the wave generation can be predicted with good accuracy. Finally, the software is also used to optimize the vessel further by means of affine transformations of the hull geometry.

Keywords: ferry design; wave generation; ship hydrodynamics

1. Introduction

1.1. Previous Studies

The principal features of the design philosophy behind the fleet of RiverCat ferries were described by Doctors, Renilson, Parker, and Hornsby [1] and Hornsby, Parker, Doctors, and Renilson [2]. These ferries were specifically designed in order to minimize the wave generation because of the requirement to limit the erosion of the banks of the Parramatta River along which these ferries operate. To this end, a total of ten different proposed designs were considered in the investigation.

As listed by Doctors, Renilson, Parker, and Hornsby [1] (Table 1), these designs consisted of three catamarans (same demihulls, but different demihull separations and displacements), a different set of three catamarans (same demihull, but with variations in the separations and displacement), and four trimarans (different subhull separations and different proportions of sidehull displacement relative to the centerhull displacement).

An elementary ship-resistance program was used in that study. This software complemented the purely experimental work of Renilson [3], in which the 1/25-scale models of the ten prospective ferries were tested in the towing tank at the Australian Maritime College (AMC). Both the physical experiments and the theoretical study suggested that the catamaran design was superior to the trimaran in terms of the wave generation, as characterized by the maximum height of the wave system at the specified speed and distance from the track of the vessel.

Two photographs of the RiverCat are shown in Figure 1. The general arrangements are shown in plan and profile in Figure 2. The extreme slenderness of the demihulls is very evident.

The particulars of the final design are listed in Table 1. It is particularly noteworthy that the demihull-beam-to-length ratio is 0.02857, which is likely to be a record for slenderness and is only possible due to the choice of a catamaran in order to solve the matter of lateral stability.

(**a**) *Dawn Fraser* during Presentation (**b**) *Shane Gould* in Operation

Figure 1. Grahame Parker Design Pty Ltd Sydney RiverCat.

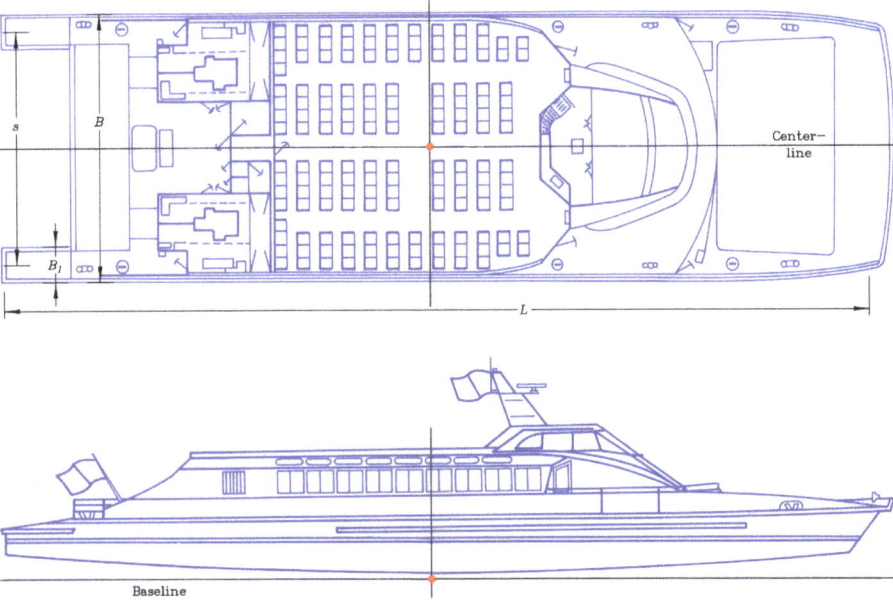

Figure 2. General arrangement of the Sydney RiverCat.

Because the key feature of the RiverCat is its very slender demihulls, it was not possible to position the propulsion engines inside them. As a consequence, the engines were placed on the deck behind the main passenger cabin. This may be seen in Figure 1b and in the plan view of Figure 2. More detail is presented in the profile view of the machinery in Figure 3 and in the section view of the machinery in Figure 4.

A negative feature of the unusual positioning of the engines is that the transmission had to be effected by means of steerable Z-drives, or azimuth thrusters, fitted with two right-angle gearboxes. These drives require more maintenance than conventional propeller shafts. On the other hand, there is a certain practical advantage of this arrangement because maintenance of the engines is made simple and very convenient.

Table 1. Particulars of the Sydney RiverCat.

Quantity	Symbol *	Value
Length on waterline	L	35.00 m
Demihull beam	B_1	1.000 m
Beam overall	B	10.06 m
Draft	T	1.226 m
Block coefficient	C_B	0.6262
Prismatic coefficient	C_P	0.6958
Slenderness ratio	$L/\nabla^{1/3}$	11.68
Transom–area ratio	A_T/A_M	0.4311
Displacement mass	Δ	55.00 t
Power	P	2 × 335 kW
Speed	U	23 kn

* Nominal loading condition.

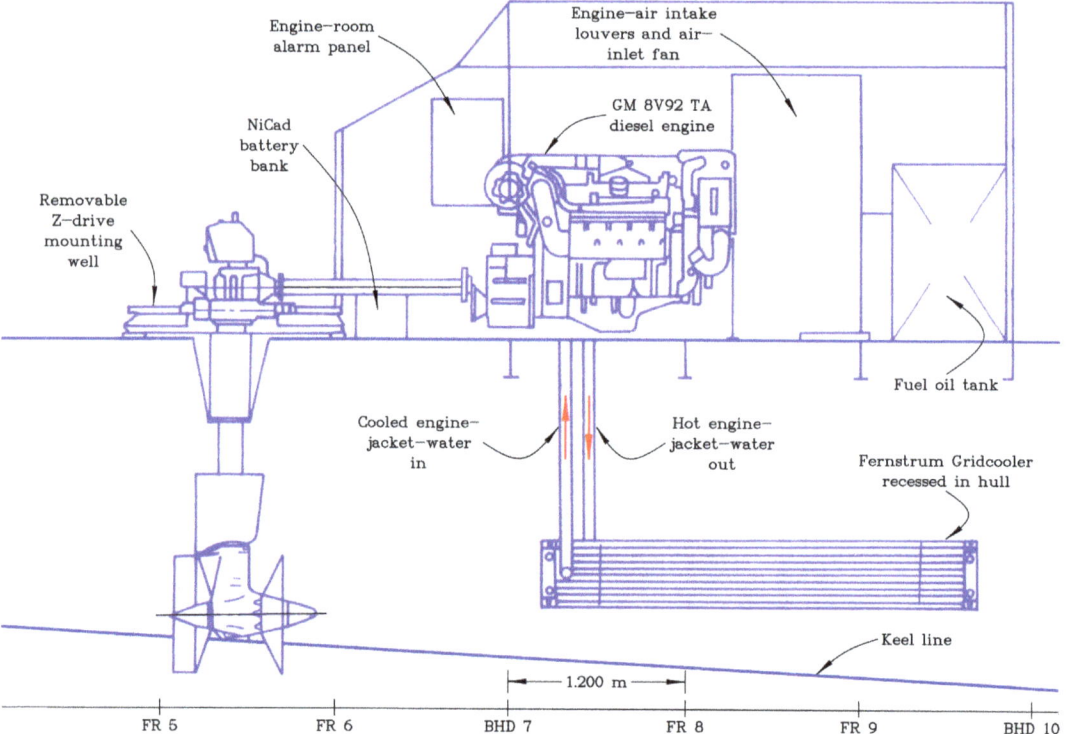

Figure 3. Profile of machinery arrangements.

Because of the low resistance of the vessel, other issues, such as cavitation and noise, have not been a practical concern. A significant design limitation was based on manning regulations. Thus, had the length of the vessel been increased slightly beyond the chosen value of 35 m, the required number of crew members would have been increased by one—leading to significantly increased operational costs.

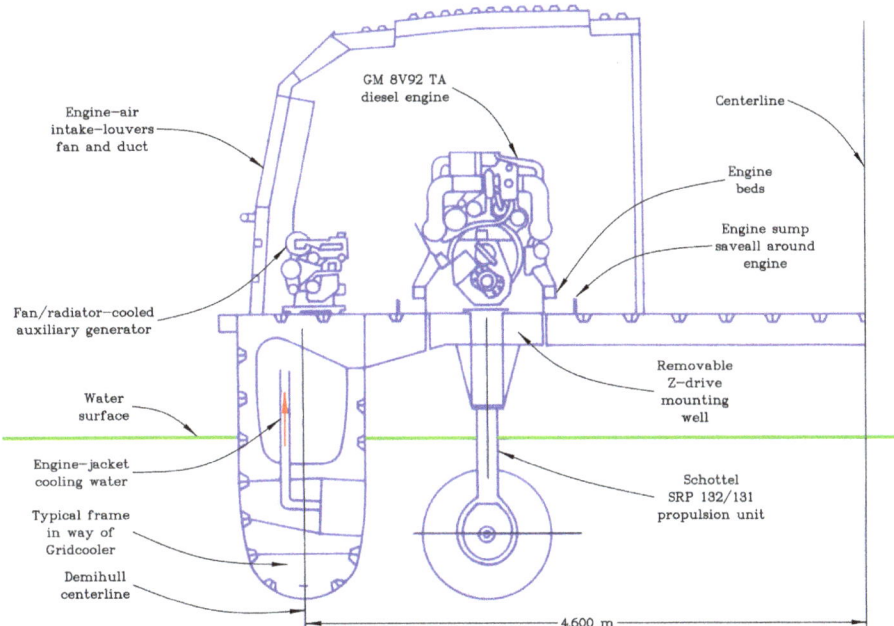

Figure 4. Section of machinery arrangements.

1.2. Current Investigation

The first purpose of the current study is to revisit the theoretical analysis supporting the design of this efficient ferry using more advanced software.

Because of the good correlation between this theory and the experimental data, the software is secondly applied to various modified geometries of the hull. These modifications will be effected by means of affine transformations of the individual demihulls.

There is a similarity between this second part of the study and the work of Doctors [4]. In that effort, vessels with one through six subhulls were analyzed. It was demonstrated theoretically that vessels with three or more subhulls experienced more wave resistance and more total resistance than the simpler catamaran. The higher total resistance of the multihulls relative to catamarans is readily explained by a consideration of their greater wetted surface, which creates more frictional resistance.

2. Hydrodynamic Theory

2.1. Decomposition of Resistance

We follow the traditional approach of decomposing the total resistance R_T into reasonably independent components, as follows:

$$R_T = R_W + R_H + f_F R_F + R_A + R_a. \qquad (1)$$

Here, R_W is the wave resistance, R_H is the hydrostatic resistance caused by the lack of hydrostatic pressure on the partly or fully ventilated transoms, R_F is the frictional resistance, R_A is the correlation allowance that accounts for roughness of the ship hull, and R_a is the aerodynamic resistance, which will be neglected in this current effort. Lastly, f_F is the frictional form factor, which accounts for the increased friction of a real vessel compared to that of a flat plate. This breakdown into components is similar to that proposed by Froude [5].

2.2. Wave Resistance

As the demihulls are considered to be thin, one may apply the classic potential-flow theory. The development of this theory can be traced to Michell [6], whose analysis applies to a monohull traveling in unrestricted water. The influence of laterally restricted water was included by Sretensky [7]. The effect of the finite depth of the water was added by Newman and Poole [8]. The extension of the analysis to a catamaran was presented by a number of researchers, including Doctors and Day [9]. A very detailed presentation of the theory was published by Doctors [10] (Section 5.2).

So, the theory accounts for the possible finite width and finite depth of the towing tank (in the case of a model) or the waterway (in the case of operation of the prototype in a river). The wave resistance is expressed as a sum of the effects of an infinite set of wave systems, each advancing at a different angle to the track of the vessel and at a different speed, such that each wave system keeps up with the vessel:

$$R_W = \frac{\rho g}{\pi} \sum_{i=0}^{\infty}{}' \epsilon \Delta k_y k k_x^2 (\mathcal{U}^2 + \mathcal{V}^2) \bigg/ \frac{df}{dk}, \tag{2}$$

$$\epsilon = \begin{cases} 1/2 & \text{for } i = 0 \\ 1 & \text{for } i \geq 1 \end{cases}. \tag{3}$$

The index i for each wave component has been omitted from most of the algebraic expressions in order to simplify the notation. The other symbols are the water density ρ, the wave number k, the longitudinal wave number k_x, and the transverse wave number k_y. The prime $'$ on the summation in Equation (2) has been used to indicate that the zeroth term, which indicates the transverse wave, is to be omitted for the supercritical case. This is relevant when the depth Froude number $F_d = U/\sqrt{gd}$ exceeds unity, where g is the acceleration due to gravity, d is the depth of the water, and U is the vessel speed.

The wave numbers are determined from the sequence of formulas:

$$\Delta k_y = 2\pi/w, \tag{4}$$
$$k_y = i\Delta k_y, \tag{5}$$
$$k_x^2 + k_y^2 = k^2, \tag{6}$$

in which w is the width of the channel and k is the solution of the transcendental equation

$$f = k^2 - k_0 k \tanh(kd) - k_y^2, \tag{7}$$

where $k_0 = g/U^2$ is the fundamental circular wave number. The solution of this implicit equation can be found in the usual way by using the Newton–Raphson iteration with the assistance of its derivative,

$$df/dk = 2k - k_0 \tanh(kd) - k_0 kd \operatorname{sech}^2(kd). \tag{8}$$

The two finite-depth-water Kochin functions in Equation (2) are defined by the formula

$$\mathcal{U} + i\mathcal{V} = 2\cos\left(\frac{1}{2}k_y s\right) \int_{S_1} b_1(x,z) \exp(ik_x x) \frac{\cosh[k(z+d)]}{\cosh(kz)} d\mathcal{S}, \tag{9}$$

in which s is the separation between the demihull centerplanes, $b_1(x,z)$ is the local demihull beam, and S_1 is the centerplane area of a demihull. A convenient location of the Cartesian coordinate origin is on the centerplane of the vessel at the stern on the undisturbed water surface, with x directed forward, y directed to port, and z directed upward. The computation of the two functions in Equation (9) is conveniently related to the two deep-water Kochin functions,

$$\mathcal{P}^{\pm} + i\mathcal{Q}^{\pm} = 2\cos(\frac{1}{2}k_y s) \int_{S_1} b_1(x,z) \exp(ik_x x \pm kz) \, d\mathcal{S}, \tag{10}$$

by means of the pair of relationships:

$$\mathcal{U} = \frac{\mathcal{P}^+ + \exp(-2kd)\mathcal{P}^-}{1 + \exp(-2kd)}, \tag{11}$$

$$\mathcal{V} = \frac{\mathcal{Q}^+ + \exp(-2kd)\mathcal{Q}^-}{1 + \exp(-2kd)}. \tag{12}$$

2.3. Hydrostatic Resistance

The water pressure acting on the transom gives rise to a negative contribution to the force in the aft direction. Figure 5a is an idealization of the flow behind the transom in the partially ventilated condition. The presence of the deadwater region suggests that we can estimate the pressure load on the face of the transom by means of simple hydrostatic considerations.

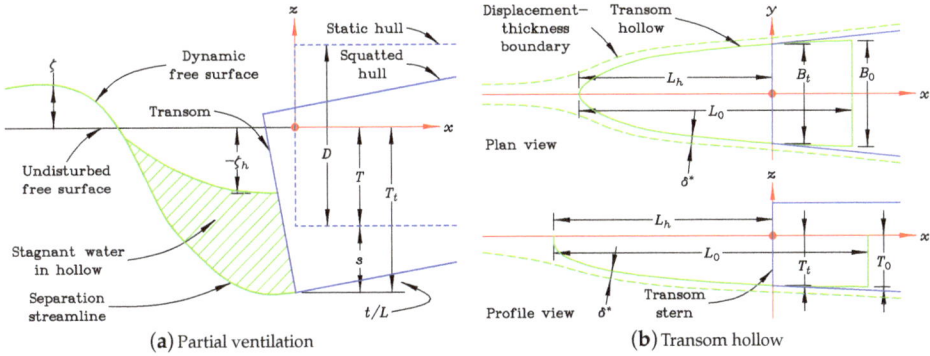

Figure 5. Hydrodynamic modeling of the transom stern.

Therefore, this load is calculated by the simple integration:

$$R_{H1} = \rho g \int_{-T_t}^{\zeta_t} b(x_t, z)(z - \zeta_t) \, dz, \tag{13}$$

where T_t is the draft at the transom and ζ_t is the local elevation of the water surface on the face of the transom, which is always negative in this context.

On the other hand, if the transom were fully wetted, as in the at-rest condition, the hydrostatic force in the aft direction would be

$$R_{H2} = -\rho g \int_{-T_t}^{0} b(x_t, z) z \, dz. \tag{14}$$

The final result for the drag is, therefore, a summation of these two contributions:

$$R_H = R_{H1} + R_{H2}. \tag{15}$$

2.4. Transom Hollow

The process of the ventilation of the transom has been researched by a number of workers. Some of the most practical design guidance was provided by Toby [11], Toby [12], and Toby [13]. As well as giving rise to the unwanted hydrostatic drag, the separation of the water flow at the stern creates a hollow in the water, which adds to the effective hydrodynamic wave-making length of the vessel. This hollow is illustrated in Figure 5b.

This occurrence of the transom hollow is generally favorable in that it reduces the wave resistance. The reader should consult Doctors [10] (Chapter 4) for an in-depth summary of the research on this question.

2.5. Frictional Resistance

The frictional resistance on the model will be computed by the use of the ITTC (1957) (International Towing-Tank Conference) formulation, described by Clements [14] (Page 374) and Lewis [15] (Section 3.5, Pages 7 to 15). This process first requires calculating the Reynolds number based on the wetted length L of the vessel:

$$R_N = UL/\nu, \tag{16}$$

in which ν is the kinematic viscosity. The coefficient of frictional resistance is estimated for extrapolation purposes as

$$\begin{aligned} C_F &= R_F \Big/ \frac{1}{2}\rho U^2 S \\ &= 0.075/[\log(R_N) - 2]^2, \end{aligned} \tag{17}$$

where S is the area of the wetted surface of the vessel.

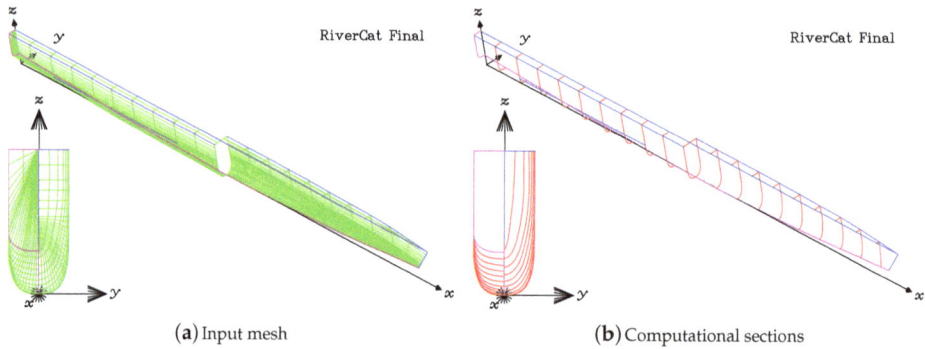

(a) Input mesh (b) Computational sections

Figure 6. Demihull of the finalized RiverCat.

3. Characteristics of the RiverCat

3.1. General Layout

Figure 2 demonstrates the very slender nature of the demihulls of the vessel. This was a primary design feature with the specific purpose of minimizing the wave generation of the ferry. The choice of a catamaran permitted the selection of such slender hulls. Had a monohull design been chosen, it would not have been possible to achieve the minimum required transverse stability.

3.2. Demihulls

The input mesh defining the geometry of the demihull is reproduced in Figure 6a. For the purpose of the computations, a regular spacing of the transverse sections is preferable. This reorganizing of the definition of the hull geometry is performed automatically by the software. The result of this process is shown in Figure 6b.

Both parts of Figure 6 show perspective split views of the demihull. Perhaps more useful are the body plans, or bow-on frontal views. These views emphasize the very rounded nature of the bilges, the very fine bow, and the transom. The pronounced rocker (rise of keel) gives as small a transom area as possible and is evident in this figure. The rocker is very noticeable in the profile view of Figure 2.

4. Numerical Computations

4.1. Finalized Vessel

Figure 7 shows the results of the towing-tank experiments and the current numerical analysis of the finalized design of the RiverCat. The calculations pertain to a model displacement of 3.434 kg. This corresponds to the prototype displacement of 55 t. The prototype demihull centerplane separation is 9.060 m in this example.

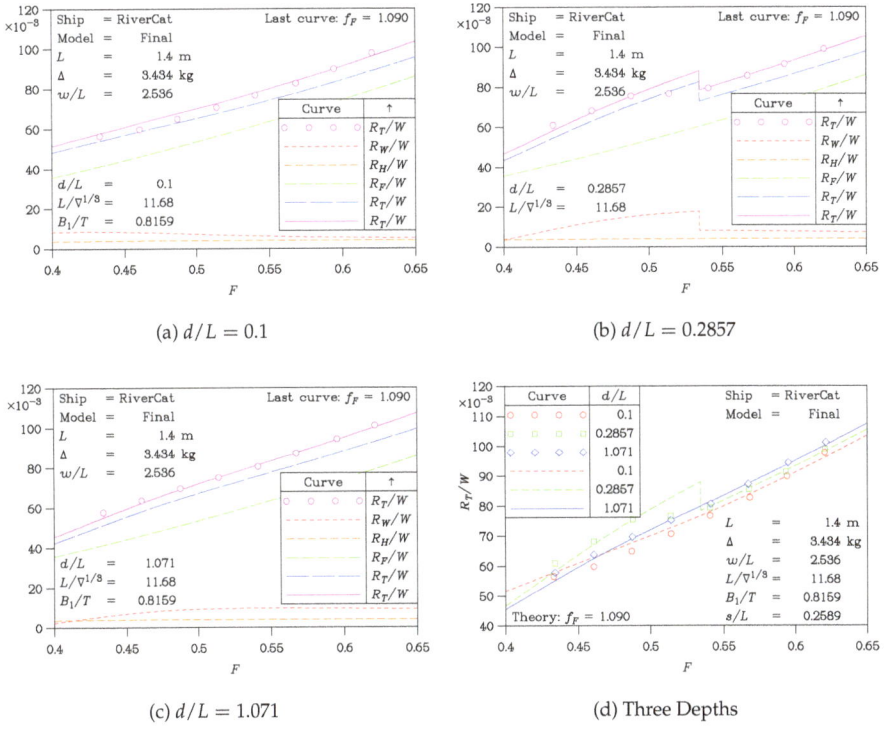

Figure 7. Experiments on the model of the Finalized RiverCat.

The RiverCat was designed to operate in very shallow water as well as in deep water. Thus, the first three plots in Figure 7 relate specifically to the depth-to-length ratios d/L of 0.1, 0.2857, and 1.071. Each of these three plots presents the theoretically computed wave resistance R_W, the hydrostatic resistance R_H, and the ITTC (1957) frictional resistance R_F. The plotted data are non-dimensionalized with respect to the displacement weight W. So, these are the specific-resistance coefficients. These components are plotted as a function of the Froude number $F = U/\sqrt{gL}$.

It is imperative to emphasize the relatively small rôle that the wave resistance plays in contributing to the total resistance budget. Even in Figure 7b, for the dimensionless depth $d/L = 0.2857$, where the speed range covered by the experiments includes the critical speed defined by a unit value of the depth Froude number F_d, the lion's share of the resistance is due to friction. On the other hand, it should be noted that in accordance with the principles of the Froude extrapolation, the resistance will be less important at the prototype scale because of the greater value of the Reynolds number.

The hydrostatic drag is evidently very small and is typically about one-half of the wave resistance—for speeds away from the critical value. The hydrostatic drag is also independent of the speed, at least for the range of speeds covered by these plots. This confirms that the transom is consistently fully ventilated in this speed range.

The plots in Figure 7 also show the total resistance R_T. Circular symbols have been used to indicate the experimental data points, and curves have been used to show the results of the computer predictions. The last dashed curve is the theoretical prediction according to the simple summation suggested by Equation (1)—with a unit value of the frictional-resistance form factor f_F.

In addition, the last and continuous curve in each case shows the total resistance, also using Equation (1), but with the more realistic choice $f_F = 1.090$. This value was deduced by means of a standard root-mean-square minimization procedure, such as that described by (de Vahl Davis [16] Section 3.10). The method was applied to Equation (1) with the one unknown f_F.

With this value of f_F, an excellent correlation between the predicted total resistance and the experimentally derived data is obtained. The most interesting case, depicted in Figure 7b, demonstrates the drop in resistance when the speed crosses the critical value. The steady-state resistance theory used here predicts a sharp drop, which is not so evident in the experiments.

The magnitude of this drop in specific resistance is given by the remarkably simple formula:

$$\Delta R_W / W = 3\nabla / 2wd^2, \qquad (18)$$

in which ∇ is the displacement volume. This formula is independent of the shape of the vessel hull.

We can verify the magnitude of the drop in Figure 7b by noting that in this case, $\Delta = 3.434$ kg, $w = 3.550$ m, and $d = 0.400$ m. These data give $\Delta R_W / W = 9.069 \times 10^{-3}$; this result is in agreement with the theoretical discontinuities in the curves for the wave resistance and for the total resistance.

The fact that the discontinuity in the experimental data is somewhat rounded off—unlike the sharp discontinuity in the theory—can be resolved by employing the unsteady wave-resistance theory published by Day, Clelland, and Doctors [17]. It was demonstrated that the time-averaging signal processing, which is used in recording ship-model-resistance data during typical tests, results in a rounding of the results in the vicinity of $F_d \approx 1$. One must, therefore, also account for the motion of the tank carriage from rest and use the true time-unsteady theory. Then, the rounded characteristic of the resistance curve can be predicted accurately. An example for a case of $w/L = 1.524$ and $d/L = 0.25$ was published by (Day, Clelland, and Doctors [17] Figure 6).

Finally, we note that only the midrange of Froude numbers was of interest in this work. It is well known that the curve of wave resistance exhibits a large number of oscillations at low values of the Froude number. The reader is referred to Doctors [10] (Figure 5.11, Page 132), where the wave-resistance coefficient is plotted for Froude numbers down to zero in value. We add here that when the more useful specific wave resistance is plotted instead, the magnitude of these oscillations is greatly reduced, and so is their significance.

4.2. Transformations of the Finalized Demihull

In this section, we will consider four different types of transformations of the demihull of the Sydney RiverCat. These are listed in Table 2.

The first transformation involves stretching the length of the demihull and reducing the local beam and local draft in equal proportions, thus maintaining a fixed value of the displacement.

The second transformation relates to changing the local sectional aspect ratio B_1/T while keeping the local sectional area and the length fixed.

The third transformation is to change the demihull separation by increasing it from the original value.

The fourth and the most challenging transformation is one in which the significance of the transom, as measured by the metric transom–area ratio A_T/A_M, is examined. In the current effort, the methodology of Doctors [4] is replicated. The first step is to create a pointed-stern version of the original demihull.

In this example, the forward half of the vessel is employed to create a pointed-stern demihull. This demihull is symmetric fore-and-aft. Of course, it is improper to directly compare the hydrodynamics of this demihull with the original transom-stern demihull because its volume is substantially lower. Consequently, the local beam and the local draft have each been increased at all stations in an affine manner by a simple factor, which is the square root of the desired ratio of volumes. The resulting pointed-stern demihull with the same displacement volume is presented in Figure 8a.

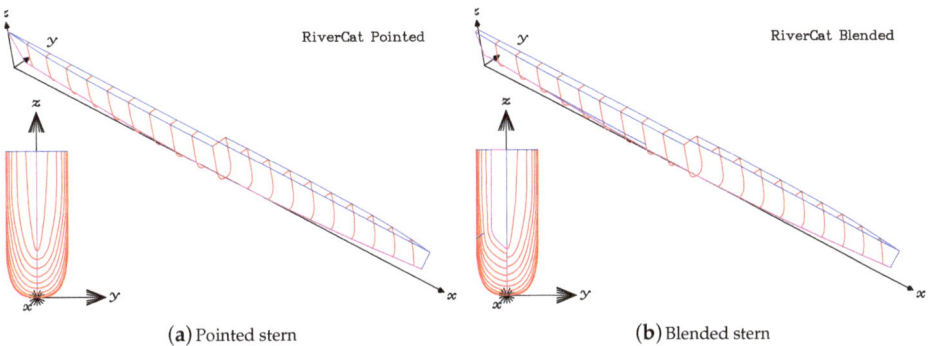

(**a**) Pointed stern   (**b**) Blended stern

Figure 8. Transformations of the finalized RiverCat demihull.

We now have two extreme cases. These are the original transom-stern demihull in Figure 6b and the pointed-stern demihull in Figure 8a. Next, a third and intermediate blended-hull demihull was created by combining or blending equal portions of these two basis demihulls, after the manner of Doctors [18]. That is, the coordinates of the points on the surface of the new and blended hull are simple averages of the original coordinates. The blended-stern vessel is presented in Figure 8b.

Table 2. Affine transformations of the demihull.

Index	Parameter	Symbol	Values		
			Affine 0 [†]	Affine 1	Affine 2
1 *	Slenderness ratio	$L/\nabla^{1/3}$	11.68	14.60	17.52
2	Beam-to-draft ratio	B_1/T	0.8159	1.275	1.836
3	Demihull separation	s/L	0.2589	0.3160	0.3731
4	Transom–area ratio	A_T/A_M	0.4311	0.2156	0

* Corresponding to the four parts in each of Figures 9–11. [†] Data for the finalized RiverCat.

4.3. Wave Resistance

The wave generation of the Sydney RiverCat was a critical performance criterion in its design. To this end, we provide a set of calculations in each of the four parts of Figure 9.

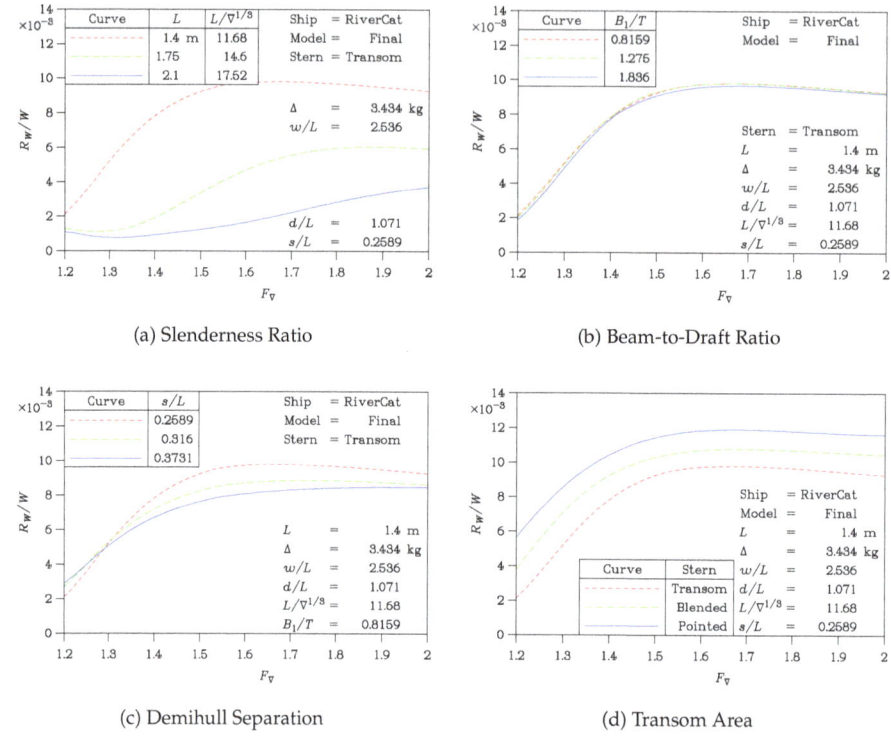

Figure 9. Influence of the affine transformations on wave resistance.

Figure 9a is a plot of the specific wave resistance R_W/W as a function of the volumetric Froude number $F_\nabla = U/\sqrt{g\nabla^{1/3}}$. This abscissa was chosen instead of the traditional Froude number F because the model length varies between the curves in the test of slenderness variation; so, it is advisable to be consistent and to use the displacement volume ∇ as the basis for rendering the speed non-dimensional.

Three different values of the slenderness ratio $L/\nabla^{1/3}$ are considered in Figure 9a, varying between 11.68 for the finalized RiverCat and 17.52 for the longest variant. As noted earlier, the cross-sectional shape has been kept constant so that B_1/T is the same for the three models. There is a significant reduction in wave generation for the slenderest vessel. This shows that there is much scope for reducing the undesired wave generation.

Modifying the section aspect ratio, as measured by the ratio B_1/T, is considered in Figure 9b. There is almost no influence of this parameter on wave generation.

The demihull–centerplane separation s is changed in Figure 9c. Increasing the separation generally reduces the wave generation, as is well known from previous research. However, the effect is not large.

Lastly the importance of the type of vessel stern is examined in Figure 9d. The original and finalized RiverCat (with the full transom) is seen to experience the lowest wave resistance. This has been explained in the past as being a result of the transom hollow creating an effectively longer hydrodynamic shape in the water, which is favorable.

4.4. Model Total Resistance

The total resistance of the model is considered in the four plots of Figure 10. The variations in the demihull modifications studied in these plots correspond to the four parts of Figure 9.

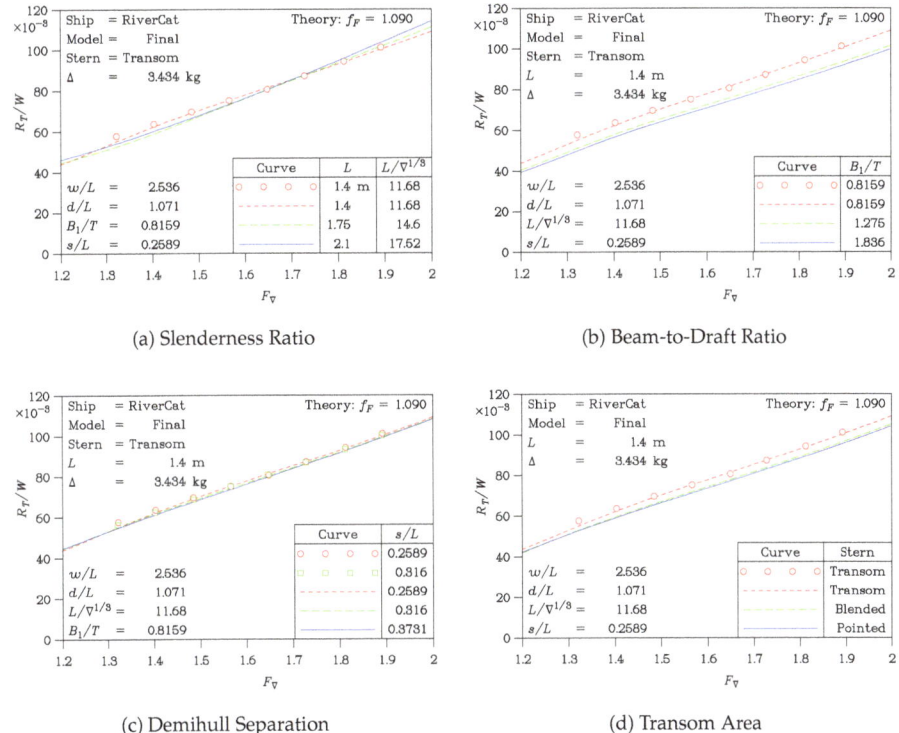

Figure 10. Influence of the affine transformations on model resistance.

The total resistance of the model was tested in some of these cases, and the deep-water model data for the finalized RiverCat were plotted already in Figure 7c.

The influence of slenderness in Figure 7a is seen to be negligible—despite its strong effect on wave resistance in Figure 9a. This outcome can be explained by the fact that slenderer hulls have a greater wetted-surface area, which is a negative attribute.

Increasing the section aspect ratio B_1/T in Figure 7b is favorable because sections closer to semicircles have a smaller perimeter, leading to less wetted surface.

The effect of increasing the demihull separation in Figure 7c is almost negligible because, while the wave resistance is noticeably reduced, the lion's share of the resistance is due to frictional resistance, which is not altered.

Lastly, the influence of transom-stern size is considered in Figure 7d. Again, the reduced wave resistance for a demihull with a larger transom is not of great importance for the total resistance because it contributes so little to the total drag budget.

4.5. Prototype Total Resistance

Lastly, we emphasize that the main purpose of performing calculations on ship models is to verify the experimental data measured on physical models and to understand the hydrodynamic phenomena. Nevertheless, we are only concerned, in a more practical sense, with the performance of the prototype vessel. These theoretical results are presented in the four graphs of Figure 11, which correspond to the four graphs of Figure 10.

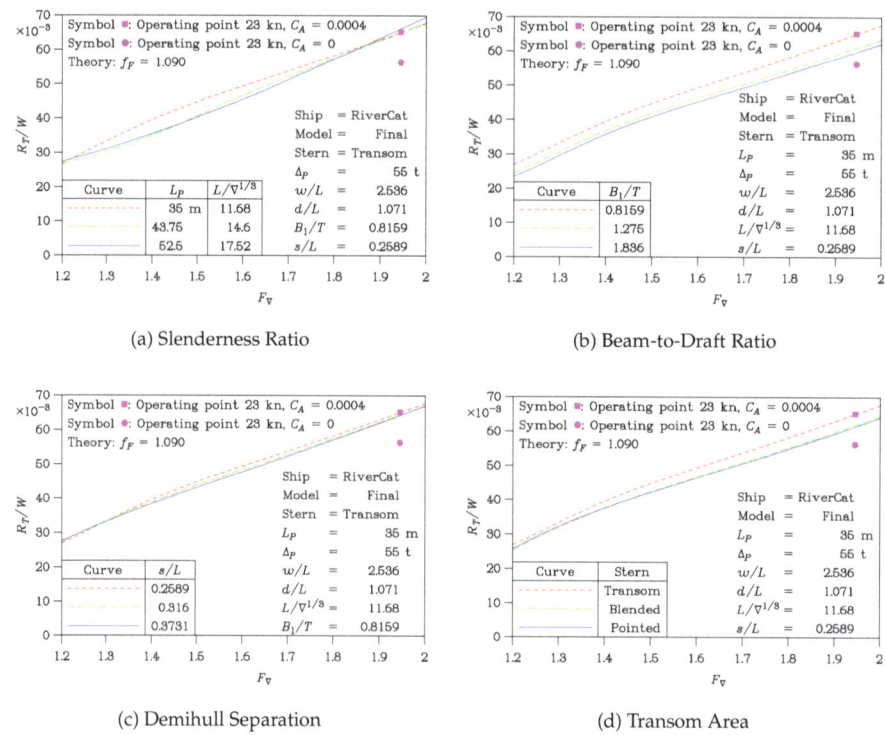

Figure 11. Influence of the affine transformations on prototype resistance.

To minimize the human effort required, these computations for the prototype were effected on the 1/25-scale hull geometry—as for the previous computations for the model. We remind the reader that the results are presented in a dimensionless manner. So, the only necessary change to the data file was to reduce the value of the kinematic viscosity ν in Equation (16) for the Reynolds number by a factor of $25^{3/2}$. In this manner, the software will calculate the correct Reynolds number at the prototype scale.

In broad terms, the plotted curves for the resistance of the prototype resemble those for the model. It is important to note that the upper limit for the vertical scale for the total resistance has been reduced from 0.12 to 0.07. That is, the values of the total resistance are almost one-half when plotted on a dimensionless basis. This substantial reduction in resistance is due to the much lower frictional-resistance coefficient, as computed using Equation (17). This is a fundamental outcome of Froude scaling. In line with standard practice, a correlation or roughness allowance $C_A = 0.0004$ has been added to the friction coefficient at prototype scale according to Equation (17).

Because the major part of the resistance is due to friction—even at the prototype scale—most of the variations in demihull geometry considered in Figure 11 are insignificant. This is particularly true at the nominal operating point, corresponding to a speed of 23 kn, which is indicated on the four plots.

4.6. Hull Finish

The transport factor, defined as

$$TF = WU/P \qquad (19)$$
$$= \eta \times (W/R_T), \qquad (20)$$

is here evaluated on the basis of the total weight W, the operational speed U, and the installed engine power P. We now use the prototype displacement of 55 t, the nominal operating speed of 23 kn, and the prototype propulsive power of 2×335 kW. So, the transport factor TF is 9.524.

We also note that at this speed, corresponding to $F_\nabla = 1.945$, we have the predicted value of the specific-resistance data at the operating point from any of the four plots in Figure 11, namely, $R_T/W = 0.0652$. This allows us to use Equation (20) and to compute the overall propulsive efficiency of the transmission and the propellers η, which is 0.621. This result is a reasonably acceptable value in naval-architecture usage.

On the other hand, if the hull finish of the vessel could be maintained at an ideal hydraulically smooth value of zero, the transport factor rises to a value of 11.03. A strong case can be made for enforcing a strict régime for cleaning the hull on a regular basis, thereby reducing fuel consumption. These results are summarized in Table 3.

Table 3. Transport factor of the Sydney RiverCat.

Assumption	Transport Factor TF *
ITTC (1957) with $C_A = 0$	11.03
ITTC (1957) with $C_A = 0.0004$	9.52
Measured	9.52

* Nominal operating point: Δ = 55 t, U = 23 kn and η = 0.621.

Further studies on the matter of hull friction and different methods of implementing the ITTC (1957) extrapolation were published by Clements [14] (Page 374) and Lewis [15] (Section 3.5, Pages 7 to 15), the ITTC (1978) extrapolation by Oosterveld [19] (Equation (1.19)) and ITTC [20], and the ITTC (2004) extrapolation by Candries and Atlar [21] (Equation (1)) and ITTC [22] (Equation (3)).

5. Conclusions

5.1. Current Investigation

The current study has demonstrated that the linearized wave-resistance theory, used in conjunction with physical models of the transom-ventilation process and the generation of the accompanying transom hollow, provides excellent predictions of the resistance of efficient river catamarans. It is important, at the same time, to have good estimates of the frictional-resistance form factor. For the Sydney RiverCat, this was determined to be 1.09.

This work has also shown that it is difficult to further improve the hull design if one wishes to reduce the total resistance. At a design operating speed of 23 kn, the only affine transformation in Figure 11 that showed promise was to increase the sectional beam-to-draft ratio, B_1/L. This increase makes the sections closely semicircular, thereby reducing the frictional resistance.

However, a principal aim of this river vessel was to reduce the wave generation. In this regard, Figure 9a demonstrated that increasing the slenderness ratio by 50% resulted in a reduction of wavemaking at 23 kn of the order of 60%.

5.2. Future Studies

Further studies should include modifications of the demihulls in various ways. Suitable modifications include embracing the concept of pure semicircular sections. These will reduce the frictional resistance to the minimum. Secondly, it is possible that modifying the sectional-area curve (the longitudinal distribution of volume) may lead to a further reduction in wave generation.

Thirdly, if wave generation is a key design consideration, one should expand the current investigation into the rôle that the slenderness ratio plays. It is understood that the consequently very long hulls might lead to some practical operational difficulties. These difficulties include maneuverability in confined waterways and manning regulations, which might impose increased salaries for a larger crew.

Funding: This research received no external funding.

Institutional Review Board Statement: Not applicable.

Informed Consent Statement: Not applicable.

Data Availability Statement: Not applicable.

Conflicts of Interest: The authors declare no conflict of interest.

List of Symbols

A_T	Transom area	T	Draft
A_M	Midship-section area	U	Ship velocity
B	Overall beam	W	Displacement weight
B_1	Demihull beam	b_1	Demihull local beam
C_A	Correlation allowance	d	Depth of water
C_B	Block coefficient	f_F	Frictional-resistance form factor
C_P	Prismatic coefficient	g	Acceleration due to gravity
F	Froude number	k	Circular wave number
F_d	Depth Froude number	k_x	Longitudinal wave number
F_∇	Volumetric Froude number	k_y	Transverse wave number
L	Length	s	Demihull centerplane separation
L_P	Prototype nominal length	w	Towing-tank or canal width
$L/\nabla^{1/3}$	Slenderness ratio	x	Longitudinal coordinate
P	Power	x_t	Longitudinal coordinate at transom
R_A	Correlation resistance	y	Transverse coordinate
R_F	Frictional resistance	z	Vertical coordinate
R_H	Hydrostatic resistance	Δ	Model displacement mass
R_T	Total resistance	Δ_P	Prototype displacement mass
R_W	Wave resistance	ζ_t	Free-surface elevation on face of transom
R_a	Aerodynamic resistance	η	Propulsion efficiency
R_N	Reynolds number	ν	Kinematic viscosity of water
S	Overall wetted-surface area	ρ	Density of water
S_1	Demihull wetted-surface area	∇	Displacement volume

References

1. Doctors, L.J.; Renilson, M.R.; Parker, G.; Hornsby, N. Waves and Wave Resistance of a High-Speed River Catamaran. In Proceedings of the First International Conference on Fast Sea Transportation (FAST '91), Norges Tekniske Høgskole, Trondheim, Norway, June 1991; Volume 1, pp 35–52. Available online: https://repository.tudelft.nl/islandora/object/uuid%3A133e695e-1a70-4a0d-b99a-17c25aeace92 (accessed on 8 February 2021)
2. Hornsby, N.; Parker, G.; Doctors, L.J.; Renilson, M.R. The Design, Development, and Construction of a 35-Metre Low-Wash Fast Catamaran River Ferry. In Proceedings of the Sixth International Maritime and Shipping Conference (IMAS '91), The University of New South Wales, Sydney, Australia, 11–13 November 1991; pp. 41–47.
3. Renilson, M.R. *Resistance Tests, Powering Estimates and Wake Wave Prediction for a 35 m Ferry*; Report 89/T/13; Australian Maritime College, AMC Search Limited: Mowbray, TAS, Australia, 1989; p. 63+i.
4. Doctors, L.J. On the Great Trimaran-Catamaran Debate. In Proceedings of the Fifth International Conference on Fast Sea Transportation (FAST '99), Seattle, WA, USA, 31 August–2 September 1999; pp. 283–296.
5. Froude, W. On Experiments with H.M.S. 'Greyhound'. *Trans. Inst. Nav. Archit.* **1874**, *15*, 36–59.
6. Michell, J.H. The Wave Resistance of a Ship. *Philos. Mag.* **1898**, *45*, 106–123. [CrossRef]
7. Sretensky, L.N. On the Wave-Making Resistance of a Ship Moving along in a Canal. *Philos. Mag.* **1936**, *22*, 1005–1013. [CrossRef]
8. Newman, J.N.; Poole, F.A.P. The Wave Resistance of a Moving Pressure Distribution in a Canal. *Schiffstechnik* **1962**, *9*, 21–26.
9. Doctors, L.J.; Day, A.H. Resistance Prediction for Transom-Stern Vessels". In Proceedings of the Fourth International Conference on Fast Sea Transportation (FAST '97), Sydney, Australia, 21–23 July 1997; Volume 2, pp. 743–750.
10. Doctors, L.J. *Hydrodynamics of High-Performance Marine Vessels*, 2nd ed.; CreateSpace: Charleston, SC, USA, 2018; Volume 1, pp. 1–421+li.
11. Toby, A.S. The Evolution of Round Bilge Fast Attack Craft Hull Forms. *Nav. Eng. J.* **1987**, *99*, 52–62. [CrossRef]
12. Toby, A.S. U.S. High Speed Destroyers, 1919–1942: Hull Proportions (to the Edge of the Possible). *Nav. Eng. J.* **1997**, *109*, 155–177. [CrossRef]

13. Toby, A.S. The Edge of the Possible: U.S. High Speed Destroyers, 1919–1942. Part 2: Secondary Hull Form Parameters. *Nav. Eng. J.* **2002**, *114*, 55–76. [CrossRef]
14. Clements, R.E. An Analysis of Ship-Model Correlation Data Using the 1957 I.T.T.C. Line. *Trans. R. Inst. Nav. Archit.* **1959**, *101*, 373–385.
15. Lewis, E.V.Q.Q. *Principles of Naval Architecture: Volume II. Resistance, Propulsion and Vibration*; Society of Naval Architects and Marine Engineers: Jersey City, NJ, USA, 1988; p. 327+vi.
16. de Vahl Davis, G. *Numerical Methods in Engineering and Science*; Allen & Unwin (Publishers) Ltd.: London, UK, 1986; pp. 286+xvi.
17. Day, A.H.; Clelland, D.; Doctors, L.J. Unsteady Finite-Depth Effects during Resistance Tests in a Towing Tank. *J. Mar. Sci. Technol.* **2009**, *14*, 387–397. [CrossRef]
18. Doctors, L.J. A Versatile Hull-Generator Program. In Proceedings of the Twenty-First Century Shipping Symposium, The University of New South Wales, Sydney, Australia, 6 November 1995; pp 140–158.
19. Oosterveld, M.W.C.Q.Q. Report of Performance Committee. In Proceedings of the Fifteenth International Towing Tank Conference, The Hague, The Netherlands, September 1978; pp. 359–404. Available online: https://repository.tudelft.nl/islandora/object/uuid%3Ad7ad3bc5-40f8-42b1-a997-be238d791b56 (accessed on 8 February 2021).
20. ITTC. Procedure 7.5-02-03-01.4—Recommended Procedures: Performance, Propulsion 1978 ITTC Performance Prediction Method. In Proceedings of the Twenty-Second International Towing Tank Conference, Seoul, Japan, Shanghai, China, 5–11 September 1999; p. 31.
21. Candries, M.; Atlar, M. On the Drag and Roughness Characteristics of Antifoulings. *Trans. R. Inst. Nav. Archit.* **2003**, *145*, 107–132.
22. ITTC. Procedure 7.5-03-03-01 Recommended Procedures and Guidelines: Practical Guidelines for Ship Self-Propulsion CFD. In Proceedings of the Twenty-Seventh International Towing Tank Conference, Copenhagen, Denmark, 31 August–5 September 2014; p. 9.

Article

Numerical Study of Hydrodynamics of Heavily Loaded Hard-Chine Hulls in Calm Water

Miles P. Wheeler [1], Konstantin I. Matveev [1,*] and Tao Xing [2]

[1] School of Mechanical and Materials Engineering, Washington State University, Pullman, WA 99164, USA; miles.wheeler@wsu.edu
[2] Department of Mechanical Engineering, University of Idaho, Moscow, ID 83844, USA; xing@uidaho.edu
* Correspondence: matveev@wsu.edu

Abstract: Hard-chine boats are usually intended for high-speed regimes where they operate in the planing mode. These boats are often designed to be relatively light, but there are special applications that may occasionally require fast boats to be heavily loaded. In this study, steady-state hydrodynamic performance of nominal-weight and overloaded hard-chine hulls in calm water is investigated with computational fluid dynamics solver program STAR-CCM+. The resistance and attitude values of a constant-deadrise reference hull and its modifications with more pronounced bows of concave and convex shapes are obtained from numerical simulations. On average, 40% heavier hulls showed about 30% larger drag over the speed range from the displacement to planing modes. Among the studied configurations, the hull with a concave bow is found to have 5–12% lower resistance than the other hulls in the semi-displacement regime and heavy loadings and 2–10% lower drag in the displacement regime and nominal loading, while this hull is also capable of achieving fast planing speeds at the nominal weight with typical available thrust. The near-hull wave patterns and hull pressure distributions for selected conditions are presented and discussed as well.

Keywords: boat hydrodynamics; hard-chine hulls; computational fluid dynamics

1. Introduction

Fast planing boats usually employ hard-chine hulls to ease water separation at high speeds from their hulls, which leads to drag reduction. Relatively small areas on hull surfaces need to stay in contact with water to provide a hydrodynamic lift sufficient for carrying the boat weight. To achieve high speeds, power requirements for fast boats with hard chines are still much greater than those of displacement boats moving at low speeds. Therefore, to keep planing boats reasonably economical, they are usually designed to be relatively light. The weight W of a fast boat can be normalized by the hull beam B, forming a beam-based loading coefficient,

$$C_B = \frac{W}{\gamma B^3}, \quad (1)$$

where γ is the specific weight of water. This coefficient is usually limited by 0.9 for planing hulls, while most fast boats are much lighter. However, hard-chine hulls operating with $C_B > 1$ also exist.

When describing different speed regimes of boats and ships, a non-dimensional Froude number is commonly used,

$$Fr_c = \frac{u}{\sqrt{gc}}, \quad (2)$$

where u is the speed, g is the gravity constant, and c is the characteristic length. Various length parameters are utilized for Froude number in ship hydrodynamics, including the hull length, L, waterline length, hull beam, and a cubic root of the volumetric dis-

placement, $\sqrt[3]{V}$. The planing regime usually corresponds to the Foude length number $Fr_L = u/\sqrt{gL} > 1\text{--}1.2$ or the volumetric Froude number $Fr_V = u/\sqrt{gV^{1/3}} > 3\text{--}4$ [1–3].

Although not very common, there are applications demanding heavily loaded fast marine transports. For example, during rescue missions, a boat may have to carry a higher payload than it was designed for. There are also a variety of special operations when boats with relatively small footprints need to transport heavy cargo at high speeds. To assess the hydrodynamic performance of boats in such conditions, namely to estimate achievable speeds (which will be of course lower than at normal loading), thrust requirements and other parameters, one needs to know the hull drag and attitude behavior in a broad range of speeds and loadings. However, the literature on the hydrodynamics of hard-chine planing hulls is essentially limited to conditions with the beam loading coefficient around 0.9 at moderate Froude numbers.

Among the approaches used for predicting hydrodynamics of usual planing hulls, empirical correlations, such as the Savitsky's method [2], are still very popular, but due to a small number of involved parameters and a broad range of possible conditions, they can be applied only for initial approximate estimations. A review of empirical methods and illustrations of hull forms intended for different high-speed regimes, including relatively heavy hard-chine hulls, is given by Almeter [4]. A variety of potential-flow modeling methods that can account for specific hull geometries have been developed in the past [5–7], but they ignore viscous effects and are often applicable only at sufficiently high Froude numbers. With the growth of available computational power, numerical methods accounting for viscosity and flow non-linearities are becoming widely used for ship hydrodynamics studies, including fast boats [8–11]. These computational fluid dynamics (CFD) tools can, therefore, be applied for modeling heavily loaded hard-chine hulls in the entire speed range. In the present study, one such CFD program (STAR-CCM+) is utilized. The authors of this paper have previously conducted a validation study of planing hulls employing a similar CFD approach [12].

The present paper has several objectives. One is to show validation of CFD for a relatively heavy constant-deadrise planning hull with $C_B \approx 0.9$, for which experimental data are available. Secondly, an overloaded (by 40%) condition of this and other hull forms is simulated to expand the knowledge base of heavy hard-chine hulls in the range of speeds from the displacement to planing states. In addition, more practical hulls with extended bow portions that have convex and concave shapes are also generated, and their hydrodynamic characteristics are quantified as well. Most simulations are conducted here with one common location of the longitudinal center of gravity (LCG), at 45% of the hull length from the transom. Several simulations of overloaded hulls in the transitional regime are also carried out with LCG = 40% to compare the performance of concave and convex bows.

Figure 1 graphically illustrates some practical motivations that guided the present study. One of the intentions is to determine a suitable hull form that would be economical at the nominal weight in the semi-displacement mode at cruise speed u_c (Figure 1), i.e., it will have reasonably low resistance R_c within $Fr_L \sim 0.5\text{--}0.8$ or $Fr_V \sim 0.7\text{--}1.7$. At the same time, this hull at the nominal weight should be able to reach a planing speed u_p ($Fr_L \sim 1.2$, $Fr_V \sim 3.0$) with full thrust typical for planing hulls T_f. The specific characteristic of most interest in the present study is the boat's ability to archive the highest speed u_h (among the considered hulls) under the same available thrust in the overloaded condition. It should be noted that only steady-state forward motion in calm water was analyzed here. Other important hydrodynamic characteristics, such as seaworthiness and maneuvering, are beyond the scope of this paper. Studies on those topics with applications to semi-displacement and planning hulls can be found in [3,9,13].

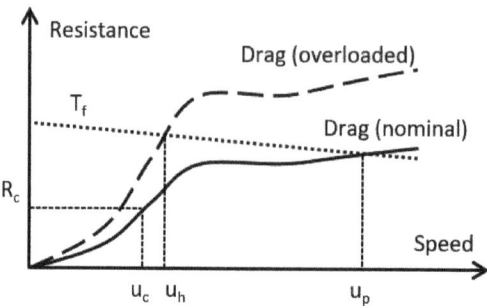

Figure 1. Typical resistance curves of nominal-weight hull (solid line) and overloaded hard-chine hull (dashed line). Approximate full thrust curve is indicated by dotted line. Symbols are explained in the text.

The novelty and contributions of this work include results for hydrodynamics of heavily loaded hard-chine hulls that are not available in the literature, so the practitioners can use these data to quickly access the performance of overloaded planing hulls. A systematic comparison of hydrodynamic performance is presented for various bow shapes, so a specific bow form can be used as a starting point in designing a heavy hull for intended speed regimes. Contributions are also made to the computational methodology for fast boat hydrodynamics. A three degree of freedom unsteady approach used to achieve steady states has been described. Comparisons are presented for different turbulence models and computational times at various speeds and loading conditions.

The rest of the paper is organized as follows. In Section 2, a description is given for the hull geometries, as well as speed and loading conditions. Section 3 elaborates on computational aspects, including the numerical domain, grid specifics, governing equations and modeling of the hull dynamics in transient regimes. The verification and validation studies, involving a comparison with experimental data, are shown in Section 4. The results of extensive parametric studies performed in this work for various hull geometries, loadings and speeds are presented and discussed in Section 5, which is followed by the conclusions.

2. Hull Geometries and Studied Conditions

The hull geometries studied in this work are relatively basic (Figures 2 and 3). Three different bow shapes were analyzed: constant-deadrise, concave, and convex hull shapes. Only two locations of the center of gravity were investigated, 40% and 45% of the hull total length from the transom. Two loading conditions were looked at, $C_B = 0.912$ and $C_B = 1.276$. The lower value corresponds to a relatively heavy planing hull, but within a common range of loadings. The higher value, obtained by increasing the hull weight by 40%, imitates an overloaded state or a special compact fast boat intended for heavy cargo.

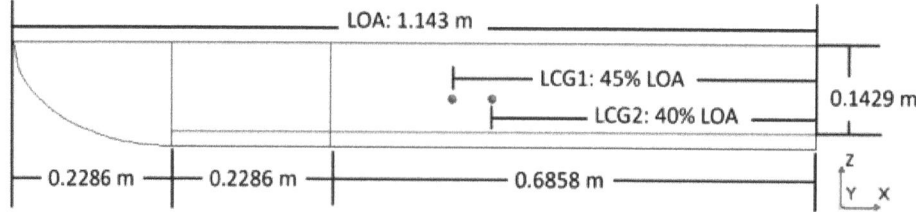

Figure 2. Key dimensions of studied hulls in the vertical plane.

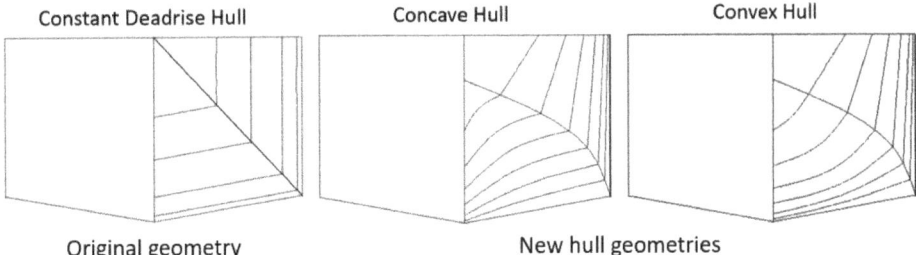

Figure 3. Transverse hull lines of three studied configurations.

The constant-deadrise basis hull with the selected loading $C_B = 0.912$ comes from the family of hulls experimentally studied by Fridsma [13]. Its length-to-beam ratio is 5, and the deadrise angle is 10°. The experimental hull was 1.143 m long, and the length of the non-prismatic bow section was 0.229 m or 20% of the hull length (Figure 2).

Two modified hull shapes were numerically generated in this study with the purpose to improve hydrodynamic characteristics in the semi-displacement regime. They have convex and concave bow shapes, shown in Figure 3, while the curved bow portion was extended to 40% of the fore part of the hull. (Initially, hulls with curved bows of 20% length overall (LOA) were also tried in this work, but due to the bow exit out of water with increasing speeds, their performance difference from the original hull was limited to a narrow range of relatively low speeds.) The aft body of hulls with curved bows was kept prismatic and identical to the constant-deadrise hull.

In most simulations, the longitudinal center of gravity (LCG) was placed at 45% of the overall length, as measured from the stern (Figure 2), although a limited set of simulations was also conducted with LCG at 40%. The vertical position of the center of gravity was 47% of the hull height or 5.9% of LOA, counted up from the keel line on the prismatic hull portion.

The lines plan for hulls of the three different bow shapes are shown in Figure 3. The geometry of hulls with curved bows was parameterized with three equally spaced transverse splines in the bow region such that the convexity and concavity were arced by the same amount, but in different directions. The maximum arc for the three splines, going from bow to stern were 4.44%, 1.11%, and 0.56% of the hull beam, respectively.

An additional hull feature in experiments of Fridsma [13] was a very small strip positioned at the chine along the prismatic portion of the hulls. This thin strip was used to prevent wetting of the hull side walls at sufficiently high speeds on the model scale that would likely not be seen if the hull was operating in full-scale conditions. This thin strip was modeled in all simulations in the same manner as described by Wheeler et al. [12].

In the parametric simulations of this study for LCG = 45% conditions, the speed range was selected between 0.25 and 1.50 of length-based Froude numbers defined by Equation (2). Given the limitation of accessible computational resources, only the model-scale hulls were investigated, corresponding to experimental dimensions [13]. The range of length-based Reynolds numbers in the chosen speed range was between 5.4×10^5 and 3.2×10^6. For the LCG = 40% conditions, the parametric calculations were carried out only in the transitional speed regime at heavy loading and in a wider speed range at nominal loading for validation purposes.

3. Computational Approach

The present study consisted of a number of simulations, and in order to ensure similar numerical accuracy for all cases, a mesh template was employed such that all geometries and simulated conditions had as similar computational grids as possible. For the reference length in the meshing procedure, a base size of $L/25$ (L being the hull length) or 4.6 cm in dimensional units was selected, and all meshing parameters were based on this length.

There were two primary fluid domains used in the computations. The first region was a larger, background region, which consisted of octree-formed hexahedral cells (Figure 4). Mesh refinements were implemented in the area near the free surface and in the areas in the wake region of the hull. The domain used an anisotropic refinement in the Z (vertical) direction of 25% of base size throughout those zones. In addition, isotropically refined cells of 25% base size were used in the areas near the hull and the area of the expected near-field wave generation. Only half of the fluid region on the port side from the hull centerplane was considered. The dimensions of the fluid domain were based on the hull length and were selected as 10 L × 4 L × 8 L (Figure 4).

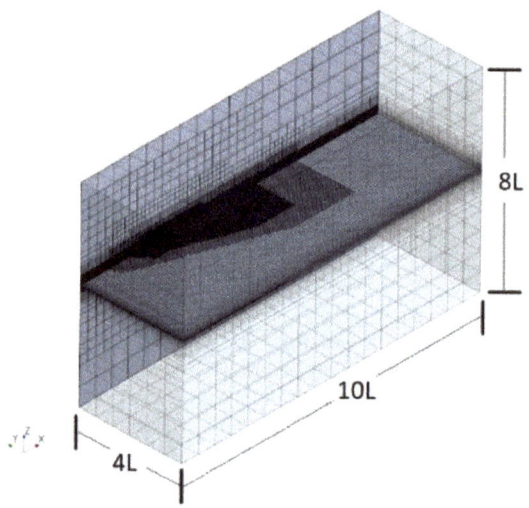

Figure 4. Mesh in the background region.

The second (overset) region encompassed the vessel and used an octree formed trimmed cell mesh with five prism layers along the surface of the hull. The size of this region was chosen by using both length and beam, with the dimensions being 2L × 2B × 2B as shown in Figure 5. Special care was used in the prism layer generation process to ensure the wall Y+ was within the acceptable range. All simulations had surface averaged Y+ values in the range between 30 and 100; therefore, empirical wall functions were used to approximate the flow turbulence. The near wall cell size of the hull was set to 6.25% of the base size. In addition, the prism layer thickness and number of layers were carefully chosen such that the cell size in the outermost layer of the prism layer mesh matched the cell size in the near-wall trimmed mesh (Figure 6). This matching produced a more uniform mesh, which, together with moderate Reynolds numbers used in this study, helped eliminate numerical ventilation. Numerical ventilation is known to be a challenging problem arising in ship hull simulations that rely on the volume-of-fluid (VOF) approach. To address this problem, a common recommendation is to employ very fine numerical grids and very small time steps which, however, may be very computationally prohibitive. Alternative methods include the artificial suppression of the ventilation, such as the phase replacement, corrections to the interface capturing scheme and more gradual transitions between the prism mesh and the volume mesh [12,14,15].

The two regions (background and overset) were then interfaced together using overset interfaces with the linear interpolation scheme. The overset region was placed 3 L behind the inlet and 4 L above the domain bottom as shown in Figure 7. A three-degree-of-freedom (3-DOF) unsteady approach was used to find the steady-state resistance, trim, and sinkage of the vessel. The three degrees of freedom of the vessel were surge, pitch, and heave, or in other words, translation in X and Z directions and rotation about the Y axis passing through of the hull's center of gravity. The background and overset region moved in surge, but only the overset region pitched and heaved. The vessel was initially at rest in a calculated hydrostatic equilibrium state. The vessel was then artificially accelerated from rest using an assumed constant acceleration (1 m/s^2) via a point force attached to the hull's center of mass. The force applied was equal to the drag of the vessel plus the mass of the hull multiplied by the assumed acceleration. This force was exerted until the vessel reached the speed of interest, at which the applied point force became simply equal to the vessel's instantaneous drag force. The reason for using this approach in lieu of the more traditional 2-DOF approach with the constant incoming flow was due to significant hull motions when the simulation was started with at-rest initial conditions. It was found that some hulls would initially exhibit severe motions due to the abrupt (unrealistic) start of the flow if the starting conditions were far from the steady-state conditions at speed. These oscillating motions would then require a long time to dampen out in the simulation. Since the final steady-state condition was not known beforehand, but it was the main objective in these simulations, the 3-DOF method was used. The hydrostatic resting position of the hull can be calculated quickly and accelerating it from rest provides a natural and more realistic evolution of the vessel's sinkage and trim. This approach proved to have a much faster turnaround for this study than the traditional 2-DOF method [10,16].

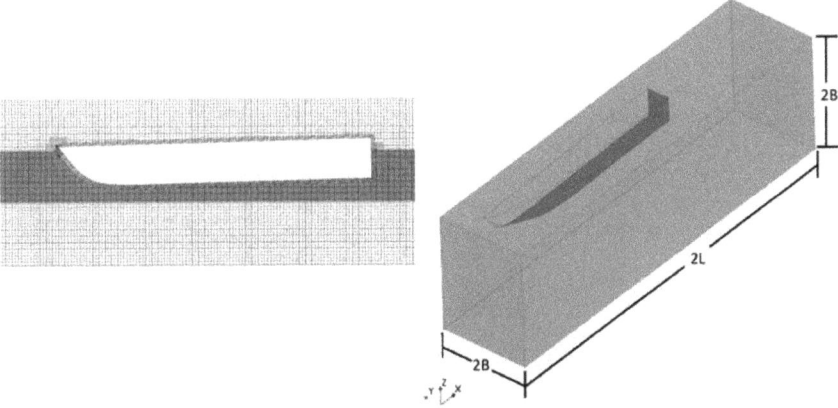

Figure 5. The overset region mesh along the hull centerplane (**left**) and dimensions of this region (**right**).

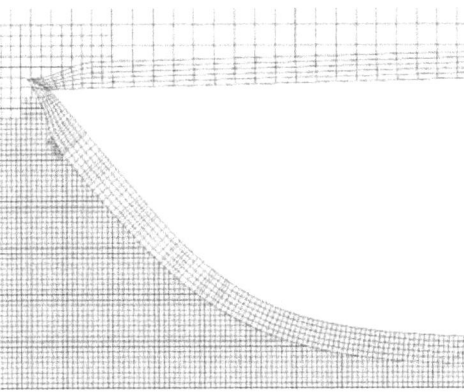

Figure 6. Zoomed-in view of the prism layer mesh near the bow along the hull centerline.

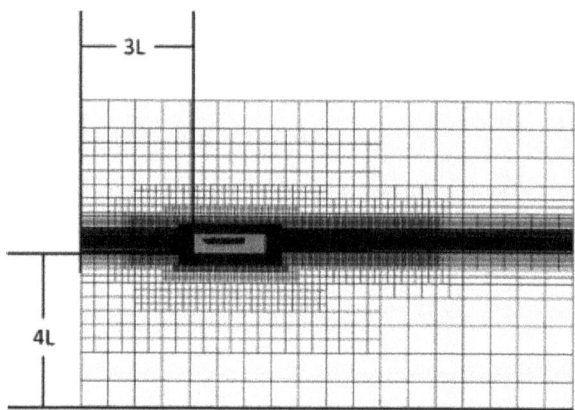

Figure 7. Location of the overset region (shown without mesh) within the background region. The background region can translate forward, and the overset region can translate/rotate in 3 degrees of freedom. The hull surface is shown in black.

The STAR-CCM+ segregated flow solver employing the SIMPLE (Semi-Implicit Method for Pressure Linked Equations) algorithm with second-order convection terms was utilized in this study. The first-order implicit stepping in time was conducted until the time-averaged flow characteristics were no longer evolving. The Eulerian multi-phase method with constant-density air and water, properties of which were consistent with experimental conditions [13]. The high-resolution interface capturing the (HRIC) approach within the volume-of-fluid (VOF) method was employed for resolving the air–water interface. The main fluid mechanics equations used by the solver include the continuity, momentum and VOF equations,

$$\nabla \cdot \bar{v} = 0, \qquad (3)$$

$$\frac{\partial}{\partial t}(\rho \bar{v}) + \nabla \cdot (\rho \bar{v} \otimes \bar{v}) = -\nabla \cdot (\bar{p}I) + \nabla \cdot (T + T_t) + f_b, \qquad (4)$$

$$\frac{\partial c}{\partial t} + \nabla \cdot (c\bar{v}) = 0, \qquad (5)$$

where v is the flow velocity vector, ρ is the density of the mixture, p is the pressure, I is the identity tensor, T is the viscous stress tensor, T_t is the Reynolds stress tensor, f_b is the gravitational body force, and c stands for the volume fraction taken by air. Then, the effective fluid density ρ and viscosity μ are found as $\rho = \rho_{air}c + \rho_{water}(1-c)$ and $\mu = \mu_{air}c + \mu_{water}(1-c)$. The overbar in Equations (3)–(5) correspond to mean flow properties. The Boussinesq hypothesis is used to model the Reynolds stresses,

$$-\overline{\rho u'_i u'_j} = \mu_t \left(\frac{\partial u_i}{\partial x_j} + \frac{\partial u_j}{\partial x_i} \right) - \frac{2}{3}\rho k \delta_{ij}, \qquad (6)$$

where k is the turbulent kinetic energy and μ_t is the turbulent eddy viscosity.

The Reynolds averaged Navier–Stokes (RANS) approach with the realizable $k - \varepsilon$ Two-Layer All-Y+ turbulence model available in Star-CCM+ was utilized [17,18]. The realizable $k - \varepsilon$ model is the most common method in CFD ship hydrodynamics [19]. Other turbulence models were also tried in several conditions (as described in the next section), but they produced results very similar to the Realizable $k - \varepsilon$ model. The governing equations of this model for the turbulent kinetic energy k and the turbulent dissipation rate ε, as well as the expression for the turbulent viscosity μ_t, are given as follows,

$$\frac{\partial(\rho k)}{\partial t} + \frac{\partial(\rho k \bar{u}_j)}{\partial x_j} = \frac{\partial}{\partial x_j}\left[\left(\mu + \frac{\mu_t}{\sigma_k}\right)\frac{\partial k}{\partial x_j}\right] + G_k - \rho \varepsilon, \qquad (7)$$

$$\frac{\partial(\rho\varepsilon)}{\partial t}+\frac{\partial(\rho\varepsilon\bar{u}_j)}{\partial x_j}=\frac{\partial}{\partial x_j}\left[\left(\mu+\frac{\mu_t}{\sigma_\varepsilon}\right)\frac{\partial\varepsilon}{\partial x_j}\right]+\rho C_{\varepsilon 1}S\varepsilon-\rho C_{\varepsilon 2}\frac{\varepsilon^2}{k+\sqrt{\nu\varepsilon}},\qquad(8)$$

$$\mu_t=\rho C_\mu\frac{k^2}{\varepsilon},\qquad(9)$$

where G_k is responsible for turbulent production, S is the magnitude of the mean strain rate tensor, ν is the kinematic viscosity, σ_k, σ_ε, $C_{\varepsilon 1}$, $C_{\varepsilon 2}$ are the model parameters [20], and C_μ depends on both the mean flow and turbulence properties [21].

Following ITTC recommendations on CFD simulations [22], near and at steady-state regimes the time step was selected as $L/(250\ U)$, where U is the hull velocity. Five inner iterations were performed at each time step during the simulations. The initial conditions included the undisturbed fluid at rest. The boundary conditions were specified as shown in Figure 8. The downstream boundary is the pressure outlet with the hydrostatic pressure gradient. The symmetry plane passed through the hull centerplane. The no-slip condition was imposed on the hull surface. Other sides of the domain were treated as the velocity inlets with zero flow condition since the entire mesh moves forward at a rate equivalent to the hull speed. The wave forcing zones of 80% of hull length were applied at the port-side boundary and the upstream and downstream boundaries. The wave forcing method involves activation of momentum sources near domain boundaries that adapt the solution to specified boundary conditions [20]. This way, one can minimize undesirable numerical wave reflections and thus use more compact (economical) numerical domains.

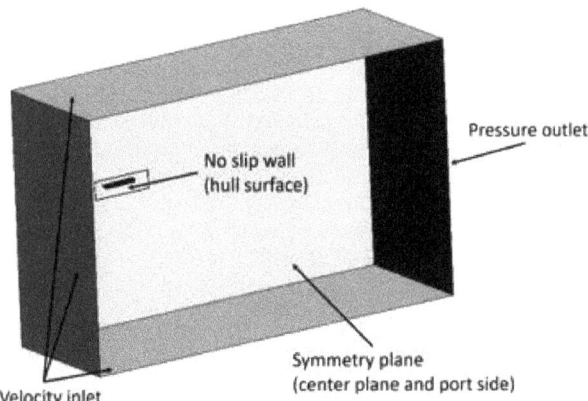

Figure 8. Boundary conditions used in simulations. The front, top, and bottom of the domain are set to velocity inlets. The rear of the domain enforces hydrostatic pressure at the outlet. The sides of the domain are modeled as symmetry planes and the hull surface is treated as a no slip wall.

4. Verification and Validation

Solution verification study was conducted at two conditions with $C_B=0.912$, for which experimental data [13] are also available. Condition (1) involves LCG = 45% and $Fr_\nabla = 1.67$ (transitional regime), whereas condition (2) corresponds to LCG = 40% and $Fr_\nabla = 2.68$ (close to planing regime). To perform the verification, solutions were obtained on three mesh levels with different characteristic cell size. The base size was changed by factors of $\sqrt{2}$ for three mesh levels. The corresponding time step was also changed by a factor of $\sqrt{2}$ as well to keep the Courant number the same between grids. As an indicator of the solution convergence, the drag coefficient based on beam [23] was used,

$$C_R=\frac{R}{0.5\rho u^2 B^2},\qquad(10)$$

where R is the total hull resistance, ρ is the water density, B is the hull beam, and u is the hull speed. Expressed in this form, the resistance coefficient becomes directly proportional to the actual resistance for hulls with the same beam, as in this work. The results for C_R obtained in the verification study are given in Table 1, showing monotonic convergence. The finest mesh had 4.3 million cells, and this mesh template was used in the rest of the study.

Table 1. Resistance coefficient obtained at three mesh levels and two operating conditions.

Mesh Type	Mesh Size	Resistance Coefficient, C_R	
		Condition (1)	Condition (2)
		LCG = 45%	LCG = 40%
		$Fr_V = 1.67$	$Fr_V = 2.68$
Fine	4.32 Million	0.110	0.0458
Medium	2.04 Million	0.111	0.0469
Coarse	1.48 Million	0.114	0.0506
Numerical uncertainty		0.006	0.006

To estimate the numerical uncertainty, first the Richardson extrapolation was used to determine the solution corrections [24],

$$\delta_{RE} = \frac{\Delta_{21}}{\beta^p - 1}, \tag{11}$$

where Δ_{21} is the difference between solutions found on fine and medium grids, $\beta = \sqrt{2}$ in this study, and p is the observed order of accuracy. Then, these corrections were multiplied by the factors of safety following one of the standard methods [25]. The numerical uncertainties came out as 5.7% and 13.6% for conditions (1) and (2), respectively. These and percentage uncertainties given below are evaluated with respect to the solution values obtained on the fine grids.

The total validation uncertainty U_V combines both the experimental U_D and numerical U_{NS} uncertainties as follows,

$$U_V = \sqrt{U_D^2 + U_{NS}^2}. \tag{12}$$

Although the experimental uncertainty was not specified, it is assumed to be about 8%, common for this type of test. Then, the validation uncertainties for the two cases become 9.8% and 15.8%, respectively. The corresponding differences between the numerically calculated and experimental values are about 4.9% and 14.9%. Since these differences are within the validation uncertainties, the CFD models can be considered as validated at these two conditions.

A comparison between numerical and experimental results in the range of speeds for two LCG values is shown in Figure 9. The agreement at transitional speeds, which are the primary interest in this study, is very good. The numerical results show somewhat higher drag than test data in the planing regimes. As stated above, the numerical and experimental uncertainties can be responsible for part of these differences. It is also noted that previous CFD simulations with planing hulls, which employed a much higher number of numerical cells than the present study, produced results demonstrating similar discrepancy with the experimental data [9,10]. Insufficiently accurate modeling of spray at high speeds is a possible cause for this discrepancy. Numerical grids of very high resolution or the development of different models for spray may be needed to address this issue.

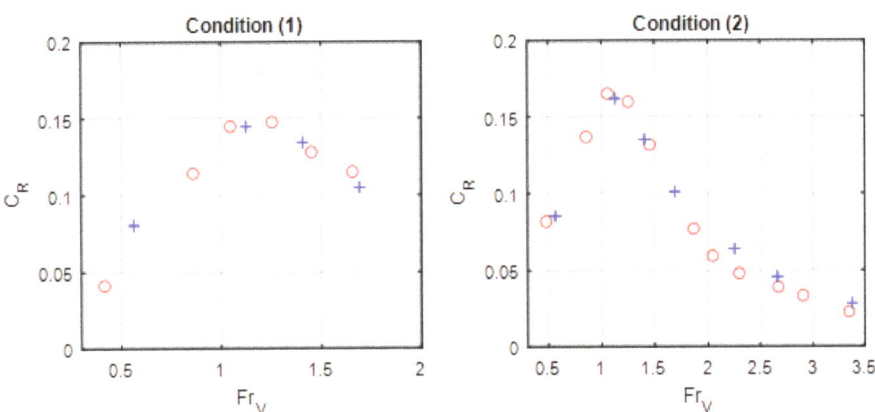

Figure 9. Experimental and numerical results for constant-deadrise hull, $C_B = 0.912$ and two LCG conditions: (**1**) LCG = 45%, (**2**) LCG = 40%. Red circles, experimental data; blue crosses, numerical results.

The computational times needed to achieve steady states for the hulls used in the validation study have been assessed as well. The central processing unit (CPU) times, defined as the actual time multiplied by the number of employed processors, is given in Figure 10 for the range of Froude numbers. The heavier hulls and intermediate Froude numbers, which correspond to semi-planing regimes, required longer CPU times.

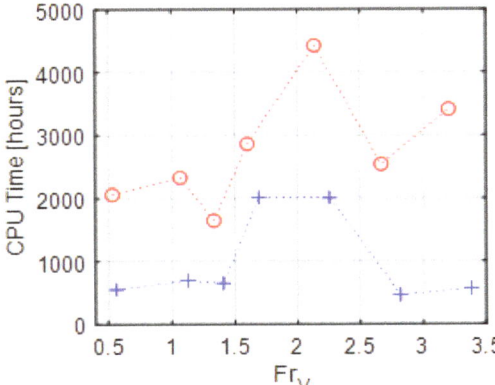

Figure 10. Central processing unit (CPU) times spent to achieve steady-state results for constant-deadrise hulls with LCG = 45% in light (blue crosses) and overloaded (red circles) conditions.

Additional simulations have been conducted here with different turbulence models on the fine mesh. These models included the $k-\omega$ SST (shear stress transport) and the Reynolds stress turbulence (RST) models. The simulations were carried out for both LCG and Froude numbers used in the verification study. Results for the resistance coefficient are summarized in Table 2. The differences in the resistance values were less than 1% for the transitional case, so accuracy of all three turbulence models is about the same at this condition. In the planing regime, the experimental value for C_R was about 0.4, so the error between the experiment and the RST model result was larger and, therefore, the RST model was not used. The realizable $k-\varepsilon$ model was chosen since it was closer to the experimental data point and showed lower oscillations in the monitored values compared to the $k-\omega$ SST values.

Table 2. Resistance coefficient obtained with different turbulence models at two operating conditions.

Turbulence Models	Resistance Coefficient, C_R	
	Condition (1) LCG = 45% $Fr_V = 1.67$	Condition (2) LCG = 40% $Fr_V = 2.68$
Realizable $k - \varepsilon$	0.110	0.0458
$k - \omega$ SST	0.110	0.0466
Reynolds Stress Turbulence	0.110	0.0485

5. Parametric Results

Since the focus of this study is on heavy hulls that perform well in the transitional speed range, the initial parametric calculations were carried for three hull geometries at the loading coefficient $C_B = 1.276$, two centers of gravity LCG = 40% and 45%, and speeds corresponding to $Fr_V = 1.0$–1.6. To present results in the non-dimensional form, three metrics are used: the resistance coefficient defined by Equation (10), the hull trim, τ, and the rise of the center of gravity (in comparison with the rest position) normalized by the hull beam, H/B.

The resistance coefficient and attitude data obtained for heavy hulls in the transitional regime are shown in Figure 11. As one can notice, the hulls with LCG = 45% consistently outperform those with LCG = 40% (Figure 11a). The trim angles of the configurations with the rearward CG are noticeably higher (by 3–4 degrees) than trims of the hull with more forward CG (Figure 11b). At moderate speeds, excessive trim angles result in larger pressure drag, while the hydrodynamic lift is not yet developed to raise these hulls to higher positions. On the contrary, the dynamic suction at these speeds increases the hull submergences. Differences between sinkages of hulls with different CG locations are not as pronounced as differences in drag and trim (Figure 11c). Thus, the hull configurations with LCG = 45% were selected for further studies in a broader speed range.

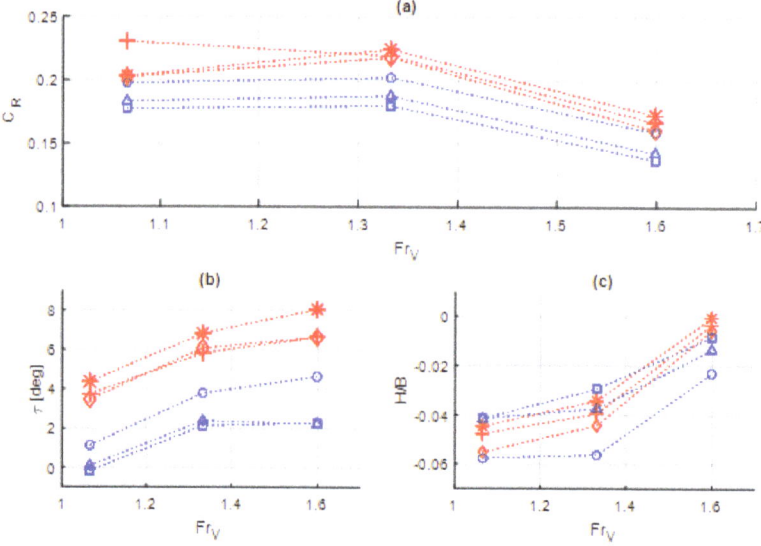

Figure 11. Comparison of hydrodynamic characteristics in the transitional regime of overloaded hulls ($C_B = 1.276$) with LCG = 45% (blue smaller symbols) and LCG = 40% (red larger symbols). (**a**) Drag coefficient, (**b**) trim angle, (**c**) normalized rise of the center of gravity. Circles and stars, constant-deadrise hull; squares and crosses, concave hull; triangles and diamonds, convex hull.

Both nominal and heavy hulls with C_B = 0.912 and 1.276 and three bow geometries were computationally simulated in Fr_V interval from 0.5 up to 3.5, covering all important regimes from the displacement to planing modes. Starting from zero speed, hulls were accelerated at 1 m/s² till their speeds reached required values. An example of time-dependent hull characteristics, demonstrating attainment of a steady-state regime, is shown in Figure 12 for one of the studied hulls. The steady-state results for the resistance coefficient, trim and CG rise for all hull configurations are summarized in Figure 13. General shapes of the resistance coefficient curves (Figure 13a) are rather common to hard-chine hulls. There is a steep drag increase at the transitional speeds, followed by the resistance coefficient peak around Fr_V = 1.2 and some reduction of resistance at the post-hump planing speeds. As expected, heavy hulls demonstrate higher drag, and the drag increase is roughly similar to the relative increase of hull displacements.

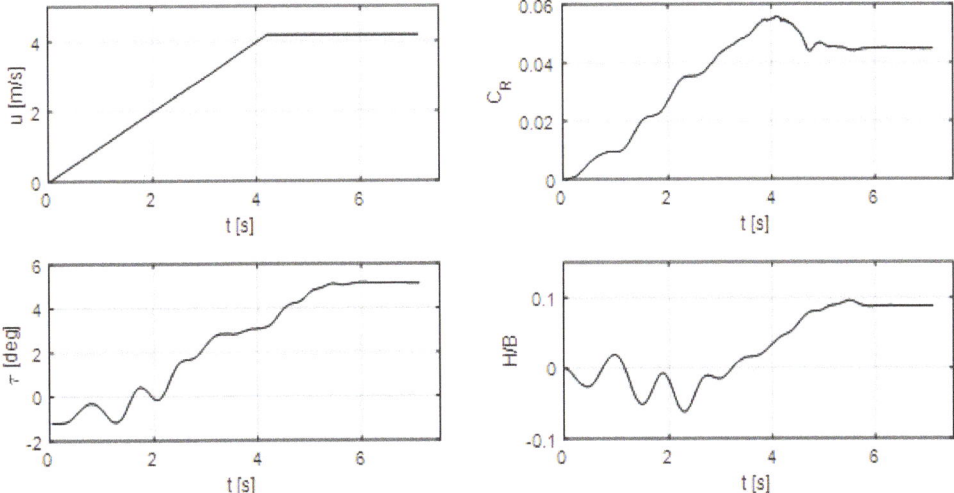

Figure 12. Time histories of hull speed, resistance coefficient, trim and relative sinkage for convex-bow hull with LCG = 45% and C_B = 0.912 at Froude number 1.25.

The trim angles generally increase with speed (Figure 13b), demonstrating faster growth at the transitional speeds and saturation at the high planing speeds. However, the hulls with curved bows (both nominal-weight and heavy) exhibit a significant drop in trim at Fr_V = 2.2. This is caused by earlier exits of finer bows at this speed (in comparison with the original constant deadrise hulls), accompanied by the loss of lift at the front portion of the hull, which results in the bow-down adjustment.

The vertical positions of the hulls' centers of gravity initially descend due to dynamic suction near Fr_V = 1.2 but, later, with increasing trim and speed, the hulls rise due to higher hydrodynamic lift (Figure 13c). Again, at Fr_V = 2.2, the hulls with finer bows do not experience significant elevation increase (as the constant-deadrise hulls) due to trim reduction and some loss of hydrodynamic lift. At higher speeds, resistance and attitude characteristics of hulls of different geometries approach each other, since the bows almost exit the water and the rear prismatic-type hull portions are identical for all three hulls studied here.

When comparing the performance of different hull forms in the overloaded condition, one can notice that the hull with concave bow has consistently lower resistance in the transitional regime, Fr_V = 1.0–1.7. Its finest bow shape (Figure 3), among those of the hulls studied here, helps this hull cut through the water more efficiently at semi-displacement speeds. At the nominal (lighter) loading and lower speeds, Fr_V < 1.0, the hull with concave bow is also superior in terms of resistance, which will allow it to operate more economically in that regime. When a high speed is needed at the nominal loading, the concave-bow hull will be able to reach planing speeds $Fr_V > 2.5$ with the available thrust-weight ratio of 0.2. Thus, the hull with the concave bow would be the best performer for the specific operational regimes of interest to this study. It should be noted that only calm-water conditions were considered here, and additional studies will be needed if operations in rough seas are taken into consideration.

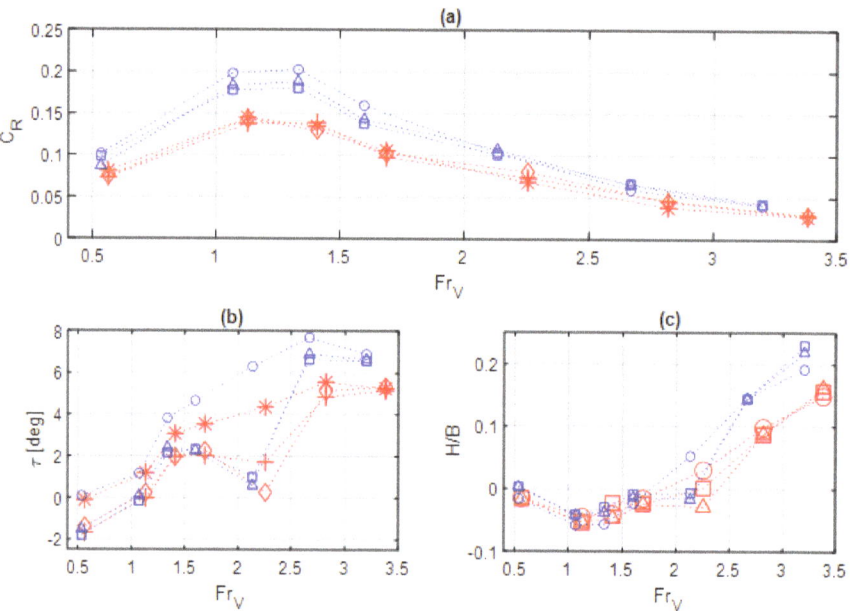

Figure 13. Steady-state results for hulls with LCG = 45%: (**a**) resistance coefficient, (**b**) trim, (**c**) normalized CG rise. Circles and stars, constant-deadrise hull; squares and crosses, concave hull; triangles and diamonds, convex hull. Red larger symbols, nominal-weight condition (C_B = 0.912); blue smaller symbols, overloaded (C_B = 1.276).

The convex-bow hull (Figure 3), while inferior to the concave counterpart in calm water, slightly outperforms the constant-deadrise hull in the transitional regimes (Figure 13). However, the hull with convex bow exhibits the largest peak of the actual drag force near $Fr_V = 2.2$ in both nominal and heavy loadings. The constant-deadrise hull is superior at the planing speeds due to its pronounced prismatic hull surfaces. Again, if operations in waves are considered, relative performances of different hull forms may change.

One of the interesting metrics of hulls intended for a broad speed range is the correspondence between pressure and friction (shear) drag components. The fraction of the pressure drag in the total hull resistance is shown in Figure 14. Obviously, heavily loaded hulls have a higher pressure drag contribution in comparison with lighter hulls. The pressure-drag fraction peaks at Froude number around 1.3. These speeds belong to the transitional regime where the hulls experience large drag but relatively low hydrodynamic lift. The secondary peak in the pressure-drag fraction is noticeable for heavy hulls at early planing speeds, $Fr_V = 2.7$. As commonly known, the frictional drag becomes more pronounced at the lowest (displacement) and highest (developed planing) speeds (Figure 14), although hulls with the finer bows also tend to have a larger frictional contribution at $Fr_V = 2.2$, when hull trim angles drop slightly (Figure 13b).

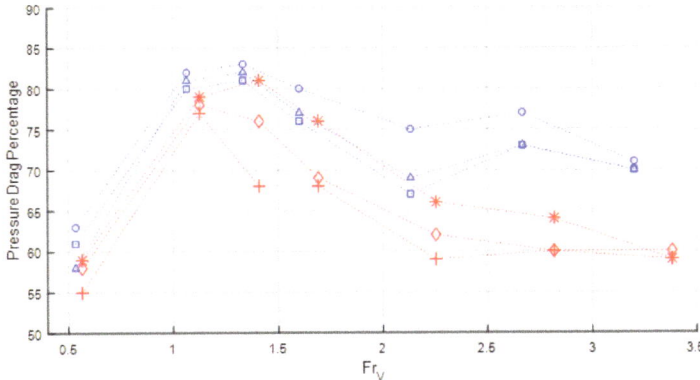

Figure 14. Comparison of the pressure drag expressed as the percentage of total drag for hulls with LCG = 45%. Circles and stars, constant-deadrise hull; squares and crosses, concave hull; triangles and diamonds, convex hull. Red larger symbols, nominal-weight condition (C_B = 0.912); blue smaller symbols, overloaded (C_B = 1.276).

More detailed insight on the flow characteristics near hulls can be gained from the distribution of pressure coefficients, C_p, on the hull bottom and the water surface deformations around hulls, which are given for selected states in Figures 15 and 16, respectively. One can notice a slightly larger wet area of a constant-deadrise hull bottom at the lower speed ($Fr_V \approx 1.3$) in comparison with other hulls that have finer bows (Figure 15). The highest pressure coefficient is observed at the water impingement zone at the bow. In the overloaded cases and lower speeds, this high-pressure zone is more pronounced for the constant-deadrise hull (Figure 15), while C_p magnitudes in this region are the lowest for the concave-bow hull. This is consistent with the resistance coefficient values shown in Figure 13a, where the constant-deadrise and concave-bow heavy hulls have the highest and lowest resistances, respectively, at $Fr_V \approx 1.3$. On the other hand, small regions with reduced pressure are visible near hull transoms, where pressure recovers back to atmospheric and, therefore, does not significantly contribute to the boat lift. At $Fr_V \approx 2.7$, pressure coefficient values are generally smaller since the flow speeds are higher, but loadings are the same. The wet area of the constant-deadrise hull is smaller at higher speed ($Fr_V \approx 2.7$) than wet areas of concave- and convex-bow hulls due to larger trim angles of the constant-deadrise hull (Figure 13b).

The near-hull water surface elevations for the same 12 cases are illustrated in Figure 16. At lower speeds ($Fr_V \approx 1.3$), significant water build-up with wave breaking features appears in front of the bow of the constant-deadrise hull, whereas the bow waves extend further along the hulls with finer bows. The concave-hull bow nose is slightly less wet than the convex-hull counterpart. The water depression at the transom and the following "rooster tail" are more pronounced for heavier hulls. At higher speed ($Fr_V \approx 2.7$), the constant deadrise-hull has a noticeably larger trim than other hulls. The "rooster tails" of convex and concave hulls are located closer to the transom than for a more prismatic hull. The divergent waves generated by heavier hulls are more pronounced, since those hulls displace more water. The wake zones of faster hulls are narrower than those behind hulls operating at lower speeds. The water depression zones behind the transom become more aligned with the hull centerline at higher speeds in comparison with wave hollows nearly split in two parts behind most hulls at lower speeds.

The full-domain wave patterns behind one of the hulls at two loadings are illustrated in Figure 17. At this high speed ($Fr_V \approx 2.7$), the well-defined divergent waves are clearly visible. The heavier hull sits deeper in the water and produces larger wave amplitudes. Near the downstream boundary on the right side of computational domains in Figure 17, the numerical forcing zone suppresses the waves, resulting in diminishing magnitudes of water surface elevations.

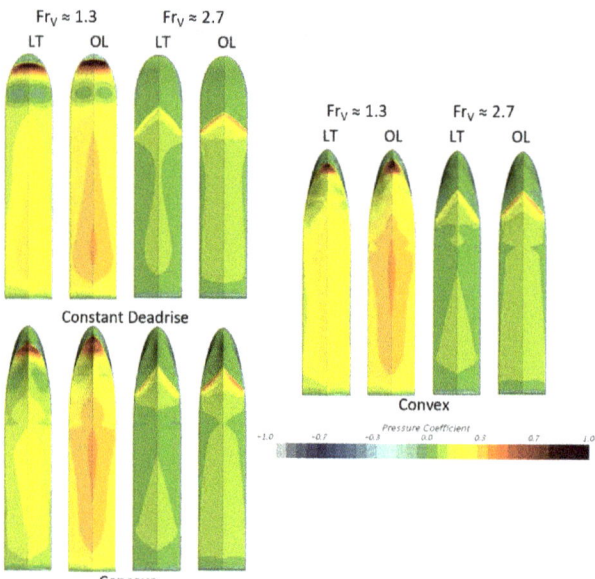

Figure 15. Comparison of pressure coefficient for each hull form at LCG = 45% and Froude numbers 1.3 and 2.7 for both lighter (LT) and overloaded (OL) configurations.

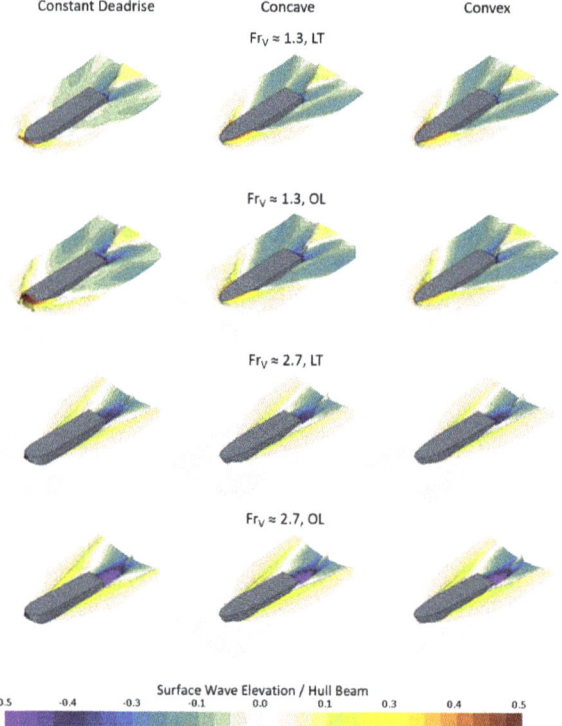

Figure 16. Comparison of near-hull waves for each hull form at LCG = 45% and Froude numbers 1.3 and 2.7 for both lighter (LT) and overloaded (OL) configurations.

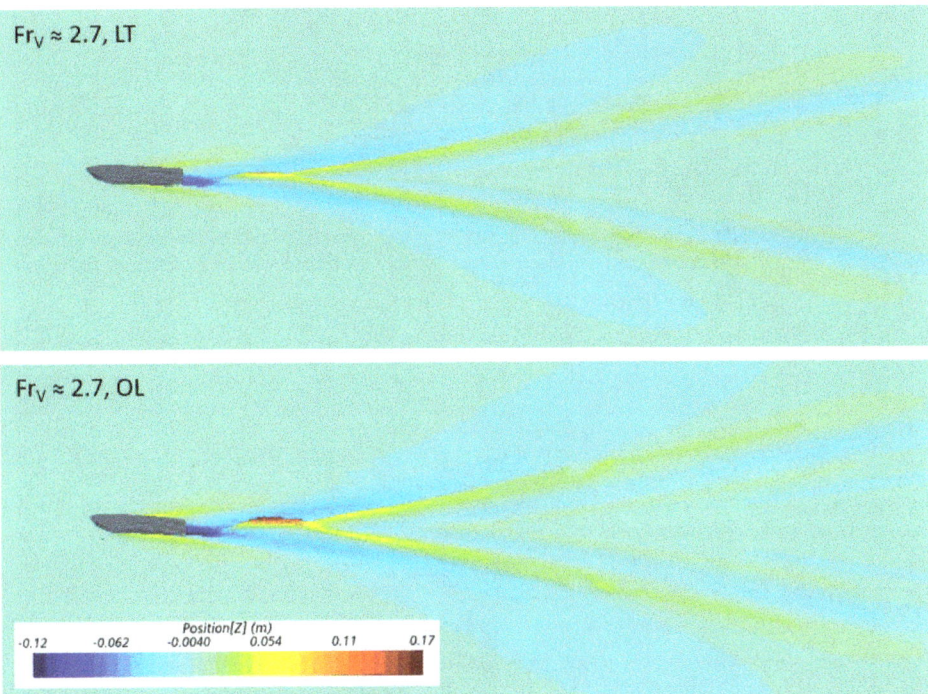

Figure 17. Far-field waves produced by convex-bow hull with LCG = 45% and at Froude number 2.7 for both lighter (LT) and overloaded (OL) configurations.

6. Conclusions

A computational study has been undertaken to evaluate the performance of basic hard-chine hull forms in heavily loaded conditions and in the broad speed range from the displacement to planing regimes. The resistance coefficient curves demonstrated a typical behavior with the peaks around the displacement Froude number of 1.2. Hulls with 40% heavier displacements manifested about 30% larger resistance than lighter hulls over the studied speed range. In the overloaded regime and transitional speeds, the concave-bow hull is found to have about 5% and 12% lower drag than the convex-bow hull and the more prismatic constant-deadrise hull, respectively. The same convex-bow hull at the nominal loading and displacement speeds showed 2–10% lower drag than the other hulls, as well as moderate resistance up to the planing speeds. The hulls with finer bows exhibited a significant 2–3° trim decrease at the hump speed. The original constant-deadrise hull with a long prismatic portion of the hull performs better at the planing speeds, demonstrating 5–15% lower drag than other hulls.

Future research directions can involve investigating the performance of heavy hulls in the presence of waves to provide recommendations for overloaded hard-chine hulls intended for variable sea conditions. The present computational approach is also suitable for determining the effects of finer geometric details on the hull surfaces, such as spray rails, steps, and appendages, and conducting hull optimization studies at the design stage.

Author Contributions: Conceptualization, M.P.W., K.I.M. and T.X.; methodology, M.P.W., K.I.M. and T.X.; software, M.P.W.; validation, M.P.W., K.I.M. and T.X.; formal analysis, M.P.W.; investigation, M.P.W.; resources, K.I.M.; data curation, M.P.W. and K.I.M.; writing—original draft preparation, M.P.W.; writing—review and editing, K.I.M. and T.X.; visualization, M.P.W.; supervision, K.I.M.; project administration, K.I.M.; funding acquisition, K.I.M. All authors have read and agreed to the published version of the manuscript.

Funding: This research was funded by the U.S. Office of Naval Research, grant number N00014-17-1-2553, and the U.S. National Science Foundation, grant number 1800135.

Data Availability Statement: The data presented in this study are available on request from the corresponding author. The data are not publicly available due to requirements from funding agencies.

Conflicts of Interest: The authors declare no conflict of interest.

References

1. Egorov, I.T.; Bunkov, M.M.; Sadovnikov, Y.M. *Propulsive Performance and Seaworthiness of Planing Boats*; Sudostroenie: Leningrad, Russia, 1978.
2. Savitsky, D. Hydrodynamic design of planing hulls. *Mar. Technol.* **1964**, *1*, 71–95.
3. Faltinsen, O.M. *Hydrodynamics of High-Speed Marine Vehicles*; Cambridge University Press: New York, NY, USA, 2005.
4. Almeter, J.M. Resistance prediction of planing hulls: State of the art. *Mar. Technol.* **1993**, *30*, 297–307.
5. Doctors, L.J. Representation of planing surfaces by finite pressure elements. In Proceedings of the 5th Australian Conference on Hydraulics and Fluid Mechanics. Proceedings, Christchurch, New Zealand, 9–13 December 1974.
6. Lai, C.; Troesch, A.W. A vortex lattice method for high-speed planing. *Int. J. Numer. Methods Fluids* **1996**, *22*, 495–513. [CrossRef]
7. Bari, G.S.; Matveev, K.I. Hydrodynamic modeling of planing catamarans with symmetric hulls. *Ocean Eng.* **2016**, *115*, 60–66. [CrossRef]
8. O'Shea, T.T.; Brucker, K.A.; Wyatt, D.; Dommermuth, D.G.; Ward, J.; Zhang, S.; Weems, K.; Lin, W.-M.; Judge, C.; Engle, A. *Validation of Numerical Predictions of the Impact Forces and Hydrodynamics of a Deep-V Planing Hull*; NSWCCD Report No. 50-TR-2012/040; Naval Surface Warfare Center Carderock Division: Bethesda, MD, USA, 2012.
9. Mousaviraad, S.M.; Wang, Z.; Stern, F. URANS studies of hydrodynamic performance and slamming loads on high-speed planing hulls in calm water and waves for deep and shallow conditions. *Appl. Ocean Res.* **2015**, *51*, 222–240. [CrossRef]
10. Sukas, O.F.; Kinaci, O.K.; Cakici, F.; Gokce, M.K. Hydrodynamic assessment of planing hulls using overset grids. *Appl. Ocean Res.* **2017**, *65*, 35–46. [CrossRef]
11. De Marco, A.; Mancini, S.; Miranda, S.; Scognamiglio, R.; Vitiello, L. Experimental and numerical hydrodynamic analysis of a stepped planing hull. *Appl. Ocean Res.* **2017**, *64*, 135–154. [CrossRef]
12. Wheeler, M.; Matveev, K.I.; Xing, T. Validation study of compact high-lift planing hulls at pre-planing speeds. In Proceedings of the 5th Joint US-European Fluids Engineering Summer Conference, Montreal, QC, Canada, 15–20 July 2018; ASME Paper FEDSM 2018-83091.
13. Fridsma, G. *A Systematic Study of the Rough-Water Performance of Planing Boats*; Report No. 1275; Davidson Laboratory, Stevens Institute of Technology: Hoboken, NJ, USA, 1969.
14. Bohm, C.; Graf, K. Advancements in Free Surface RANSE Simulations for Sailing Yacht Applications. *Ocean Eng.* **2014**, *90*, 11–20. [CrossRef]
15. Gray-Stephens, A.; Tezdogan, T.; Day, S. Strategies to minimise numerical ventilation in CFD simulations of high-speed planing hulls. In Proceedings of the ASME 38th International Conference on Ocean, Offshore and Arctic Engineering, Glasgow, UK, 9–14 June 2019.
16. Frisk, D.; Tegehall, L. Prediction of High-Speed Planing Hull Resistance and Running Attitude; X-15/320. Master's Thesis, Chalmers University, Gothenburg, Sweden, 2015.
17. Shih, T.-H.; Liou, W.W.; Shabbir, A.; Yang, Z.; Zhu, J. A new k-ϵ eddy viscosity model for high Reynolds number turbulent flows. *Comput. Fluids* **1995**, *24*, 227–238. [CrossRef]
18. Rodi, W. Experience with two-layer models combining the k-ϵ model with a one-equation model near the wall. In Proceedings of the 29th Aerospace Sciences Meeting, Reno, NV, USA, 7–10 January 1991; AIAA Paper 91-0216.
19. De Luca, F.; Mancini, S.; Miranda, S.; Pensa, C. An extended verification and validation study of CFD simulations for planing hulls. *J. Ship Res.* **2016**, *60*, 101–118. [CrossRef]
20. STAR-CCM+ Manual 2020. Available online: https://www.plm.automation.siemens.com/global/en/products/simcenter/STAR-CCM.html (accessed on 10 December 2020).
21. Mulvany, N.; Tu, J.Y.; Chen, L.; Anderson, B. Assessment of two-equation modeling for high Reynolds number hydrofoil flows. *Int. J. Numer. Methods Fluids* **2004**, *45*, 275–299. [CrossRef]
22. International Towing Tank Conference. *Practical Guidelines for Ship CFD Aplications*; Publication 7.5-03-02-03; ITTC Association: Zürich, Switzerland, 2011.

23. Chambliss, D.B.; Boyd, G.M. *The Planing Characteristics of Two V-Shaped Prismatic Surfaces Having Angles of Deadrise of 20° and 40°*; NACA Technical Note 2876; Langley Aeronautical Laboratory, National Advisory Committee for Aeronautics: Washington, DC, USA, 1953.
24. Ferziger, J.H.; Peric, M. *Computational Methods for Fluid Dynamics*; Springer: Berlin/Heidelberg, Germany, 1999.
25. Xing, T.; Stern, F. Factors of safety for Richardson extrapolation. *ASME J. Fluids Eng.* **2010**, *132*, 061403. [CrossRef]

Article

Bubble Sweep-Down of Research Vessels Based on the Coupled Eulerian-Lagrangian Method

Wei Wang, Guobin Cai, Yongjie Pang, Chunyu Guo, Yang Han and Guangli Zhou *

College of Shipbuilding Engineering, Harbin Engineering University, Harbin 150001, China; wwei@cssc.net.cn (W.W.); caiguobin@hrbeu.edu.cn (G.C.); pangyongjie@hrbeu.edu.cn (Y.P.); guochunyu@hrbeu.edu.cn (C.G.); hanyang@hrbeu.edu.cn (Y.H.)
* Correspondence: zhouguangli@hrbeu.edu.cn; Tel.: +86-1884-508-1827

Received: 4 November 2020; Accepted: 15 December 2020; Published: 21 December 2020

Abstract: To explore the reason for the bubble sweep-down phenomenon of research vessels and its effect on the position of the stern sonar of a research vessel, the use of a fairing was investigated as a defoaming appendage. The separation vortex turbulence model was selected for simulation, and the coupled Eulerian-Lagrangian method was adopted to study the characteristics of the bubble sweep-down motion, captured using a discrete element model. The interaction between the bubbles, water, air, and hull was defined via a multiphase interaction method. The bubble point position and bubble layer were calculated separately. The spatial movement characteristics of the bubbles were extracted from bubble trajectories. It was demonstrated that the bubble sweep-down phenomenon is closely related to the distribution of the bow pressure field and that the bubble motion characteristics is related to the speed and initial bubble position. When the initial bubble position is between the water surface and the ship bottom, the impact on the middle of the ship bottom is greater and increases further with increasing speed. A deflector forces the bubbles to both sides through physical shielding, strengthening the local vortex structure and keeping bubbles away from the middle of the ship bottom.

Keywords: bubble sweep-down; detached eddy simulation; Coupled Eulerian-Lagrangian method; distinct element method; multiphase interaction method; bubble point position and bubble layer; motion track

1. Introduction

Exploring and understanding the ocean are prerequisites for the development of marine resources and the protection of the marine ecological environment. The marine research vessel is a type of ship extremely suitable for this. The research vessel is dedicated to scientific investigations of the sea, with the purpose of obtaining comprehensive marine geology, biology, and ecology survey information of the atmosphere, for example. As the "eye" of the research vessel, the sonar equipment, specifically its performance, plays a vital role in the accuracy of the research vessel's detection results.

Karafiath [1] analyzed the occurrence of the phenomenon of bubble sweep-down and believed that, under actual sea conditions, owing to the strong fluidity of seawater, strong sea breeze, and the effects of wave breaking and rainwater impact, the seawater near the water surface has a certain air content, within a certain water depth range. A layer of suspended bubbles is formed in this water layer.

Deane and Stokes [2] measured the bubble size distribution in breaking waves in the laboratory and on the high seas, provided a quantitative description of the bubble formation mechanism in the laboratory, and analyzed the dependence of scale on bubble generation and propagation and the mechanism of breaking wave conditions. Thorpe [3] described the fact that small bubbles with a radius of less than 1 mm are stabilized by surface tension, whereas bubbles with a larger radius are broken by the shear stress in the turbulent motion caused by the collapse event. Smaller bubbles rise very slowly; hence, they persist in the water column and flow at greater depths. When the research vessel sails in this bubble layer, the bubbles move along the hull surface owing to hull wakes. Moving down to the bottom of the ship, the phenomenon of bubble downward scanning occurs, which affects the performance of the sonar at the stern position and affects the detection function of the research vessel. Sebastian and Caruthers [4] recorded the impact on the operation of a multibeam sonar.

For some ship types, an excellent inlet design can eliminate the down-sweep phenomenon of bubbles, but the phenomenon still appears after the speed increases. Rolland [5] shows that direct installation of a defoaming attachment inevitably brings a certain increase in resistance, sometimes even up to 20%. Therefore, it is a better direction to first study the bubble motion characteristics of the transducer surface.

The current research on ship performance widely uses a ship model pool, and the scale effect is unavoidable. It is difficult to form a uniform microbubble layer in the water. It can only be supplemented by a bubble generator to generate bubbles in real time. To make matters worse, large physical pools are often left for a long time, resulting in pools containing far less air, so that bubbles dissolve in water more quickly than under actual sea conditions. Therefore, this study combines research on computational fluid dynamics (CFD) methods, which was conducted to verify, e.g., conventional resistance research, and single or limited location bubble generation research.

Delacroix [6] quantified the backscattered signal on the bubble cloud image with an echo sounder, and studied the influence of wind speed under navigation conditions on the characteristics of the bubble sweep-down. Mallat [7] used the particle image velocimetry (PIV) test method, using the bow longitudinal section and in the form of streamline, to study the 3D characteristics of the sweep-down of the bubble.

Many people, including Han [8], have used CFD for many years for hull shape optimization. Delacroix used it for a characteristic study of bubble sweep-down and Palaniappan and Subramanian [9] worked out the hydrodynamic design for bubble sweep-down.

In order to make up for the deficiencies of the bubble experiment in the pool, this study used the Eulerian-Lagrangian method to model the bubbles, as described in Section 2, and the CFD method was adopted to calculate the bubble point and bubble layer. By ignoring the dissolution and breaking of bubbles, the phenomenon of bubble sweep-down was studied. In Section 3, the accuracy of the CFD calculation method is verified through the resistance and single-bubble point towing tank test. In Section 4, we show more comprehensive calculation results, including further calculation and extraction results of bubble characteristics and flow field details such as velocity field.

2. Numerical Method

2.1. Governing Equations

Energy exchange is not involved in the research, and the continuity equation and momentum equation are, respectively, as follows:

$$\frac{\partial}{\partial t}(a_q \rho_q) + \nabla \cdot (a_q \rho_q v_q) = 0 \tag{1}$$

$$\frac{\partial}{\partial t}(\rho v) + (\rho v v) = -\nabla p + \nabla \cdot \left[\mu(\nabla v + \nabla v^T)\right] + \rho g + F \tag{2}$$

where a_q is the volume fraction of phase q, ρ_q is the density of phase q, ρ is the mixed phase density, v_q is the velocity of phase q, μ is the sum of turbulent viscosity and molecular mixing viscosity, g is the acceleration, and F is the external force.

2.2. Turbulence Model and Coupled Eulerian-Lagrangian Method

In the research of microbubble scale, it is considered that the bubble and water are both interacting and represent two relatively independent phases. The bubble phase is located in the water phase but is not soluble in water. When the volume fraction is used to express the volume fraction of the phase, the volume fraction function of the two phases of bubbles and water is continuous in time and space and the sum is 1. The volume of fluid (VOF) method is used to track the interface between the two phases of water and air, and the bubbles ejected from the bow of the ship are used as discrete bubble-phase particles distributed in the continuous fluid domain, and the motion model of the bubble particles is established by the discrete element method (DEM) method. In the study, the bubble diameter was set to 1 mm, and the bubble spacing was 16.8 mm. Therefore, the interaction between the bubble particles is relatively weak, and the influence on the continuous fluid domain can be ignored.

Maxwell [10] used the DEM model to study the interaction force between bubbles and particles and the sliding of particles. Bérard [11] summarized the progress of a CFD-DEM calculation of solid-liquid coupling in chemical engineering. Based on the coupled Eulerian-Lagrangian method, Xinhong Li [12] solved the trajectory of the discrete-phase bubble particles, so that the force is balanced during the movement as follows:

$$\frac{du_p}{dt} = F_d(u - u_p) + \frac{g(\rho_p - \rho)}{\rho_p} + F \qquad (3)$$

where u is the towing speed of the ship, u_p is the bubble particle velocity, F_d is the drag force, measured by experiment, ρ is the continuous phase density, and ρ_q is the bubble particle density.

The sweep-down of bubbles involves capturing motion near the wall of the hull. The shear stress transmission (SST) k-omega detached eddy simulation (DES) turbulence model is used for simulation to close the equation.

DES is a hybrid modeling method that uses time-averaged Reynolds-averaged Navier-Stokes (RANS) to solve near the wall boundary layer, while the turbulence is away from the wall. The area is solved by transient large eddy simulation (LES), which balances calculation accuracy and calculation cost. Zhang and Ahmadi [13] used the Eulerian-Lagrangian calculation model to simulate gas-liquid-solid three-phase flow. Watson [14] used a delayed separation vortex simulation to calculate the unsteady flow of the ship hull. Home and Lightstone [15] used DES-SST to study the flow of interstitial vortices. Jee and Shariff [16] proposed the v2-f DES model and calculated the cylindrical flow around and the turbulence phenomena. Although DES reduces the requirements for computing grids compared to LES, the requirements for grid quality are still higher.

The SST model was used to simulate the inverse pressure gradient near the wall. With respect to the unsteady flow, the finite volume method was used to solve the problem, and the coupling solver of VOF and DEM was used.

2.3. Numerical Scheme in CFD

CFD calculation software STAR-CCM+ was used for numerical calculation, and the grid scheme was designed according to the above calculation model method. The geometry of the hull and the boundary conditions established in CFD are shown in Figure 1.

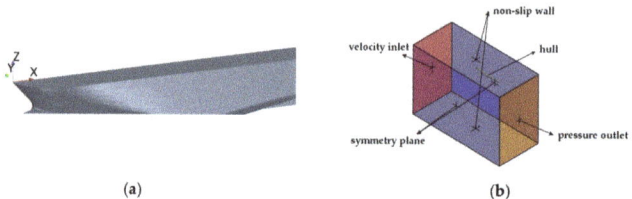

Figure 1. Computational domain: (**a**) hull model of research vessel; (**b**) boundary conditions.

The main parameters of the hull are shown in Table 1.

Table 1. Main parameters of the research vessel.

Item	Symbol	Real Ship	Model	Unit
Waterline length	L_{WL}	90.2	3.608	m
Breadth	B	16.8	0.672	m
Fore draft	T_F	5	0.2	m
Aft draft	T_A	5	0.2	m
Displaced volume	∇	3844.6	0.2461	m^3
Wet surface area	S_0	1663	2.661	m^2
Longitudinal center of buoyancy	L_{CB}	41.008	1.64	m
Block coefficient	C_B	0.5074		
Scale ratio	λ	1	25	

Since the hull is symmetrical, calculations were performed on the half hull to save computing resources. Considering that a considerable part of the bubbles were entrapped by the bow vortex to the bottom of the ship, additional structured grid discretization of the bow grid and free surface was required in the CFD calculation, as shown in Figure 2.

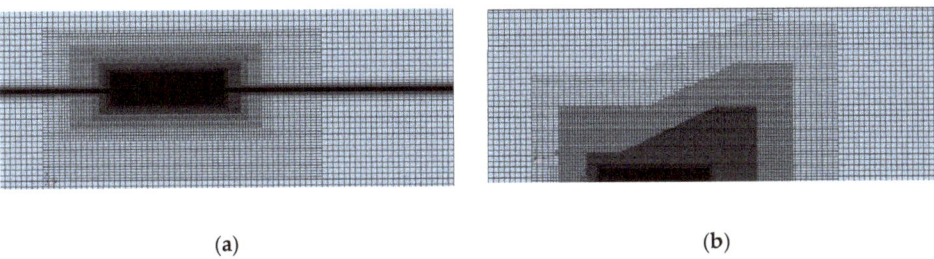

Figure 2. Mesh discretization: (**a**) discretization with free surface; (**b**) mesh discretization.

To simulate the navigation state under natural conditions and facilitate the analysis of the movement characteristics of a single bubble, two schemes were set up for calculation. The structured grid discretization of the wall is shown in Figure 3b. The final number of grid cells generated was 11.84 million.

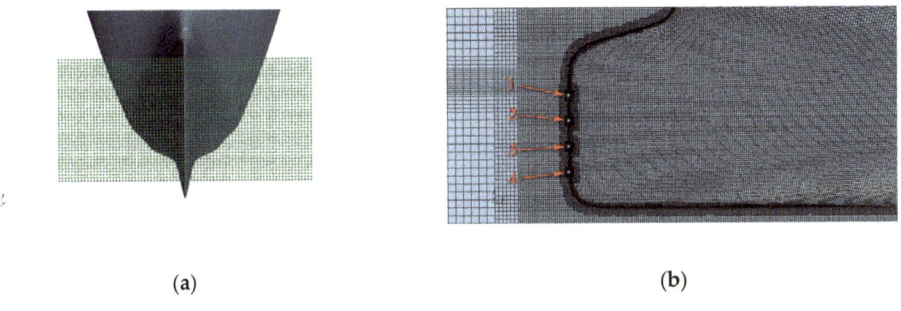

Figure 3. Bubble generation scheme settings in CFD: (**a**) bubble layer settings; (**b**) bubble point setting and grid discretization.

The following two schemes were used to calculate the movement of bubbles:

Scheme 1. For natural conditions, the bubbles in the water exist in the form of a suspended relatively stable bubble layer; hence, it is necessary to calculate the movement characteristics of the bubble layer under this condition. To make the bubble development more complete, we set the bubble layer in the bow of the ship (0.042 LWL in the front) and can observe that 40 bubble points in the vertical direction cover a certain draft range and 80 bubble points in the width direction of the ship are larger than the width of the ship. The relative position of the bow and the bubble layer is shown in Figure 3a.

Scheme 2. The position of the bubble point is the same as the test, as shown in Figure 3b.

As shown in Figure 4, the Y+ value of the final generated grid scheme is within 1.

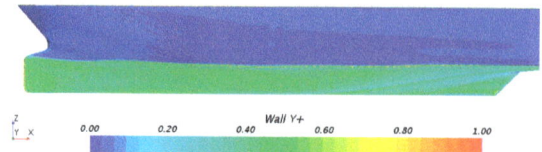

Figure 4. The value distribution of Y+ on the hull surface.

According to the International Towing Tank Conference (ITTC) 's convergence judgment criterion, when a CFD calculation is used, three sets of grid schemes with different grid sizes are obtained by changing the basic size of the grid according to the fixed fineness ratio $\sqrt{2}$. The uncertainty analysis of the resistance calculation results of the scheme is then carried out.

According to the principle that the convective Courant number is less than 1, the time step of non-steady state calculation is set to 0.001. The grid schemes compared are shown in the Table 2.

Table 2. Number of grids and simulated resistance.

Grid Scheme	Base Size (M)	Number of Grids (Million)	Simulated Drag Coefficient ×10^3	Test Resistance Coefficient ×10^3
1	0.1	6.8	4.745	
2	0.071	9.7	4.782	4.827
3	0.05	11.8	4.790	

According to the data in Table 2, in the process of gradually increasing the number of grids from scheme 1 to scheme 3, we can judge according to the grid convergence rate as follows:

$$\varepsilon_{G21} = R_2 - R_1$$
$$\varepsilon_{G32} = R_3 - R_2 \qquad (4)$$

$$R_G = \frac{\varepsilon_{G32}}{\varepsilon_{G21}} = 0.22 \qquad (5)$$

The grid convergence rate, R_G, is a positive number less than 1, which conforms to the ITTC's grid convergence criteria. It shows that the calculation scheme is convergent. The smaller the value, the faster the convergence speed. The uncertainty is further verified according to the ITTC standard. The results for the correlation coefficient of the uncertainty analysis are shown in Table 3.

Table 3. Uncertainty verification of resistance.

C_{tm1}	R_G	P_G	C_G	U_G	δ_G^*	U_{Gc}
4.790	0.216	1.531	0.70	0.0114	0.010	0.003

The order estimate P_G is greater than 1, indicating that the resistance calculation accuracy is high. From the correction factor C_G value near 1, it can be concluded that the resistance calculation result is near the asymptotic value after convergence. The error δ_G^* with correction factor and the uncertainty U_{Gc} of the correction value can then be estimated.

At the same time, in the calculation of the bubble, in order to reflect the state of the ship passing the bubble at a constant speed and the stability of the calculation, the conventional resistance is calculated first, and the bubble jet is activated after the resistance calculation is stable. It takes at least 20 s of physical time for resistance calculation to stabilize, and at least 40 s for bubble calculation.

3. Experimental Method

The resistance test adopts the method of fixing the posture of the model, and the test is completed in the towing tank of the ship model of Harbin Engineering University, Harbin, China. The static state of the ship model after installation is shown in Figure 5.

Figure 5. Initial state of the ship model test with fixed model attitude.

For large towing pools, the air content is usually much lower than for real sea conditions. It is extremely difficult to generate a suspended bubble layer similar to real sea conditions. Therefore, a bubble generator is used to generate bubbles at a vertical spacing of 0.04 m below the bow water surface. The four vent holes are connected to the bubble generator to observe the position of the individual bubble generation point. The opening position is shown in Figure 6.

Figure 6. Schematic diagram of the bubble generation location.

The experiment uses the PIV method to obtain the spatial movement information of the bubbles by using the white stripe positioning on the surface and laser irradiation.

As the main diameter of the bubble sweep-down is 500–1000 μm, it is necessary to generate 20–40 μm bubbles. The bubble generator used in the test produces bubbles in the range of 10–50 μm, which meets the test requirements. The working conditions of the resistance test are shown in Table 4, and numerical calculations were carried out under the same working conditions.

Table 4. Test conditions.

Actual Ship Speed V_s (Kn)	Froude Number F_r	Model Speed V_m (m/s)	Trim Angle (°)	Wave
12	0.225	1.234	0	Static water
16	0.277	1.646	0	Static water

A sweep-down test of air bubbles was carried out with a working speed of 1.234 m/s in a pool.

4. Results and Analysis

4.1. Spatial Movement Characteristics of Bubbles Under Sweeping

The results of the resistance test were verified. The test results of bubble motion at a multibeam operating speed of 1.234 m/s are shown in Figure 7.

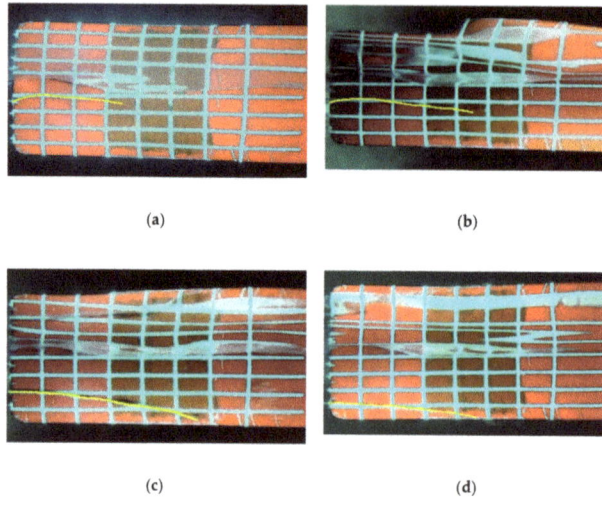

(a) (b)

(c) (d)

Figure 7. *Cont.*

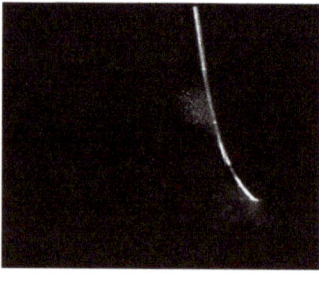

(e)

Figure 7. Bubble motion test at 1.234 m/s: (**a**) bubble point 1; (**b**) bubble point 2; (**c**) bubble point 3; (**d**) bubble point 4, and (**e**) bubble cloud diagram of a cross-section.

In Figure 7, the trajectory of the bubbles is highlighted and extracted in yellow by the bubble point. Figure 7e is a cross-sectional bubble cloud diagram obtained by PIV. The space movement position of the bubble cluster center is obtained by the white grid line in Figure 7 and this cloud image. The base point is located in front of the bow of the ship, with the length of the ship as the X-axis and the stern direction as the positive direction. The trajectories in the longitudinal plane (X-Z direction) and the horizontal plane (X-Y direction) result in line graphs, expressed in dimensionless form, by dividing by T and L_{wl}, respectively. Here, T represents the average draft.

In the test, the size of the bubbles is extremely small because of the scale, which makes them difficult to observe. It is speculated that the bubbles generated by the bubble generator in the experiment have different sizes, and the density of the bubbles generated in the experiment is larger, resulting in stronger interactions such as fusion. Therefore, the lift force of the bubble is increased to a certain extent, and the density of the bubble in the actual sea state is relatively low, and the numerical results are considered to be relatively consistent. Although the bubbles sweep down near the bow of the ship, they dissolve in the water quickly. Therefore, it is necessary to carry out CFD calculations and tests at multibeam working speed. The trajectory comparison between the experiment test results and the CFD calculation results is shown in Figure 8.

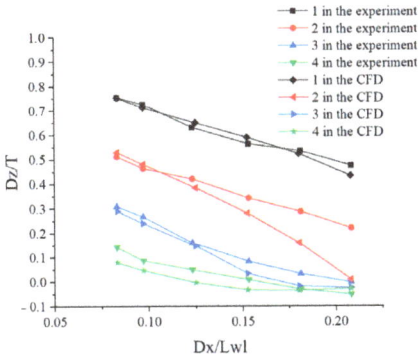

Figure 8. Comparison of bubble sweep-down trajectories between test and CFD calculation at 1.234 m/s.

It can be seen that the bubble motion trajectories of the four bubble generation points are close, but the bubble sweep-down calculated by CFD at point 2 is faster. According to Figure 7, the sweep-down trend of bubble points 3 and 4 is obvious, and bubble point 1 is the most affected by wave making. Bubble point 2 is in the area where the wave making is obviously weakened, but there is still a sweep-down trend under the dual effects of wave making and relatively steady flow. However, the relatively deeper dip of bubble points 3 and 4 is not as obvious here.

Combining the test and calculation results, the CFD method is feasible to calculate bubble motion characteristics. The calculation of bubble motion and flow field details was carried out. The calculation results of bubble layer and bubble point under multibeam working speed and research vessel design speed are shown in Figures 9 and 10.

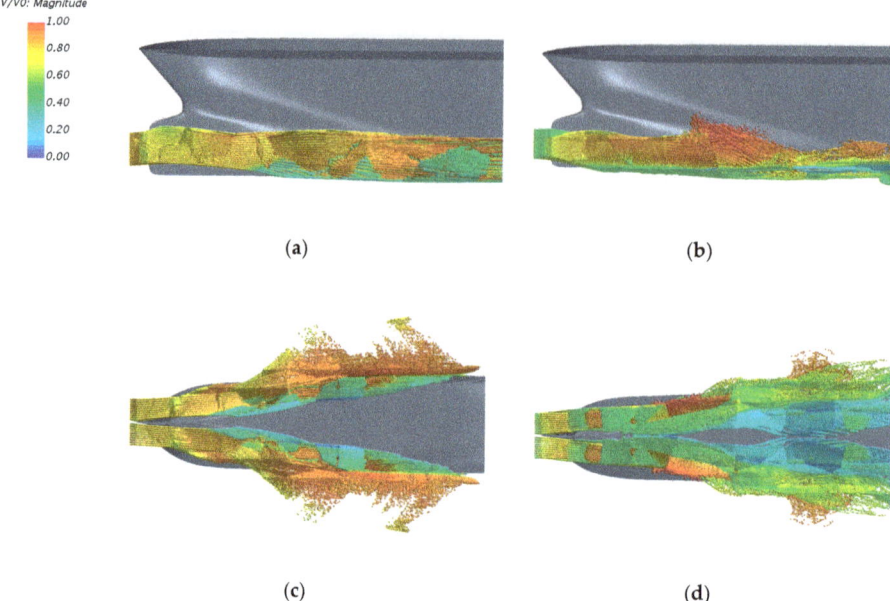

Figure 9. Bubble movement for the bubble layer setting: (**a**) speed 1.234 m/s from the side; (**b**) speed 1.646 m/s from the side; (**c**) speed 1.234 m/s at the bottom; and (**d**) speed 1.646 m/s at the bottom.

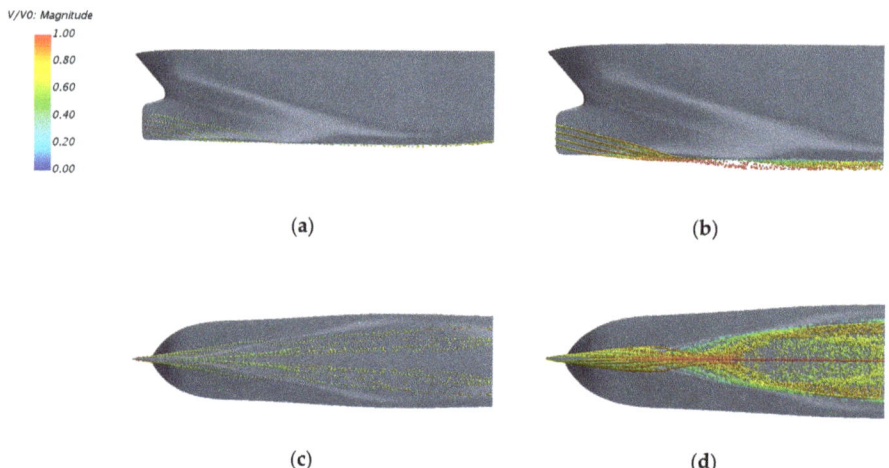

Figure 10. Bubble movement for the bubble point setting: (**a**) speed 1.234 m/s from the side; (**b**) speed 1.646 m/s from the side; (**c**) speed 1.234 m/s under the bottom; and (**d**) speed 1.646 m/s under the bottom.

It can be seen that, when the research vessel is sailing, the bubbles located in the front of the hull sweep under the surface of the hull to the bottom of the ship, causing a large number of bubbles to accumulate at the bottom of the ship. It is particularly obvious at the higher of the calculated speeds. This has a significant influence on the position of the sonar at the stern of the ship. From the calculation of the bubble layer in Figure 9, it can also be seen that, when sailing under natural conditions, in addition to the bubbles that produce the sweep-down phenomenon, a large number of bubbles move with the hull. Regular analysis becomes difficult; hence, the fixed bubble point positions are calculated.

In the comparison between the bubble point in Figure 10 and the calculation result of the bubble layer setting in Figure 9, the first bubble layer tilts down with the water surface, which has the characteristics of sweep-down. A large number of bubbles gather near the center line of the bottom surface of the ship and the bubble layer trajectory covers the trajectory of the bubble point. However, when the bubble point is set at 1.234 m/s, the sweep-down trend of the bubble point is more obvious than that of the bubble layer. It is speculated that the bubble layer has a higher density, relatively low speed, and relatively turbulent wave making, which causes more frequent interactions between the bubbles. Under actual sea conditions, as the size of the bubble becomes larger but the relative density becomes smaller, the effect of the bubbles on the surface of the hull is relatively weakened. From the image results, the above calculation of individual bubble points can better reflect the movement of the bubble layer state. The results at different speeds are shown in Figure 11.

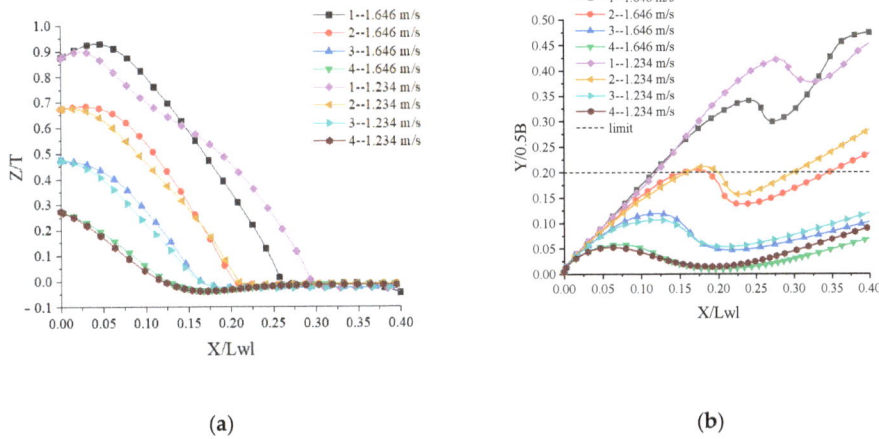

Figure 11. Spatial trajectories of bubbles: (**a**) X-Z direction; (**b**) X-Y direction.

Extracting the bubble space movement trajectory, the bubble X-Z spatial movement characteristics (Figure 11a) are obtained. The bubble initially floats up for a period of time, and then it moves down to the bottom of the ship, and finally moves to the rear of the ship under the bottom of the ship. The starting position of this final stage can be set as the down-sweep point. The lower the position of the bubble, the closer to the bow of the ship when it sweeps down to the bottom, and the lower the speed (absolute value of the tangential slope of the curve) during the sweep-down. With increasing speed, both the initial floating process of the bubbles and the sweep-down part are enhanced. The bubble at the same position sweeps down to the bottom of the boat and moves toward the bow, and the velocity at which the bubble sweeps down increases. According to the bubble movement trajectory curve in the X-Z direction, the lower sweep position of the bubble movement can be divided into the following two movement processes: before and after the lower sweep. For the speed of 1.234 m/s, the lower sweep point is 0.27 L and, for 1.646 m/s, it is 0.26 L.

The bubble movement curve in the width direction of the ship is used to analyze the influence of bubble movement on the position of the center line of the hull, and then to measure its influence on the sonar at the stern, as shown in Figure 11b. The bubble movement curve in the X-Y direction has two obvious curvature optimum points, that is, the peak appears first and then the valley. Combined with the hull model, the peak is the position that sweeps down to the outside of the bottom of the ship. The valley is the position closest to the mid-longitudinal section of the hull where the bubbles can continue to move from the outside of the bottom. The movement between peaks and valleys is the sweep-down process, and the sweep point of the bubble is also located in this process. The lower the initial bubble position, the closer the valley position after the sweep-down is to the middle of the ship. As the speed increases, this position moves toward the bow. After that, the bubble movement gradually moves away from the midship position, but it still has an impact on the position of the sonar at the stern. The dotted line in Figure 11b represents the safety limit line. The bubbles at positions 3 and 4 have a great impact on the multibeam.

4.2. Flow Around the Bow

The next step is to calculate the distribution of the pressure field near the bow and analyze the relationship between the bubble sweep-down phenomenon and the pressure field.

The pressure coefficient is defined by $C_p = (P - \rho g h)/0.5\rho V_0^2$, where V_0 represents the speed, and we can obtain the dimensionless pressure difference relative to the hydrostatic pressure. The streamline set by the bubble point is represented by a gray-white line. The calculation result is shown in Figure 12.

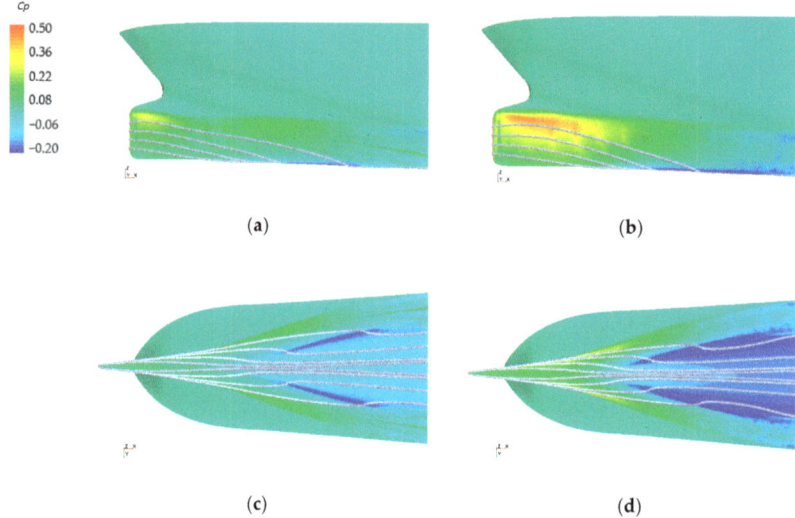

Figure 12. Distribution of dimensionless pressure difference coefficients on the hull surface: (**a**) speed 1.234 m/s from the side; (**b**) speed 1.646 m/s from the side; (**c**) speed 1.234 m/s under the bottom; and (**d**) speed 1.646 m/s under the bottom.

From the surface pressure distribution of the hull at different speeds in Figure 12, it can be concluded that the bow pressure is relatively high, a maximum value appears at the peak of the bow wave surface uplift, and the pressure gradually decreases toward the back of the ship and the bottom of the ship. Therefore, it is easy to force the microbubbles in the water to move during the navigation. The bubbles migrate from the high-pressure position to the low-pressure position under the action of the pressure difference. The bottom of the ship has a minimum pressure value, and the bubbles continue to the middle of the bottom of the ship owing to the inertial movement, and then move backward with the ship sailing. Because the shape of the bow of the research vessel in this study is relatively flat, the bubbles migrate to the bottom of the ship under the dual effects of the bow wake and the pressure difference, resulting in the phenomenon of bubble sweep-down. The pressure distribution changes with speed. At the higher speeds, the pressure coefficient of the side and bottom of the ship is smaller than that at lower speeds; hence, bubble sweep-down is more likely to occur at high speeds.

The velocity field distribution of multiple cross-sections at the bow of the hull was calculated, where the zero point of the cross-section corresponds to the foremost end of the deck, and the direction to the stern is the positive x direction and the vertical upward direction is the positive z direction. The cross-sectional positions 0.45, 0.55, 0.65, 0.75, and 0.85 m are shown in Figure 13.

Figure 13. Schematic diagram of the extracted cross-sectional velocity field position.

Through calculation, the vertical velocity distribution on the cross-section is obtained, and the vertical velocity is expressed in dimensionless V_Z/V_0, where V_Z is the vertical velocity component, $V_0 = V_m$, which is the towing speed of the ship model. Furthermore, the position of the bubble and the position of the free water surface (the horizontal black line) on the cross-section are marked, as shown in Figures 14 and 15.

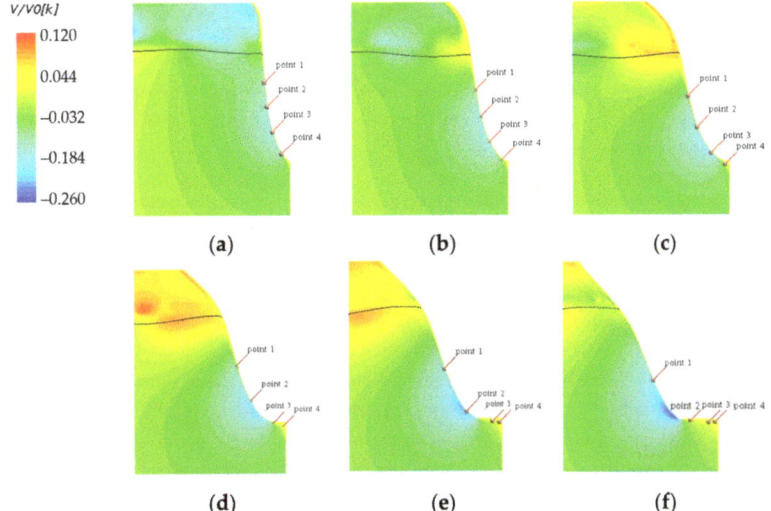

Figure 14. Vertical velocity distribution of the cross-section at a speed of 1.234 m/s: (**a**) 0.35 m cross-section; (**b**) 0.45 m cross-section; (**c**) 0.55 m cross-section; (**d**) 0.65 m cross-section; (**e**) 0.75 m cross-section; and (**f**) 0.85 m cross-section.

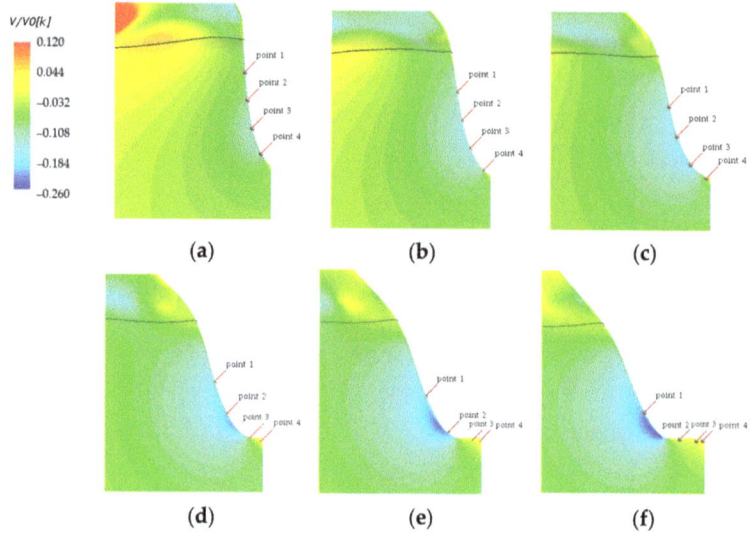

Figure 15. Vertical velocity distribution of the cross-section at a speed of 1.646 m/s: (**a**) 0.35 m cross-section; (**b**) 0.45 m cross-section; (**c**) 0.55 m cross-section; (**d**) 0.65 m cross-section; (**e**) 0.75 m cross-section; and (**f**) 0.85 m cross-section.

According to the vertical dimensionless velocity distribution of multiple cross-sections in Figures 14 and 15, it can be concluded that the vertical velocity component near the hull wall below the water surface is only upwards in the part closest to the water surface. This upward velocity core area moves back with the section position at 1.234 m/s, first close to the hull and then away from the hull, but keeps away from the hull at 1.646 m/s. This is related to the influence of the hull's wave making. The vertical downward velocity component near the wall gradually increases. The downward velocity core area formed increases with the increase in speed, reaches the maximum value near the bottom of the ship, and then gradually decreases. The range of motion is in this wall area. According to the change of the bubble position, it can be seen that the trajectory of the bubble before it moves to the bottom of the ship is near the vertical downward velocity core area. This provides a certain amount for the sweep-down of the bubble. When the bubble moves to the bottom of the ship, a vertical upward velocity component is generated, so that the bubble moves upward to the bottom of the ship while moving toward the stern. This is also one of the reasons why the sweep-down of the bubble affects the position of the stern sonar.

4.3. Shroud

A diversion cover was used as a defoaming appendage, and the principle of action was calculated and analyzed. The geometric shape and the installation position of the diversion cover are shown in Figure 16. Some parameters of the shroud are shown in Table 5.

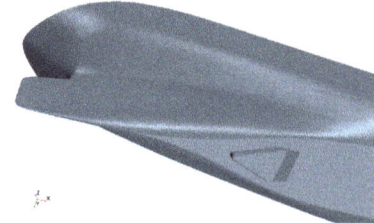

Figure 16. Installation position of the air deflector.

Table 5. Some parameters of the shroud.

Item	Symbol	Model	Unit
Displaced volume	∇	0.0002	m^3
Wet surface area	S	0.0416	m^2
Installed surface area	S	0.0295	m^2

According to the shroud's parameters, the installation of the deflector has minimal impact on the parameters of the bare hull. Among them, the displacement is increased by 0.08%, and the wet surface area is increased by 0.4%. The draft is increased by less than 0.22 mm, and its impact is almost negligible. The comparison of the vorticity distribution at the bottom of the ship is shown in the Figure 17.

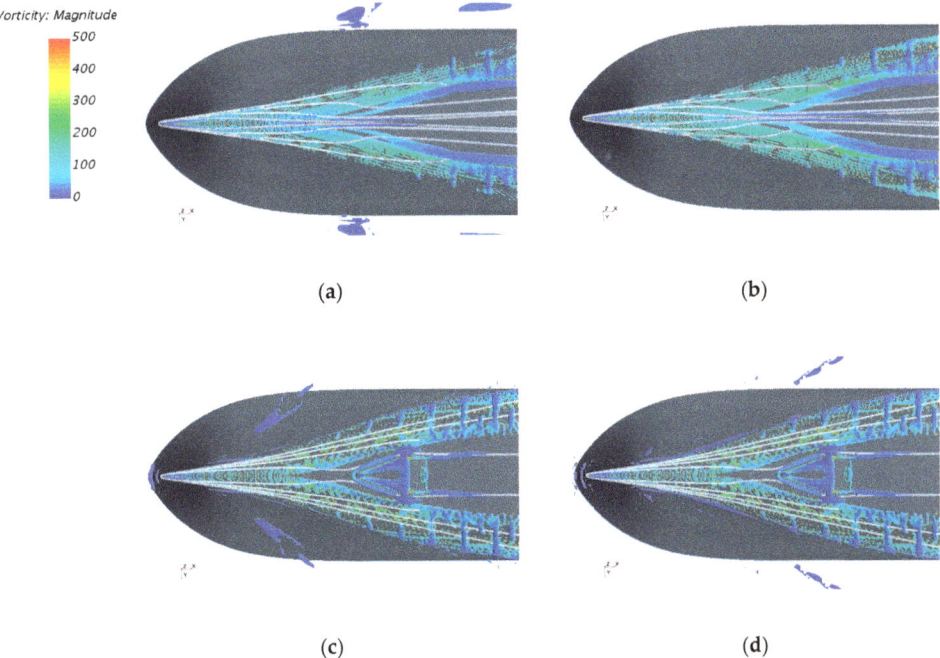

Figure 17. Vorticity distribution of the bottom of the bare hull and the ship with a deflector: (**a**) naked hull at 1.234 m/s; (**b**) naked hull at 1.646 m/s; (**c**) with deflector at 1.234 m/s; and (**d**) with deflector at 1.646 m/s.

According to the calculation results in Figure 17, the structure of the vortex system at the bottom of the ship is relatively simple. There are two parts of the vortex system distributed on the side of the ship and the bottom of the bow. The deflector effect of the vortex system on the ship side is minimal, but the ship bottom vortex structure is dispersed, changing the vortex structure in the middle of the bottom of the ship. First, the position of the deflector is located at the position of the vortex structure on the first center line of the hull, which eliminates or greatly reduces the effect of this vortex structure in bringing bubbles into the ship. The installation of the diversion cover adds a vortex structure away from the ship, which makes the bubbles move outward along the edge of the diversion cover from both the physical shielding and the vortex structure guidance. Based on this, the diversion cover has a certain defoaming performance.

The changes in the total resistance coefficient and the remaining resistance coefficient are shown in Figure 18. The influence of the dome on the resistance performance is mainly reflected in the increase in the residual resistance of the hull within a certain low speed range, and the maximum increase in the total resistance can reach 5.1%. However, the influence gradually decreases as the speed increases. For the two calculated speeds, the drag increase in the dome is extremely small.

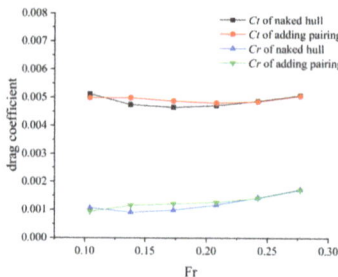

Figure 18. Change in drag coefficient before and after installation of the shroud.

Based on the comparison of the defoaming effect before and after the installation of the diversion cover, shown in Figure 17, and the resistance increase, shown in Figure 18, the diversion cover can produce a small increase in resistance and at the same time reduce the effect of bubble sweep-down on the stern.

5. Conclusions

The results of the research on the phenomenon of sweep-down of bubbles in research vessels provide the following conclusions.

1) It is feasible to use the Eulerian-Lagrangian method to calculate the bubble sweep-down phenomenon.
2) The phenomenon of bubble sweep-down is related to the shape of the bow of the ship and the distribution of the pressure field. The pressure difference caused by the decrease in the hull surface pressure with the increase in water depth and the vertical downward velocity component near the wall forces the bubbles to sweep down.
3) The movement characteristics of the bubble sweep-down space are related to the initial position and speed of the bubble. When the bow is closer to the bottom of the ship, the sweeping position of the bubble is closer to the bow, and the position after the bottom of the ship is closer to the center line of the bottom of the ship. Therefore, the influence on the position of the stern sonar is greater, and the degree of this influence increases with the increase in speed.
4) After the bubble moves through the sweep-down point, it moves to the center line of the bottom of the ship to strengthen the influence on the sonar position. From the perspective of hydrodynamics, the installation of the diversion cover plays the role of physical shielding and strengthens the guidance of the lateral vortex system, so that the bubbles move to the side of the ship with the vortex to achieve the purpose of defoaming.
5) The installation of the shroud produces a certain increase in resistance while achieving a certain defoaming effect, which is especially obvious at low speeds but is already extremely small at working speeds. This occurs because viscous resistance dominates at lower speeds, whereas pressure resistance is more important at higher speeds.

The spatial motion of the bubble sweep-down and the fine characteristics of the partial flow field were studied and the above results were obtained. The research results have some reference value for the law of bubble sweep-down of research vessels and for the reduction of its influence on sonar equipment spatially. However, the work was carried out only from the hydrodynamic point of view, without considering the noise characteristics under the actual sonar work. In addition, the dissolution and breakage of bubbles were ignored. This method is beneficial to the study of bubble movement, but the adverse effect on the study of noise cannot be ignored. The follow-up work will study the influence of the bubble sweep-down phenomenon based on the noise characteristics of sonar work, such as frequency.

Author Contributions: This paper was drafted and written by W.W. and all authors worked on the towing tank test and simulation results. G.C. contributed to the query, determination, and calculation of the simulation program. Y.P. and C.G. provided guidance for the overall research ideas and plans. Y.H. and G.Z. provided guidance for the formulation and implementation of test methods. Figure 7 in the text was provided by Y.H. All authors have read and agreed to the published version of the manuscript.

Funding: This research received no external funding.

Acknowledgments: We wish to thank the National Natural Science Foundation of China (NSFC) program "Research on Prediction Method of Coupling Effect of Pod Propeller and Ice-Water Mixed Media" (Fund number: 51809055) for support. We thank Harbin Engineering University, Harbin, China, for the use of the model towing tank for the experimental conditions. Finally, we thank the reviewers for their insights, which played a major role in the improvement and final completion of this article.

Conflicts of Interest: The authors declare no conflict of interest.

References

1. Karafiath, G.; Hotaling, J.M.; Meehan, J.M. Fisheries Research Vessel hydrodynamic design minimizing bubble sweep-down. In Proceedings of the Oceans 2001 MTS/IEEE Conference and Exhibition, Washington, DC, USA, 5–8 November 2001; Volume 2, pp. 1212–1223.
2. Deane, G.B.; Stokes, M.D. Scale dependence of bubble creation mechanisms in breaking waves. *Nat. Cell Biol.* **2002**, *418*, 839–844. [CrossRef] [PubMed]
3. Thorpe, S.A. *The Turbulent Ocean*; Cambridge University Press: Cambridge, UK, 2005; pp. 240–243.
4. Sebastian, S.M.; Caruthers, J.W. Effects of naturally occurring bubbles on multibeam sonar operations. In Proceedings of the Oceans 2001 MTS/IEEE Conference and Exhibition, Washington, DC, USA, 5–8 November 2001; Volume 2, pp. 1241–1247.
5. Rolland, D.; Clark, P. Reducing bubble sweep-down effects on research vessels. In Proceedings of the Oceans 2001 MTS/IEEE Conference and Exhibition, Washington, DC, USA, 5–8 November 2001; pp. 1–7.
6. Delacroix, S.; Germain, G.; Berger, L.; Billard, J.-Y. Bubble sweep-down occurrence characterization on Research Vessels. *Ocean. Eng.* **2016**, *111*, 34–42. [CrossRef]
7. Mallat, B.; Germain, G.; Billard, J.-Y.; Gaurier, B. A 3D study of the bubble sweep-down phenomenon around a 1/30 scale ship model. *Europ. J. Mech.* **2018**, *72*, 471–484. [CrossRef]
8. Han, S.; Lee, Y.-S.; Choi, Y.B. Hydrodynamic hull form optimization using parametric models. *J. Mar. Sci. Technol.* **2012**, *17*, 1–17. [CrossRef]
9. Palaniappan, M.; Subramanian, V.A. Hydrodynamic design for mitigation of bubble sweep down in sonar mounted research vessels. *Int. Shipbuild. Prog.* **2017**, *64*, 101–126. [CrossRef]
10. Maxwell, R.; Ata, S.; Wanless, E.; Moreno-Atanasio, R. Computer simulations of particle–bubble interactions and particle sliding using Discrete Element Method. *J. Colloid Interface Sci.* **2012**, *381*, 1–10. [CrossRef]
11. Bérard, A.; Patience, G.S.; Blais, B. Experimental methods in chemical engineering: Unresolved CFD-DEM. *Can. J. Chem. Eng.* **2020**, *98*, 424–440. [CrossRef]
12. Li, X.; Chen, G.; Zhu, H. Modelling and assessment of accidental oil release from damaged subsea pipelines. *Mar. Pollut. Bull.* **2017**, *123*, 133–141. [CrossRef]
13. Zhang, X.; Ahmadi, G. Eulerian-Lagrangian simulations of liquid-gas-solid flows in three-phase slurry reactors. *Chem. Eng. Sci.* **2005**, *60*, 5089–5104. [CrossRef]
14. Watson, N.A.; Kelly, M.; Owen, I.; Hodge, S.; White, M. Computational and experimental modelling study of the unsteady airflow over the aircraft carrier HMS Queen Elizabeth. *Ocean. Eng.* **2019**, *172*, 562–574. [CrossRef]

15. Home, D.; Lightstone, M. Numerical investigation of quasi-periodic flow and vortex structure in a twin rectangular subchannel geometry using detached eddy simulation. *Nucl. Eng. Des.* **2014**, *270*, 1–20. [CrossRef]
16. Jee, S.; Shariff, K. Detached-eddy simulation based on the v 2–f model. *Int. J. Heat Fluid Flow* **2014**, *46*, 84–101. [CrossRef]

Publisher's Note: MDPI stays neutral with regard to jurisdictional claims in published maps and institutional affiliations.

 © 2020 by the authors. Licensee MDPI, Basel, Switzerland. This article is an open access article distributed under the terms and conditions of the Creative Commons Attribution (CC BY) license (http://creativecommons.org/licenses/by/4.0/).

Article

Direct Tracking of the Pareto Front of a Multi-Objective Optimization Problem

Daniele Peri †

Istituto per le Applicazioni del Calcolo "Mauro Picone", Consiglio Nazionale delle Ricerche (IAC-CNR), 00185 Roma, Italy; d.peri@iac.cnr.it
† Current address: Via dei Taurini 19, 00185 Roma, Italy.

Received: 12 August 2020; Accepted: 7 September 2020; Published: 9 September 2020

Abstract: In this paper, some methodologies aimed at the identification of the Pareto front of a multi-objective optimization problem are presented and applied. Three different approaches are presented: local sampling, Pareto front resampling and Normal Boundary Intersection (NBI). A first approximation of the Pareto front is obtained by a regular sampling of the design space, and then the Pareto front is improved and enriched using the other two above mentioned techniques. A detailed Pareto front is obtained for an optimization problem where algebraic objective functions are applied, also in comparison with standard techniques. Encouraging results are also obtained for two different ship design problems. The use of the algebraic functions allows for a comparison with the real Pareto front, correctly detected. The variety of the ship design problems allows for a generalization of the applicability of the methodology.

Keywords: multi-objective optimization; pareto front; normal boundary intersection; ship design optimization

MSC: 78M50; 90C29; 58E17; 62K05; 49Q10; 46N10

1. Introduction

Most of the real-life optimization problems are multi-objective. It is quite uncommon that a single objective is sufficient for the determination of the optimal qualities of a design. As a consequence, tools for the determination and the analysis of the alternatives coming from a multi-objective design optimization problem are of great importance.

In the current literature, recently reviewed in [1], a very large number of optimization algorithms have been proposed to solve this task. Some of them are single-objective optimizers, applied to a single objective function where the different objective functions are recasted into a single objective function using some weights for the different objectives, being these weights static or dynamic. Another opportunity is the definition of a *goal programming* problem, where a specific value of each objective function is required, and the objective function to be minimized is represented by the Euclidean distance of the current solution and the target solution in the objective functions space. Other methods are adaptations of optimization algorithms formulated for a single objective: the different objectives are considered without an aggregation, and the concept of Pareto optimality is adopted. A typical example is the multi-objective version of the popular Genetic Algorithm (GA), namely Multi-Objective Genetic Algorithm (MOGA). The kernel of MOGA is exactly the same as GA, but only the fitness function, that is, the function providing the ranking of the different solutions, is different. Summarizing, none of the above algorithms are, in general, tracking directly the Pareto front (an exception is represented by the Normal Boundary Intersection method—NBI [2],) and the Pareto front is typically obtained by a recombination of the actual solutions.

In this paper, three different approaches aimed at the determination and enrichment of the resolution of the Pareto front are presented. A regular sampling of the Design Variables Space (DVS) is producing a first approximation of the Pareto front, and each successive step is producing an improved approximation of the Pareto front. Application to algebraic and industrial problems give positive indications about the efficiency of the approach.

2. Materials and Methods

The formulation of a multi-objective optimization problem requires the definition of *optimal point* in a multi-objective contest. In fact, it is absolutely uncommon to find a single solution that presents the minimum (or maximum) values for all the objectives at the same time. A widely accepted definition is the following:

Definition: The vector $F(x)$ is said to dominate another vector $F(y)$, x and $y \in C$, denoted $F(x) \prec F(y)$, if and only if $f_i(x) \leq f_i(y)$ for all $i \in \{1, 2, ..., n\}$ (where n is the number of criteria) and $f_j(x) < f_j(y)$ for at least one $j \in \{1, 2, ..., n\}$. A point $x \in C$ is said to be globally Pareto optimal or a globally efficient point if and only if there does not exist $y \in C$ satisfying $F(y) \prec F(x)$. F(x) is then called globally non dominated or non inferior.

In this case, we have not a single optimal solution, but a variety of Pareto-optimal solutions, distributed in the so-called *Pareto front*, that is, the locus of the Pareto optimal points. The determination of the Pareto front represents the solution of a multi-objective optimization problem. It is an hard task, mainly because the Pareto front is defined in the objective-function space, and the relationship between the DVS and the objective-function space is typically not trivial. For a single-objective minimization problem we can easily find a search direction in the DVS, i.e., using the local gradient of the objective function, so we can move along this direction in order to detect an improved value of the objective function. This opportunity is not explicit for a multi-objective problem, because the definition of the Pareto front cannot be translated to the DVS. Furthermore, a single direction able to improve simultaneously all the objectives rarely exists: consequently, it is hard to determine a search direction driving us to a better Pareto point.

As already recalled, the algorithms for the solution of a multi-objective optimization problem are often adaptations of algorithms developed for a single-objective problem. One of the more representative examples is the classical approach for the adaptation of the GA to the Multi-Objective GA (MOGA) [3]. The Pareto front is evaluated using the current set of solutions, and the value of 1 is assigned to the fitness function for all the points belonging to the Pareto front. After that, a second level of the Pareto front can be determined excluding the previously detected Pareto points, and a value of 2 is assigned to the fitness function for these points (second level of the Pareto front). The procedure continues until a value of the fitness function has been assigned to every available solution. Since then, the typical operations of the GA are performed using a single valued objective function, as usual. In other words, the kernel of the algorithm is not changed, while the objective function is reformulated in such a way that the original multi-objective problem can be treated as a single-objective problem.

A more ambitious strategy could be to produce a search algorithm for the direct identification of new Pareto points, starting from a first approximation of the Pareto front. In this case, the search is not performed on the full DVS, but in selected regions where the new Pareto points are supposed to lie. We are here following three main guidelines:

1. **Local sampling**—Perform a search in a small region around each actual Pareto point.
2. **Interpolation**—Try to identify a curve in the DVS composed by the actual Pareto points and then compute new solutions refining and regularizing the resolution of this curve.
3. Normal Boundary Intersection method (NBI)—Identify a direction in the DVS, aligned with the local normal of the Pareto front, able to improve the current Pareto point.

Local sampling is very intuitive: we can find a true Pareto point in the neighbor (referring to the DVS) of an approximation of a Pareto set. We are hypothesizing a local continuity of the involved objective functions, and this hypothesis could be reasonable.

Interpolation is a little more reckless and simplistic: we cannot generally think that the continuity of the Pareto front represents a guarantee of a continuity for the Pareto points in the DVS. This statement is strongly different than the previous one, since we can have the same value of the objective function in different points in the DVS, not located side-by-side. Anyway, in this study we are going to test also this opportunity.

NBI is the much more solid strategy for the determination of new Pareto points. The original formulation can be found in [2]. The idea is that an approximation of the Pareto set can be improved if we are able to move the point along a direction normal to the Pareto front in the objective function space. This requires a sort of inverse map of the objective function, from the objective function space to the DVS, while we usually have the opposite.

More details for the different phases are provided hereafter.

2.1. Initial Sampling

A first approximation of the Pareto set is obtained by an uniform sampling of the DVS. Since we have not preliminary information about the location of the Pareto front, every point in the full DVS could be, at the start, a Pareto point. In other terms, the probability to find a Pareto point is uniform over the full DVS. As a consequence, in order to have a regular sampling of the DVS, we are using an Uniformly Distributed Sequence (UDS), that is, a point distribution in which all the outcomes are equally likely. Among the different strategies, we can use Sobol sequence [4], Latin Hypercube Sampling (LHS) [5] or D-Optimal design [6]. In this examples, we are using the Sobol sequence [1].

2.2. Local Sampling

Once a first approximation of the Pareto front is available, a small UDS is placed around every Pareto point, performing a local search. A different Sobol distribution is generated for every point, and at each iteration a permutation of the design variables is applied, in order not to recompute the same points if the Pareto candidate is not improved.

2.3. Interpolation

The actual Pareto set is ordered in the DVS with respect to the first objective function, and then a curve passing thru these points is traced in the DVS. A gaussian filter is also applied in order to regularize the curve, since the resulting points are possibly poorly aligned. The curve is divided into a number of regularly spaced intervals, and the resulting points are computed.

2.4. Normal Boundary Intersercion—NBI

In the original formulation of NBI, the new tentative Pareto points are obtained by the solution of a suite of *goal programming* sub-problems, where the starting point are the actual approximations of the Pareto front and each with a constraint forcing the solution along a vector normal to the Pareto front. As a consequence, we are looking for an improvement of each point of the current approximation of the Pareto front, but along a precise direction. This could represent a limitation, since the real Pareto point could lie in the vicinity of the starting point, but in a direction, in the objective function space, not coincident with the local normal. Also for this reason, in this particular implementation we are not computing the normal to the Pareto front, and the new point is obtained, also in this case, by solving a *goal programming* problem. The target point is represented by a point in the objective function space whose values of the objective function are selected by following two different criteria[2]:

---

[1] the related software can be actually found in https://people.sc.fsu.edu/~jburkardt/src/sobol/sobol.html.
[2] we are considering here a minimization problem for all the objectives: in case of a maximization, indications should be reversed.

1. **Direct tracking:** the target point is defined by **decreasing** the objective function values of the actual Pareto point by a small amount, say 1%.
2. **Indirect tracking:** the target point is defined by **increasing** the objective function values of the actual Pareto point by a small amount. The new point is then mirrored with respect to the reference Pareto point in order to have an improvement for all the objectives.

Direct tracking is similar to the original NBI formulation with the exclusion of the constraint of the solution to stay along the normal of the Pareto front. The indirect tracking has been introduced because we are not sure about the existence of the point generated for the direct tracking option, for which the objectives are all improved. On the contrary, it is more reasonable to assume the existence of a point deteriorating all the objectives. The mirroring operation is performed hypothesizing a linear behavior of all the objective functions in a small neighborhood of the point, so that the deterioration along one direction indicates an improvement on the opposite direction: it can be reasonably applied only for small variations of the objective functions.

The solution of these *goal programming* problems could be very expensive, depending on the complexity of the objectives: if we are managing CPU hungry objective functions, this phase could take too long. For this reason, a multidimensional spline [7] is adopted in order to produce an algebraic approximation/interpolation of the objective function to be applied during NBI. The actual evaluations of the objective function are used as training set for the metamodel, so that the NBI takes a moderate amount of time. The solution of each sub-problem is finally computed by using the true expression of the objective functions.

3. Algebraic Test Functions

As a first test, a commonly used set of test functions has been adopted, namely the Kursawe functions. Here we have two objectives:

$$f_1(x) = \sum_{i=1}^{n-1} \left[-10 exp\left(0.2\sqrt{x_i^2 + x_{i+1}^2}\right) \right]$$

$$f_2(x) = \sum_{i=1}^{n} \left[|x_i|^{0.8} + 5sin(x_i^3) \right]$$

The number of design variables n is 2, and both the design variables can assume any value $\in [-5:5]$. No further constraints are applied.

A first test has been performed by applying the three methodologies separately. Results are reported in Figure 1: from top to bottom, we can observe the results obtained by applying only local resampling (top), only Pareto front interpolation (middle) and only NBI (bottom). Here is evident that Pareto front interpolation cannot be applied without another enrichment method: the Pareto front obtained at the first iteration is quite well developed, but it is obviously no further improved since the added points are always on the same line. Local resampling gives a variety of points and it is able to improve at each iteration, but it appears to be pretty slow in this case, and a branch of the curve is not obtained: this is probably connected with the quality of the initial sampling. NBI gives the more interesting results, with a continuous improvement, iteration by iteration.

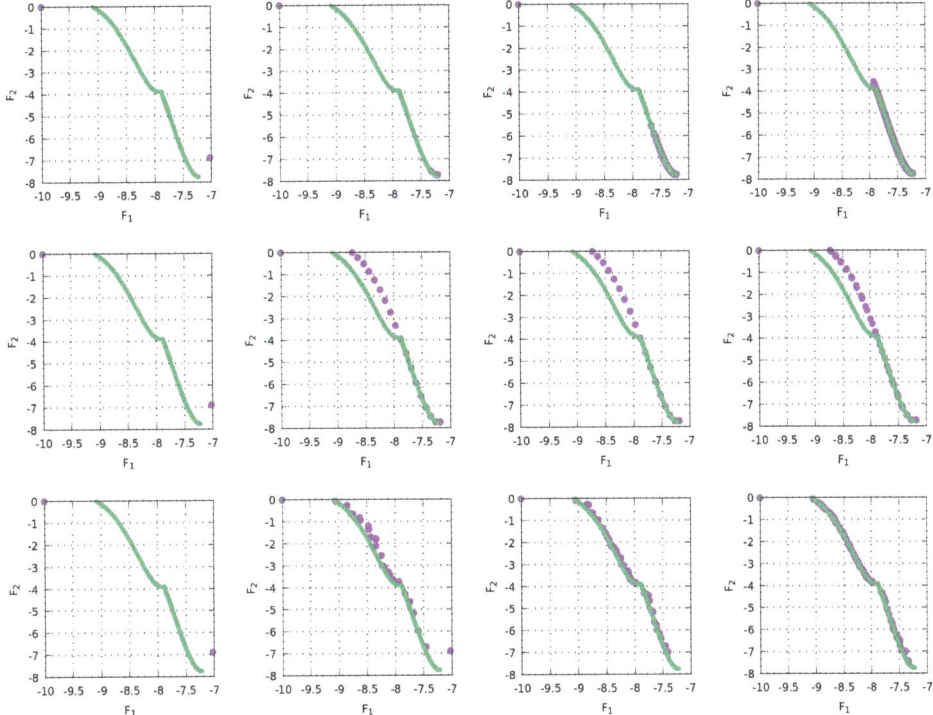

Figure 1. Pareto front approximation obtained by using the three methodologies separately. From top to bottom, the results obtained by applying only local resampling (**top**), Pareto front interpolation (**middle**) and NBI (**bottom**). From left to right, four different stages of evolution of the methodology. Green dots are the true Pareto front.

Completely different results are obtained if the three methodologies are applied in sequence, with the same order as indicated previously: resampling, interpolation and lastly NBI. The effect of the combination of the three methodologies is reported in Figure 2. In the same picture, the results obtained by a standard MOGA implementation, namely NSGA-II [8], and a Multi-Objective Particle Swarm Optimization (MOPSO) algorithm [9] are also presented in order to produce a comparison with other standard methodologies. The same total number of objective function evaluations is fixed for all the algorithms, in order to produce a fair comparison. The first graph on the left of Figure 2 reports the comparison of the results obtained by NBI and the true Pareto front. Real Pareto front is produced by using a recursive intensive regular sampling of the design space. We have a perfect agreement of NBI with the true Pareto front, with an impressive uniformity of the distribution of points along the front. The other two pictures in Figure 2 report a comparison of NBI with other two standard multi-objective algorithms: NSGA-II in the center plot and MOPSO in the plot at the right end side of the picture. Both NSGA-II and MOPSO are producing good approximation of the Pareto front, but with a lack of precision with respect to NBI. NSGA-II is missing the best point for the first objective function, and in general both the methods do not present the same uniformity the NBI is able to provide. We can conclude that the NBI algorithm is much more efficient than the two selected algorithms, at least for this example.

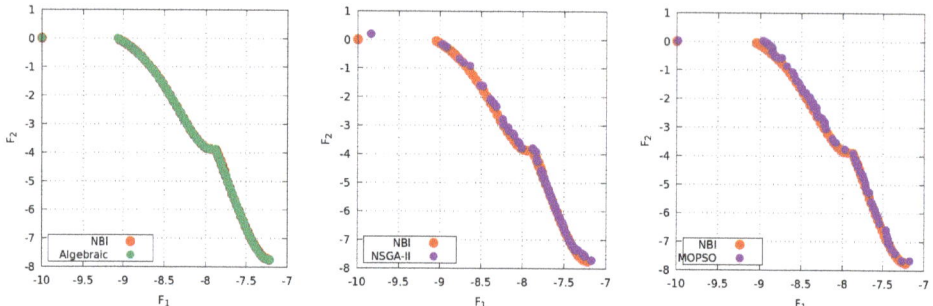

Figure 2. Pareto front approximation obtained by using the three methodologies sequentially, in comparison with a different optimization algorithm (NSGA-II and MOPSO). In the first picture at the left, NBI is compared with the true Pareto front. The other two pictures show the comparison of the results obtained by NBI versus NSGA-II and MOPSO respectively.

4. Results and Discussion

Two ship design applications are presented in the following, in order to give further elements demonstrating the efficiency of the approach. In the first example, a powering reduction problem for two different speeds is formulated and solved: a relatively simple but reliable Computational Fluid Dynamics (CFD) tool, based on the potential flow theory [10], is utilized. In the second example, the problem of the improving the quality of the flow at the propeller disk of a single-skrew ship is solved by using a more sophisticated CFD solver. Different parameterization techniques are adopted in order to present different alternatives for the hull geometry modification.

4.1. Ship Design Application: Powering

In order to further demonstrate the efficiency of the methodology, a second test case has been designed. For this test, the effective powers required by a ship advancing at two different speeds are the objectives of the multi-objective optimization problem. The two selected speeds are characterized by a nondimensional quantity (Froude number—Fr): this is to emphasize the different flow regimes at the two different speeds. With $Fr = 0.3$ we are in the displacement regime, where the lifting actions of the flow on the ship are negligible. With $Fr = 0.45$, we are in the semi-displacement regime, and the lifting actions are more intense, of the same order of magnitude of the buoyancy actions. Although the adopted flow solver is modeling the lifting forces in a simplified way, the results provided by the solver are reliable in both the two regimes. Furthermore, the use of two different flow regimes is reflected on the optimal shape, different for the range of application: typically, semi-displacement ships have finer bow entry and flatter aft with respect to displacement ships. For this reason, we are expecting different optimal shapes for the two objectives, and than a large Pareto front.

The parent hull to be modified and optimized is taken from the NPL series [11]. The 5A hull has been scaled up to an overall length 36 m, maximum width 6 m, moulded depth 4 m, mean draft when fully loaded 1.75 m, tonnage 200 tonnes. The hull surface has been replicated using a single Non-Uniform Rational Basis-Spline (NURBS—see the left image in Figure 3), and the control points represent the design parameters of the optimization problem. Every control point can move in every direction: the only fixed nodes are the ones at the bridge (locked in the vertical direction), at the stern (locked in the longitudinal direction) and the fore point on the bridge (locked in every direction). We have a total Number of Design Variables (NDV) of 21.

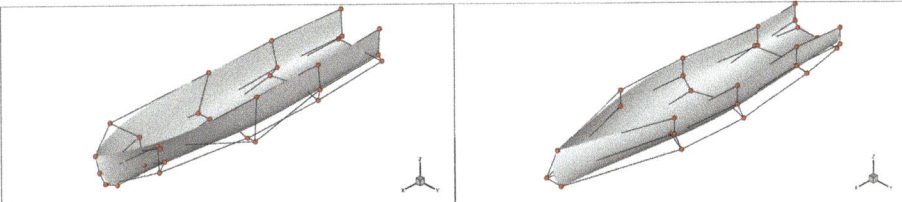

Figure 3. Representation of the hull surface adopted in the effective power optimization problem by means of a single NURBS: original shape (on **left**) and optimal shape (on **right**) obtained with a shift of the control points of the NURBS.

40 iterations of the full NBI procedure have been performed, evaluating a total number of 20,000 different configurations (around $1000 \times NDV$): this number can be assumed as representative of a medium effort for a single-objective optimization problem. Results are reported in Figure 4. The top left picture of Figure 4 reports the initial approximation of the Pareto front after the initial sampling phase: it is obtained by using $16 \times NDV$ solutions. Since then, in every picture the black dots are indicating the newly added solutions at every iteration. The central solution of the Pareto front is reported in the right part of Figure 3, and a comparison of the section views, original versus optimal, is reported in Figure 5. From this last picture we can observe the finer bow entry and flatter aft, as expected. We can see also in Figure 4 how the method is concentrating the newly added points around the current approximation of the true Pareto front: it seems that all the new solutions are not dispersed in a useless area, but they are all concentrated around the real Pareto front, improving its resolution at each iteration. This is an indication of the efficiency of the method.

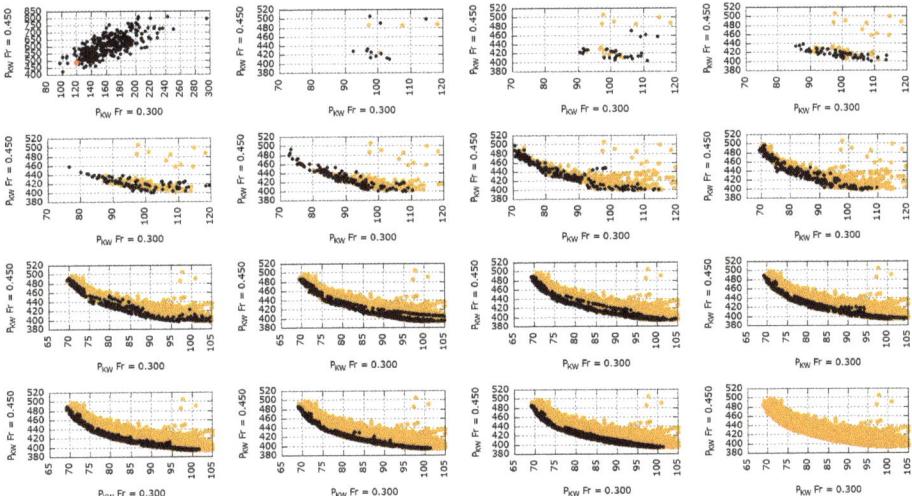

Figure 4. Evolution of the Pareto front during the iterations of the multi-objective optimization procedure of the effective power. Black dots are the newly added points after one full iteration (all the three methodologies).

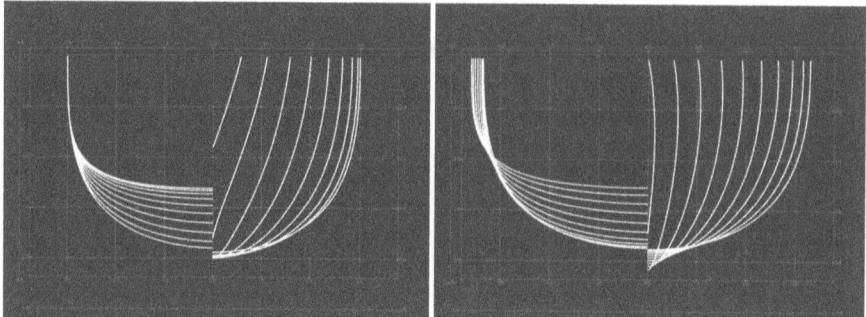

Figure 5. Comparison of the section views for the original (**left**) and the optimal (**right**) hull shapes for the effective power optimization problem.

4.2. Ship Design Application: Propulsion

A second realistic ship design optimization problem has been solved using the proposed methodology. In this case, the flow at the propeller disk for a bulk carrier has been optimized. The parent hull form is represented by the Japan Bulk Carrier (JBC), whose geometry has been adopted as test case for several workshops[3]. Full loaded condition as reported in the workshop tests has been considered.

In order to obtain a more regular flow at the propeller disk, possibly improving the working conditions of the propeller, two quantities have been selected to be minimized: the alignment of the local flow with the advancing direction and the variance of the local speed vector module. The first objective function is measured by the average value of the scalar vector between the local speed and the advancing direction: since they are pointing in opposite directions, the value we hope to get is -1, so that also this objective is to be minimized.

Local flow is computed by using the suite `OpenFoam v7`: the solver is based on the the unsteady Reynolds-Averaged Navier-Stokes Equations (RANSE) model, implemented in the solver (`interFoam`) with k-Omega$_{SST}$ turbulence model. Simulation is stopped at 30 s of physical time: we have observed that solution becomes stable after around 25 s.

The modification of the hull has been restricted to the bossing area and the surrounding part. A portion of the aftbody has been included in order to facilitate the fairing of the resulting surfaces. The modification methodology is here the Free Form Deformation (FFD) [12,13]. An FFD with $7 \times 2 \times 7$ subdivisions along the X, Y and Z axis respectively, has been placed in the area in front the propeller. In order to preserve the fairing of the hull surface, some of the nodes of the FFD are locked at the original position: the total number of design variables is 9, since only lateral movements are allowed in this case. A perspective view of the hull and the FFD box is reported in Figure 6. The method have produced 1350 alternatives ($150 \times NDV$), a smaller number with respect to the previous example, since the computational effort is considerably higher. The resulting Pareto front is reported in Figure 7. Also in this case, the Pareto front (indicated with black dots) is rich and well distributed. For a comparison of the alternatives from the geometrical point of view, three shapes have been selected: the best solutions for each objective and an intermediate one, representing the best trade-off between the objectives. This last shape has been selected normalizing the objectives in between 0 and 1, with a proportional scaling, and than selecting from the Pareto front the solution closer to the bisector of the objective function space. The selected geometries are reported in Figure 8.

---

[3] for details, see https://t2015.nmri.go.jp/jbc_gc.html.

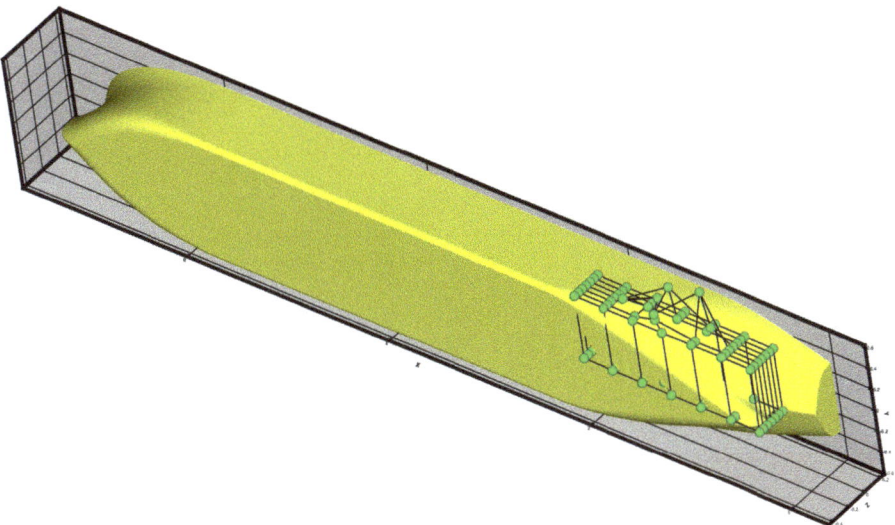

Figure 6. Perspective view of he hull surface (JBC) together with the FFD box for this hull parameterization. Green dots are representing the nodes of the FFD, and only 9 of them can move trasversally.

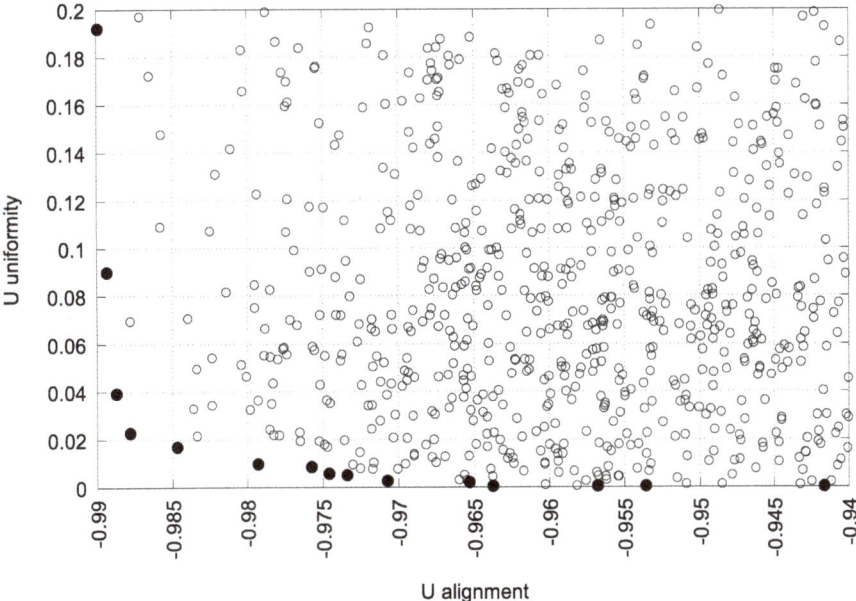

Figure 7. Pareto front for the multi-objective optimization problem of the JBC containership. Black dots are representing the computed Pareto front, white dots are indicating other tentative solutions.

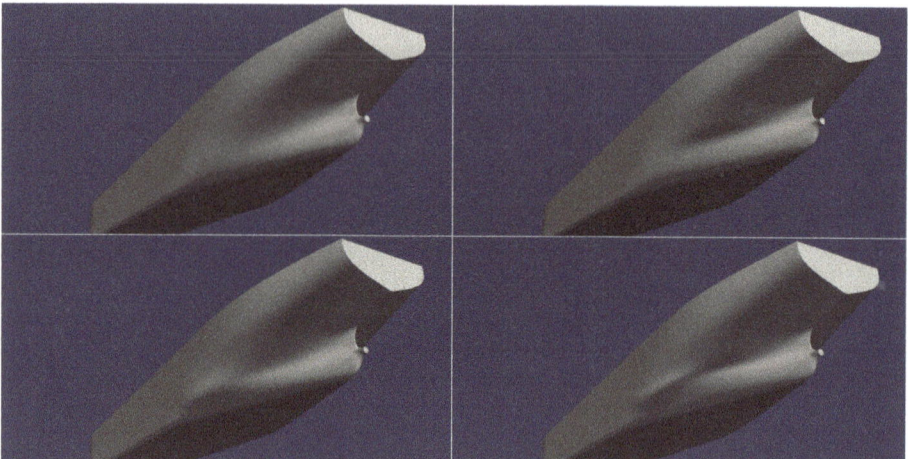

Figure 8. Comparison of the original hull shape (**top left**) and three Pareto optimal solutions: the best for the first objective function (**top right**), a solution with the best balance between the two objectives (**bottom left**) and the best for the second objective (**bottom right**).

We can observe how the modification of the three alternative hulls is very similar, being the amount of the modification the only difference. The optimizer finds that the most promising area producing an appreciable variation of the flow at the propeller disk is the region of the aftbody just in front of the restriction of the hull generating the bossing area. A bump is generated in this part of the hull, and the height of the bump (and its sharpness) represents the main difference between the three hulls.

In Figure 9 we can observe the flow at the propeller disk for the original configuration and for the three alternatives. The corresponding values of the objectives are reported in Table 1. Here we can observe how clearly the geometry with the best value of the local speed alignment is also increasing the average value of the local flow (looking at the longitudinal component), while the intermediate solution (the second from right in Figure 9) has a quite uniform value of the longitudinal component of the local speed. A different situation is arising for the third shape, where the decrease of the variance is obtained at the expense of the average value, strongly reduced. Furthermore, we can also observe the occurrence of a negative side effect, that is, a strong vortex in the lower part of the propeller disk. From the methodological standpoint this is not representing a problem, since the Pareto front has been correctly identified, but practically this is representing an failure. The problem should be probably reformulated, including the average value of the local speed into the definition of the second objective function.

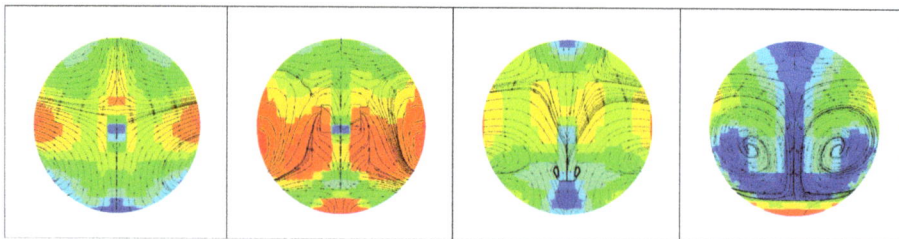

Figure 9. Flow at the propeller disk for four different configurations. From left to right, the original shape, the best shape for the first objective function, the configuration with the best balance between the two objectives and the best configuration for the second objective. Colors are representing the axial velocity of the flow, while the transverse component is indirectly reported by the streamtraces.

Table 1. Objective function values for some selected solutions.

	$\bar{u}$	σ_u	$\bar{D}$	σ_D
Original	−0.737169	0.126924	−0.951803	0.001652
Best $\bar{D}$	−0.822615	0.192103	−0.989913	0.000016
Intermediate	−0.734401	0.001688	−0.984661	0.000876
Best σ_u	−0.603269	0.000019	−0.941700	0.000520

5. Conclusions

A new methodology for the determination of the Pareto front in a multi-objective optimization problem has been formulated. Although the three components of the full methodology are not able to provide good results if applied in isolation, the combined use has been demonstrated to be very efficient. Application to algebraic functions and realistic applications gives the measure of the efficiency of the methodology: a dense and regular Pareto front is provided with an expense similar to the one of the solution of a single-objective optimization problem. The comparison with a standard MOGA highlights the good performances of the algorithm, being the results of the combined NBI procedure outperforming the MOGA. The solution of the two ship design applications is also giving clear indications about the ability of the proposed methodology to correctly identify the full Pareto front at the expense of a moderate computational effort, also in this case similar to the one required by a standard single objective optimization process.

One of the selected applications show some shortcomings in the definition of one of the objective functions, an it could be interesting to repeat the computations in order to observe the effects on the final configuration. Also the comparison with a single different methodology is not enough, and a more extensive and complete comparison with more similar algorithms has to come

New design applications could also be formulated and solved in order to generalize the actual results, also increasing the number of objectives.

Funding: This research was funded by Italian Minister of Instruction, University and Research (MIUR) to support this research with funds coming from PRIN Project 2017 (No. 2017KKJP4X entitled "Innovative numerical methods for evolutionary partial differential equations and applications").

Conflicts of Interest: The author declare no conflict of interest.

References

1. Cui, Y.; Geng, Z.Q.; Zhu, Q.X.; Han, Y.M. Review: Multi-objective optimization methods and application in energy saving. *Energy* **2017**, *125*, 681–704. [CrossRef]
2. Das, I.; Dennis, J.E. A closer look at drawbacks of minimizing weighted sums of objectives for Pareto set generation in multicriteria optimization problems. *Struct. Optim.* **1997**, *14*, 63–69. [CrossRef]
3. Goldberg, D.E. Genetic Algorithms in Search. In *Optimization and Machine Learning*; Addison-Wesley: Reading, MA, USA, 1989.
4. Sobol, I.M. Distribution of points in a cube and approximate evaluation of integrals *Zh. Vych. Mat. Mat. Fiz.* **1967**, *7*, 784–802. (In Russian)
5. Beachkofski, B.; Grandhi, R. *Improved Distributed Hypercube Sampling*; Paper 1274; American Institute of Aeronautics and Astronautics: Reston, VA, USA, 2002.
6. Vandenberghe, L.; Boyd, S.; Wu, S.P. Determinant maximization with linear matrix inequality constraints. *SIAM J. Matrix Anal. Appl.* **1998**, *19*, 499–533. [CrossRef]
7. Peri, D. Easy-to-implement multidimensional spline interpolation with application to ship design optimisation. *Ship Technol. Res.* **2018**, *65*, 32–46. [CrossRef]
8. Kasat, R.B.; Gupta, S.K. Multi-Objective Optimisation of an Industrial Fluidized-Bed Catalytic Cracking Unit (FCCU) Using Genetic Algorithm (GA) with the Jumping Genes Operator. *Comput. Chem. Eng.* **2003**, *27*, 1785–1800. [CrossRef]
9. Pinto, A.; Peri, D.; Campana, E.F. Multiobjective Optimization of a Containership Using Deterministic Particle Swarm Optimization. *J. Ship Res.* **2007**, *51*, 217–228.

10. Gadd George, E. A method for calculating the flow over ship hulls. *Trans. Inst. Nav. Archit.* **1970**, *112*, 335–345.
11. Bailey, D. The NPL High Speed Round Bilge. In *Displacement Hull Series: Resistance, Propulsion, Manoeuvring and Seakeeping Data*; Royal Institution of Naval Architects: London, UK, 1967.
12. Sederberg, T.W.; Parry, S.R. Free-form deformation of solid geometric models. In Proceedings of the 13st Annual Conference on Computer Graphics and Interactive Techniques, SIGGRAPH 1986, Dallas, TX, USA, 18–22 August 1986; Volume 20, pp 151–160.
13. Tahara, Y.; Peri, D.; Campana, E.F.; Stern, F. Computational fluid dynamics-based multi-objective optimization of a surface combatant using a global optimization method. *J. Mar. Sci. Technol.* **2008**, *13*, 95–116. [CrossRef]

© 2020 by the authors. Licensee MDPI, Basel, Switzerland. This article is an open access article distributed under the terms and conditions of the Creative Commons Attribution (CC BY) license (http://creativecommons.org/licenses/by/4.0/).

Article

Numerical and Experimental Optimization Study on a Fast, Zero Emission Catamaran

Apostolos Papanikolaou [1,*], Yan Xing-Kaeding [1], Johannes Strobel [1], Aphrodite Kanellopoulou [2], George Zaraphonitis [2] and Edmund Tolo [3]

1. Hamburgische Schiffbau-Versuchsanstalt, 22305 Hamburg, Germany; Xing-Kaeding@hsva.de (Y.X.-K.); Strobel@hsva.de (J.S.)
2. School of Naval Architecture and Marine Engineering, National Technical University of Athens, 15780 Zografou, Greece; afroditi@deslab.ntua.gr (A.K.); zar@deslab.ntua.gr (G.Z.)
3. Fjellstrand AS, 5632 Omastranda, Norway; etolo@fjellstrand.no
* Correspondence: papanikolaou@hsva.de

Received: 16 July 2020; Accepted: 25 August 2020; Published: 26 August 2020

Abstract: The present study focuses on the hydrodynamic hull form optimization of a zero emission, battery driven, fast catamaran vessel. A two-stage optimization procedure was implemented to identify in the first stage (global optimization) the optimum combination of a ship's main dimensions and later on in the second stage (local optimization) the optimal ship hull form, minimizing the required propulsion power for the set operational specifications and design constraints. Numerical results of speed-power performance for a prototype catamaran, intended for operation in the Stavanger area (Norway), were verified by model experiments at Hamburgische Schiffbau Versuchsanstalt (HSVA), proving the feasibility of this innovative, zero emissions, waterborne urban transportation concept.

Keywords: fast catamaran; zero emission; battery propulsion; hydrodynamic ship design; parametric design; multi-objective optimization; model experiments

1. Introduction

The herein presented work is conducted in the frame of the Horizon 2020 European Research project "TrAM—Transport: Advanced and Modular", which is a joint effort of 13 stakeholders (Kolumbus/Rogaland Prefecture, Norway (coordinator); Fjellstrand Yard, Norway; Hamburgische Schiffbau Versuchsanstalt (HSVA), Germany; Fraunhofer IEM, Germany; Hydro, Sweden; Leirvik, Norway; Nat. Tech. Univ. Athens (NTUA), Greece; Univ. of Strathclyde, United Kingdom; MBNA Thames Clippers, United Kingdom; Servogear, Norway; De Vlaamse Waterweg, Belgium; Wärtsilä Netherlands V.S., Netherlands; NCE Maritime CleanTech, Norway) of the European maritime industry [1]. The aim of this project is to develop zero emission fast passenger vessels through advanced modular production, with the main focus on electrically powered vessels operating in coastal areas and inland waterways. The project is innovative for the introduced zero emission technology, the design and manufacturing methods, while it should prove that electric-powered vessels can be fast and competitive in terms of offered services, of the environmental impact and the life-cycle cost.

In the frame of this project, intensive research was carried out on the hydrodynamic optimization of a battery-driven catamaran's hull form, in order to minimize power requirements and energy consumption, while introducing new propulsion and hull solutions related to the concept of electrically driven fast vessels. It should be noted that present replications of the battery-driven waterborne concept are limited to the lower and medium speed range. In that respect, hydrodynamic (and structural design) optimization is imperative for fast vessels and even more for battery driven vessels with a limited range of operation for the installed battery capacity.

Main authors of this paper have a long-standing experience in the optimization of fast twin-hull vessels—Papanikolaou et al., 1991 [2] and 1996 [3]; Zaraphonitis et al., 2003 [4]; Skoupas et al., 2019 [5]. The state of the art in the field is, however, longstanding and widely represented. Even though it is not the purpose of this paper to elaborate on this, some other representative and useful works dealing with the optimization of fast twin-hull vessels and their experimental verification can be found in the listed references—Insel, 1990 [6]; Molland et al., 1994 [7]; Brizzolara, 2004 [8]; Bertram et al., 2017 [9]. This list is not claimed to be exhaustive nor fully balanced.

The objective of this paper is to briefly present parts of the conducted numerical and experimental optimization study on the development of the hull form of the "Stavanger demonstrator". Numerical studies were conducted by use of the Computer Aided Design (CAD) software platform CAESES® and included the development of surrogate models for the ship's resistance based on calculations for a large number of design variants using HSVA's panel and Computational Fluid Dynamics (CFD) codes and the multi-objective global and local optimization by use of genetic algorithms. Obtained numerical results were verified by systematic model tests at the large towing tank of HSVA. A demonstrator of the presently studied catamaran concept will be built and also start operations in a multistop commuter route in the Stavanger area, Norway, before the end of the project in 2022 (https://tramproject.eu/).

2. The Parametric Model

2.1. Background

Based on the preliminary lines plan of a reference vessel, a parametric model for the demihulls of the Stavanger demonstrator was developed by use of the CAESES® software platform of Friendship Systems, Potsdam, Germany [10,11]. The developed parametric model offers the designer the possibility to control/specify the main particulars of the demihull along with the hull form details within a reasonable range of variation of the defined design variables, while at the same time, adequate quality (fairness) of the hull is ensured. The designer is enabled to explore the huge design space of automatically generated hull forms and decide on the most favorable ones on the basis of rational, holistic criteria [12].

The external dimensions of the vessel providing the required passengers transport capacity were set equal to 31.0 m length overall by 9.0 m beam overall (Figure 1). The vessel should be able to carry up to 147 passengers with a maximum operating speed of about 23–25 kn, depending on the loading condition and installed power of the propulsion e-motors. The overall length of each demihull was set equal to 30.6 m. Because of uncertainties in the weight calculations inherent in the early design stage of a prototype, it was decided that calm water predictions should be carried out for three different displacements (Δ1, Δ2 and Δ3).

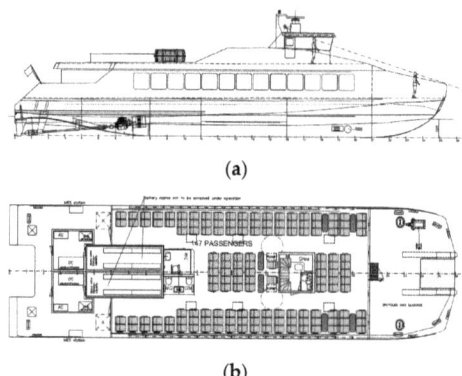

Figure 1. Preliminary general arrangement of the Stavanger Demonstrator. (**a**) side profile (**b**) main deck plan.

2.2. Parametric Model

A set of 20 design variables was first specified, defining the main dimensions, as well as local hull details, such as the width, immersion and shape of the transom and the shape bow area of the vessel. From the set of 20 design parameters, the four most important referring to the catamaran's main dimensions and the transom width were selected as design variables during the optimization studies (Table 1). The overall beam of the catamaran is herein kept constant due to design/construction reasons (yard's specification of deck superstructure module). It is of course acknowledged that increasing the separation distance of the demihulls would lead to lower wave resistance at some speeds, but the increase in lightweight and production cost is expected to outweigh this resistance benefit. It is also noted that the vessel's operational Froude number will be close to 0.70, thus far beyond the last hump of wave resistance; thus, viscous resistance will be dominant at the catamaran's service speed. The remaining 16 design parameters are kept constant at their default values. For the definition of the stern region, the most important parameters are the transom height at centerline and the height difference from centerline to the chine at transom. Negative values of the latter parameter indicate designs with the chine located lower than centerline, forming the propeller tunnel area. Another important parameter for stern definition is the x-coordinate of the maximum height of the tunnel at centerline. There is also a set of parameters for the definition of the bow area, such as the waterline fullness at the design draught and at deck height. The effect of the local design parameters is mainly of interest during the local hull form optimization.

Table 1. Optimization variables.

L_{WL}	Definition waterline length
HB_{DES}	Demihull's definition half beam
T_{INIT}	Definition draught
$CHINEY\_ATS0$	Transom width definition

Based on the specified values of the optimization variables, and the default values for the remaining design parameters, a grid of parametrically defined curves was created.

At first a set of primary definition curves was generated, such as the deck line, transom and centerline. Then a set of parametrically defined sections and diagonals was gradually added until the grid was completed. The creation of the primary definition curves was entirely based on the values of the optimization variables and the remaining design parameters. Subsequent curves were created using mostly reference points (making reference to previously defined curves) in order to ensure the consistency of the grid and to keep the necessary information for the elaboration of the hull form to the minimum. The sections were created using Interpolation Curves definition available in CAESES® software, while the diagonals using 3D curves definition. A view of the parametrically defined grid is presented in Figure 2a. Subsequently, a series of metasurfaces and lofted surfaces was generated, as presented in Figure 2b. The metasurfaces were generated using a set of features, developed in CAESES® software, exploiting the built-in scripting language.

After the initial hull definition, the hydrostatic values were calculated. A Lackenby transformation was then applied in order to obtain a demihull with a longitudinal center of buoyancy close to the expected longitudinal center of gravity. Additionally, the prismatic coefficient was adjusted in order to achieve a displacement close to the desired value. The Lackenby transformation parameters were considered constant during the optimization studies. The resulting hull form is illustrated in Figure 2c.

It should be noted that the design variables were used for the definition of the initial hull form. The actual waterline length, beam and draught values of the demihull were calculated afterwards based on the final hull resulting from the Lackenby transformation for the immersed volume(s) specified by the user. This also included the hydrostatics for the initial draught and the set displacement volume. A tank top was assumed fitted at a certain height to protect the ship against raking damage, according to the definition of the areas vulnerable to raking damage provided in Chapter 2 of the

International Maritime Organisation (IMO)'s High Speed Craft Code, for the maximum assumed displacement (Δ3). To ensure the availability of ample space for the installation and maintenance of the battery racks, the dimensions of the corresponding compartments were checked against the specified requirements/constraints.

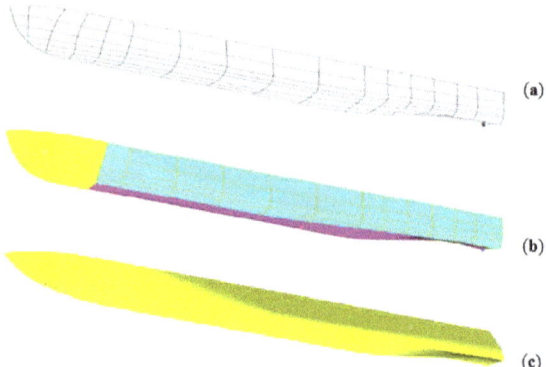

Figure 2. Phases of development of the catamaran's demihull form by CAESES—(**a**) definition grid, (**b**) resulting meta- and lofted surfaces and (**c**) final demihull after Lackenby transformation (**c**).

The hydrodynamic assessment of each design alternative was based on HSVA's in-house hydrodynamic tools, i.e., the panel code for wave resistance v-SHALLO [13] and the RANSE code FreSCo$^+$ [14]. Since these tools require considerable computing resources, it was decided to explore the possibility provided by CAESES® to precompute data for later usage. To this end, a series of so-called Design of Experiments (DoE) were carried out, to obtain the resistance of a sufficiently large number (about 1000) of alternative hull forms. Based on the collected precomputed data, surrogate models were developed, enabling the sufficiently accurate estimation of the hydrodynamic quantities of interest during the optimization study in practically zero time (in our case, the calm water resistance of each design variant at various displacements and service speeds). Apart from drastically reducing the calculation time, surrogate models increased the robustness of the whole process by avoiding the need for remote computing.

3. Potential Flow Calculations

3.1. Theoritical Background

HSVA's panel code v-SHALLO is a fully nonlinear, free surface potential CFD method computing the inviscid flow around a ship hull moving on the free water surface. The code is based on a superposition of a given free stream velocity with the flow induced by a number of 3D Rankine point sources on the ship's hull and the free surface. v-SHALLO treats the nonlinear free surface boundary condition iteratively by a collocation method and uses a patch method for dealing with the body boundary condition and pressure integration [13,15]. The hull and the free surface were discretized by means of triangular and/or rectangular panels, and the individual source strengths were determined by solving a linear equation system resulting from the discretization of a Fredholm integral equation.

The applied panel mesh for the demihull of the catamaran is shown in Figure 3 as an example. Trim and sinkage were estimated based on the vertical forces, and the body grid was moved accordingly. The wave elevation at the collocation points was computed from Bernoulli's equation. A typical wave pattern computed for the Stavanger demonstrator at a speed of 23 kn is shown in Figure 4.

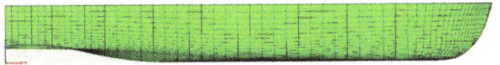

Figure 3. Discretized demihull of the Stavanger demonstrator.

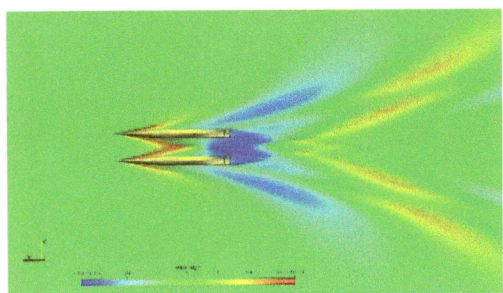

Figure 4. Wave Pattern of the Stavanger demonstrator at 23 kts.

3.2. Integration with the Parametric Model in CAESES®

v-SHALLO was the first of HSVA's simulation tools that was fully integrated into the CAESES® platform in the frame of the H2020 HOLISHIP project [16]. The CAESES® SoftwareConnector provides a platform for the integration of external simulation software in a user friendly and fast way. The panel mesh was generated directly in CAESES® and saved as a pan file. The SoftwareConnector was directly connected with the parametric model of the catamaran's demihulls in CAESES®, which means that every geometrical change in the parametric model would be processed by v-SHALLO. Additionally, by keeping herein the overall beam of catamaran constant, the separation distance of the demihulls would be calculated automatically, while considering the maximum demihull beam, and v-SHALLO computed the interference between the wave systems of the two demihulls, which was especially important for the value of the wave resistance.

By configuring the optimization engine in CAESES®, different optimization strategies can be applied. Once the optimization strategy, the design variables as well as the objective functions and constraints have been decided, either the Design of Experiment (DoE) or other methods of design exploration/optimization can be carried out in CAESES® fully automatically. Results can be directly visualized and checked within CAESES®. In the meantime, a design table in ASCII format containing all the results as well as a folder including all the information related to each design is created. This design table can be read by other tools, such as Excel, for further analysis and processing of the results.

3.3. Design of Experiments (DoE)

Based on the operational profile of the Stavanger demonstrator, three speeds (21, 23 and 25 kn) and three displacements (Δ1, Δ2 and Δ3) were specified for the DoEs, which resulted in 9 design conditions in total. The SOBOL random sequence generator was selected to explore the design space, in which the user can specify a fixed number of design variants, and the SOBOL algorithm creates the distribution of the design variables to satisfactorily cover the entire design space. The required number of design variants was tested in the beginning, showing that 200 design variants per design condition seemed to be a good compromise with respect to accuracy and efficiency of the design exploration. The actual computing time on a desktop computer with four parallel processors took only a few hours, which was acceptable for the early design phase. While the results obtained with the potential theory code v-SHALLO were generally sufficient for the ranking of the explored designs, the values of selected designs were later finetuned by use of the more sophisticated RANSE code FreSCo$^+$ of HSVA, which gave more accurate resistance levels.

4. Surrogate Models

Hydrodynamic performance calculation codes are usually time consuming and require increased memory and storage capabilities. Therefore, it was considered more efficient to replace CFD tools during the optimization process with the so-called surrogate models, providing a sufficiently accurate approximation of the hydrodynamic performance of each design variant at practically zero computing time. A surrogate model was created on the basis of a series of simulations, typically produced by means of a design-of-experiment (DoE) for a predefined set of free variables. This approach seeks to determine a relationship between the input variables and the response of the objective function. Based on the conducted DOE by use of v-SHALLO by HSVA, a series of response surfaces was created in CAESES® for the calculation of calm water resistance at three different speeds (21, 23 and 25 kn) and three displacements (Δ1, Δ2 and Δ3).

Various surrogate models were created and tested using software tools available in CAESES®. The most accurate methods were MARS (Multivariate Adaptive Regression Splines) [17] and Neural Networks [18,19]. The MARS method presented smaller differences with the v-SHALLO results and proved to be more accurate and faster (approximately 25%) than neural networks. Therefore, the response surfaces used during the optimization studies were developed using the MARS method with a training dataset of 200 samples. In Figure 5, representative results obtained with the response surfaces are compared with those obtained with v-SHALLO. Almost all designs are between the dashed lines corresponding to 0.5% error. Thus, very good correlation between the calculated and the estimated results was observed in all cases, indicating that the developed surrogate models can safely replace the potential flow calculations during the optimization studies.

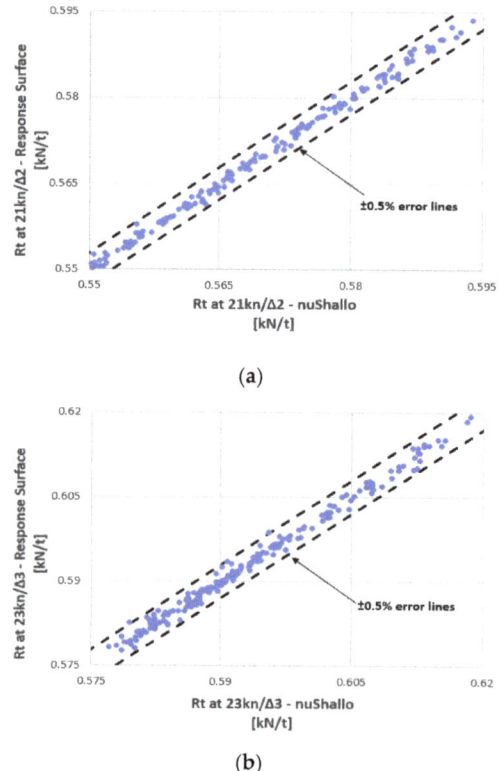

Figure 5. Comparison of calm water resistance per ton of displacement obtained with the response surfaces and v-SHALLO for (**a**) speed 21 kn at displacement Δ2 and (**b**) speed 23 kn at displacement Δ3.

5. Global Optimization Studies

As a next step, a global optimization study was performed in order to identify the optimum combination of the selected design variables for the Stavanger demonstrator.

5.1. Optimization Settings

For the optimization studies, the catamaran's length overall and beam overall were herein assumed to remain fixed at 31.0 and 9.0 m, respectively, by yard's specification. The selected design variables are listed in Table 1.

The range of variation of the design variables was as follows: length at waterline, from 29.0 to 30.2 m; halfbeam of demihull, from 1.0 to 1.3 m; initial draught, from 1.2 to 1.6 m; and transom width at chine, from 0.8 to 0.85 (nondimensional). The optimization study was carried out by employing the NSGAII algorithm [20], already integrated in the CAESES® environment. NSGAII has already been successfully applied by NTUA in various other similar studies during the last ten years (e.g., [5]), and the experience gained from these studies verified its efficiency and suitability for the problem at hand. The population and generation size, mutation and crossover probabilities used during the optimization are presented in Table 2.

Table 2. Optimization Settings.

Genetic Algorithm	NSGA-II
Generations	10
Population size	100
Mutation probability	0.01
Crossover probability	0.9

The objective of the present study was to minimize the calm water resistance of the bare hull in a range of displacements and speeds, which mostly represent the operational profile of the Stavanger demonstrator. Therefore, it was decided to evaluate the calm water resistance of each design alternative at 21, 23 and 25 kn and at three displacements $\Delta 1$, $\Delta 2$ and $\Delta 3$ and based on the results, to evaluate a weighted average of the calm water resistance. The resulting weighted objective function used during the optimization studies is presented in the following:

$$R_T = 0.12 R_{T\_21kn, \Delta 1} + 0.20 R_{T\_23kn, \Delta 1} + 0.08 R_{T\_25kn, \Delta 1} + 0.09 R_{T\_21kn, \Delta 2} + 0.15 R_{T\_23kn, \Delta 2} + 0.06 R_{T\_25kn, \Delta 2} + 0.09 R_{T\_21kn, \Delta 3} + 0.15 R_{T\_23kn, \Delta 3} + 0.06 R_{T\_25kn, \Delta 3}$$

A set of constraints was also applied in order to verify that each feasible design alternative provided sufficient space for the installation of the battery racks in the demihulls and large diameter propellers. The minimum demihull width and height around amidships for the batteries' installation and the minimum height of transom stern tunnel at its centerline for the propeller installation were specified by the yard and controlled during optimization for compliance. Additionally, the minimum draft at the lightest displacement $\Delta 1$ should be greater than the tunnel height, so that the tunnel is fully immersed when the vessel is at rest. In a second design alternative with the battery racks place on deck, the minimum demihull width constraint was removed.

5.2. Global Optimization Settings

A large number of optimization studies were carried out as more knowledge and data for the problem at hand were gradually collected during the course of the project [21]. In Figures 6 and 7, some representative results of the final optimization study are illustrated. Out of 1000 produced designs, 824 were feasible, whereas 176 violated at least one of the constraints. The optimum design is marked in the figures with a green circle. Based on the results, the overall optimum design had a very slender hull form with a length at the waterline (WL) close to the maximum, a beam close to the minimum and increased draught.

This section may be divided by subheadings. It should provide a concise and precise description of the experimental results, their interpretation as well as the experimental conclusions that can be drawn.

In addition, as can be observed from the figures, for the lightest displacement Δ1, hull forms with the smaller beam at WL were those with a lower calm water resistance. This is also true for the intermediate displacement Δ2 at 21 and 23 kn. At 25 kn however, hull forms with an approximately 10% larger beam at WL exhibited a lower calm water resistance. At the same speed and the highest displacement Δ3, this tendency became more pronounced. The designs with the best performance at higher displacement and higher speed had an increased beam of demihull, and their calm water resistance at lower displacement and lower speed increased by approximately 2% to 2.5% in comparison with the overall optimum. It should be noted that operational conditions characterized by higher displacement and higher speed were of particular importance because they determined the required power of the selected propulsion unit and the capacity of the batteries.

The characteristics of the finally promoted design along with its estimated calm resistance are illustrated in Table 3. This design was further improved during the local hull form optimization study by use of the RANSE code FreSCO+ and was tank tested by HSVA in December 2019.

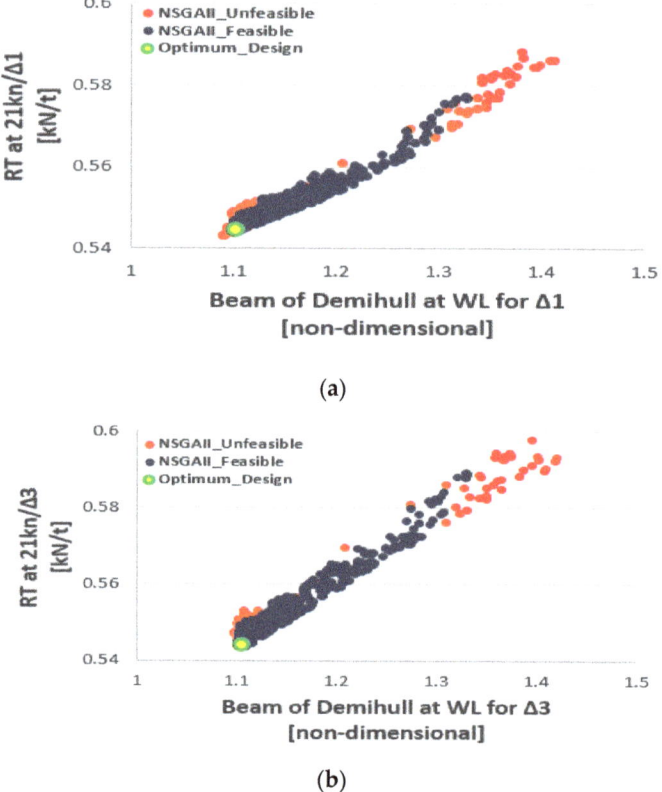

Figure 6. Calm water resistance per ton of displacement against nondimensional beam at WL at 21 kn for (**a**) Δ1 and (**b**) Δ3.

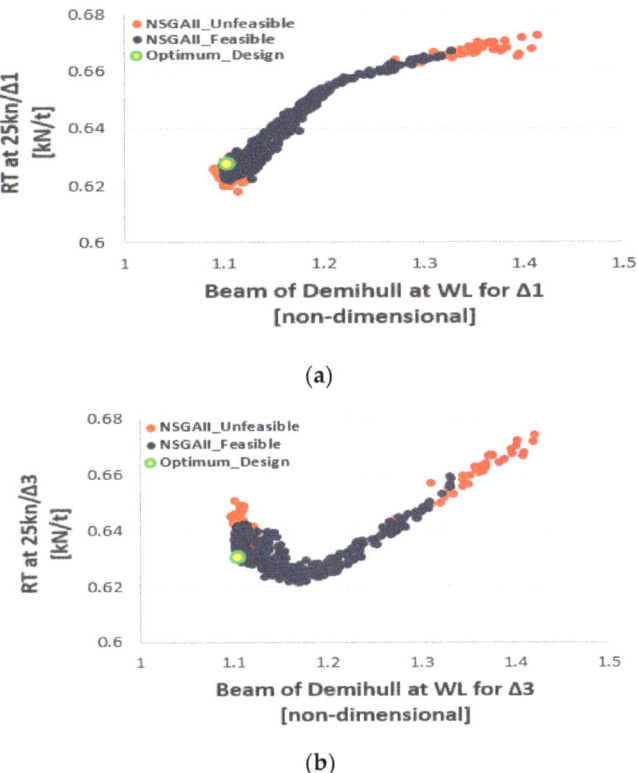

Figure 7. Calm water resistance per ton of displacement against nondimensional beam at WL at 25 kn for (**a**) Δ1 and (**b**) Δ3.

Table 3. Obtained results for the selected design (battery racks in the demihulls).

	at Δ1	at Δ2	at Δ3
LWL	29.29	29.34	29.39
Beam at WL	2.442	2.442	2.444
Draft	1.226	1.272	1.317
Rt_21 kn/Δ	0.558	0.599	0.642
Rt_23 kn/Δ	0.588	0.620	0.663
Rt_25 kn/Δ	0.658	0.684	0.717

It should be noted that both due to safety reasons related to the storage of the batteries and for easier installation, inspection and maintenance of the battery racks, an alternative design option was developed by the shipbuilder (Fjellstrand), in which the battery racks were placed on deck (see Figure 1). This allowed the removal of constraints on the demihull beam and led to a new hull form that proved significantly better than the originally optimized one, namely by more than 6% at 23 knots and even more than 10% at intermediate speeds, as elaborated in later sections (see later Figure 17).

Finally, it should be pointed out that during the global optimization, both single-objective and multi-objective optimizations were carried out. In the former case, the minimization of the weighted average of the total resistance, as defined earlier, was used as the objective function. In the latter case, nine optimization functions were introduced, each one corresponding to the calm water resistance at one combination of speed and displacement. Both studies resulted in practically identical results; therefore, the results of the single-objective optimization were used and are presented in this paper. This facilitated the quick assessment/visibility of the impact of the selection of alternative weights (for displacement and speed combinations) on the resistance (and powering) by the partners (end-users: yard and operator) and the identification of the overall optimal design alternative for the Stavanger scenario case.

6. Numerical Methods for the Local Optimization Study

The computational method applied to the local optimization of the globally optimized hull form for resistance and propulsion was the RANS method FreSCo$^+$ code of HSVA. Though modeling the propeller in RANS is nowadays possible, the more practical approach is to simulate the propeller effect through a body-force model, as this allows the quick evaluation of a large number of design variants. The body forces were obtained from a Propeller Vortex Lattice Method (QCM), which was coupled iteratively to the RANS method to enable numerical self-propulsion simulations. More details of these methods are explained below.

6.1. RANSE Method

The HSVA in-house code FreSCo$^+$ is a finite volume fluid flow solver developed in cooperation with the Institute of Fluid Dynamics and Ship Theory (FDS) of the Hamburg University of Technology (TUHH) and the Hamburg Ship Model Basin (HSVA). The FreSCo+ code solves the incompressible, unsteady Navier-Stokes-equations (RANSE). The method is applied to fully unstructured grids using arbitrary polyhedral cells or hanging nodes. Additionally, features such as sliding interface or overlapping grid techniques have been implemented into the code [14].

Various turbulence-closure models are available with respect to statistical (RANS), such as k-ε (Standard, RNG, Chen), k-ω (Standard, BSL, SST), Menter's One Equation model and the Spalart-Allmaras turbulence model. In this paper, the k-ω SST was mainly used.

6.2. Propeller Vortex Lattice Method QCM

The HSVA QCM method is similar to the approaches published by [22–24]. The chord wise arrangement of corner-points of the vortex–lattice is set up by the "Cosine-Spacing" as originally recommended by [25]. The results for the loading distribution become identical with the exact solutions of the continuous theory for 2-dimensional thin profiles. Due to this property, the method was named the "Quasi-Continuous Method" (QCM). By using QCM, [24] calculated open water characteristics of various propellers that were in good agreement with experimental results, and established a method for estimating open-water characteristics of unconventional propellers, e.g., contra-rotating, controllable pitch and tandem propellers. A typical vortex structure in the propeller wake as simulated in QCM is illustrated in Figure 8.

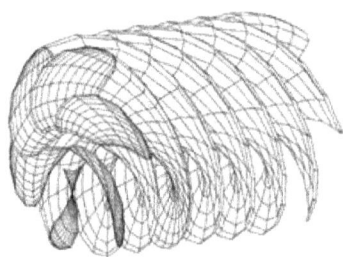

Figure 8. Vortex structure in the propeller wake in QCM for a typical propeller in a homogeneous inflow.

6.3. Numerical Self-Propulsion Using RANS-QCM Coupling

The present local hull form optimization studies aiming at very high propulsive efficiency have been performed using the RANS-QCM coupling approach of HSVA to numerically simulate the self-propulsion test. To this end, the code FreSCo+ is coupled with QCM for propeller analysis in an iterative fashion as sketched in Figure 9.

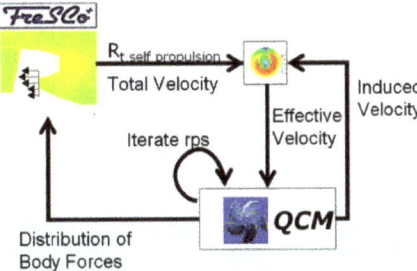

Figure 9. Numerical self-propulsion test scheme.

At the start of the simulation, a nominal wake distribution is extracted from the converged RANS solution without the propeller effect. This velocity distribution and an estimated turning rate are used as an input for the QCM code to compute the forces on the propeller blades (thrust and torque). The turning rate is adjusted until the propeller thrust required to overcome the ship resistance (in propulsion mode) is obtained. The hydrodynamic forces of the propeller are converted in the form of 3D body forces (source terms) assigned to cells which are representing the propeller disk. More details of this method can be found in [26].

7. Local Optimization Studies

While the global optimization referred to the determination of the main dimensions and integrated hull form characteristics minimizing the calm water resistance, the local hull form optimization takes into account not only the calm water resistance, but also the propulsive speed-power performance. In this respect, the local optimization study focused on the optimization of the stern tunnel area and the propulsive efficiency. The stern hull form area was mathematically captured by six local form parameters and in addition, four parameters were related to the main propeller characteristics, such as propeller diameter, position and shaft inclination. The ten local optimization variables are listed in Table 4. The Dakota Optimization Toolkit of Sandia National Laboratories disposed in CAESES® was utilized for the optimization. This toolkit allows a comprehensive exploration of the multiparametric design space by use of proper sampling methods, such as Latin hypercube sampling, orthogonal arrays and Box-Behnken designs, simultaneously allowing the implementation of the design constraints. Due to the fact that the ship will be actually built, a large number of design constraints related to the fitting of the propeller, its shaft and brackets in the tunnel area need to be considered as well. These constraints are mainly related either to class specification (induced vibrations) or to limitations from the practical and construction point of view. The defined design constraints are given in Table 5. The benefit of such constraints integrated into the design optimization process is that only the designs fulfilling all the requirements and constraints will be further kept in the optimization process. The design and optimization process as such had two stages. In the first stage, the whole design space was explored (only limited to the available computational resource and available design time) for valid designs, which satisfied all the design requirements/constraints. Due to the very strict design constraints in this design case, only a relatively small number (around 2 percent) of the generated designs proved to be feasible. Figure 10 shows some of the generated transom stern designs, which fulfill the design constraints as an example. In the second stage, the resulted valid designs were reviewed by experienced designers and propeller manufacturers to make a final check and select the final designs before performing the time-consuming RANS-QCM computations for the resistance and propulsion power prediction.

Table 4. Local optimization variables.

CLZ_{ATS0}	Z Coordinate of the Centerline at Transom
CLZ_{ATX1_7m}	Z Coordinate of the Centerline at Tunnel
$CHINEDZ\_{ATS0}$	Z Coordinate of the Chine relative to Centerline at Transom
$CHINEZ\_{AST2}$	Z Coordinate of the Chine relative to Centerline at Tunnel
$CHINEY\_{ATS0}$	Transom width definition
$CHINEY\_{ATS2}$	Tunnel width definition
$D\_{Propeller}$	Propeller Diameter
$X\_{Propeller}$	X Coordinate of the Propeller Position
$Z\_{Propeller}$	Z Coordinate of the Propeller Position
$Beta\_{Propeller}$	Propeller Inclination angle

Table 5. Design constraints in local optimization stage.

Propeller tip clearance	Greater than 20% of diameter of propeller
Propeller shaft forward end above the hull	Greater than certain value to guarantee the propeller shaft, gearbox and El Motor installation
Propeller shaft entry above the hull	Greater than certain value to guarantee the propeller shaft, gearbox and El Motor installation
Propeller shaft inclination	Less than certain degrees value
Z_max_prop	Propeller submergence at smallest displacement
LWL at largest displacement	Less than certain value
Height of shaft bracket lead	Less than certain value to guarantee the propeller shaft installation
Height of shaft bracket tail	Less than certain value to guarantee the propeller shaft installation
Longitudinal position of El. Motor forward end	Less than certain value to guarantee the space for electromotors installation

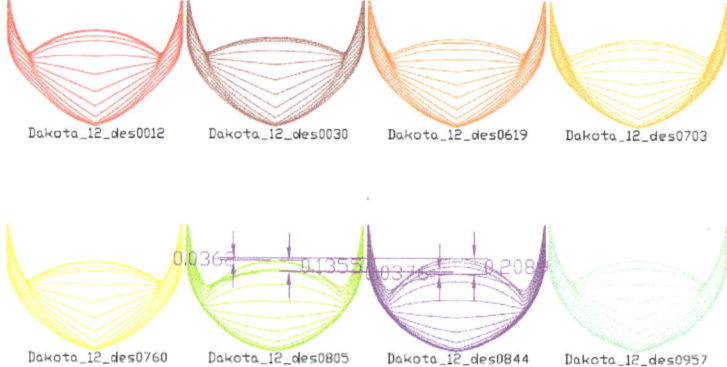

Figure 10. Transom stern design alternatives generated by local optimization.

The most promising designs were then evaluated by use of the HSVA's RANS-QCM coupled method as previously described. In this procedure, the free surface, free sinkage and trim of the catamaran were considered as well. The numerical mesh applied had around 5.3 million cells in total, including a refinement around the free surface field, the bow thruster tunnel area and the propeller/ship transom stern region, as shown in Figure 11.

The identified best design with respect to the required delivered horsepower (DHP) was further fine-tuned to minimize the risk of air suction in the propeller tunnel. For the selection of the best hull form, a range of displacements and speeds was evaluated to assess the performance of the hull variants at various off-design conditions. Figure 12 shows the computed Cp distribution. The propeller was simulated via a body force method, where the three-dimensional blade forces coming from the panel code QCM (RANS-BEM coupling) were incorporated, as can be better observed in Figure 13, together with the streamlines passing through the propeller discs.

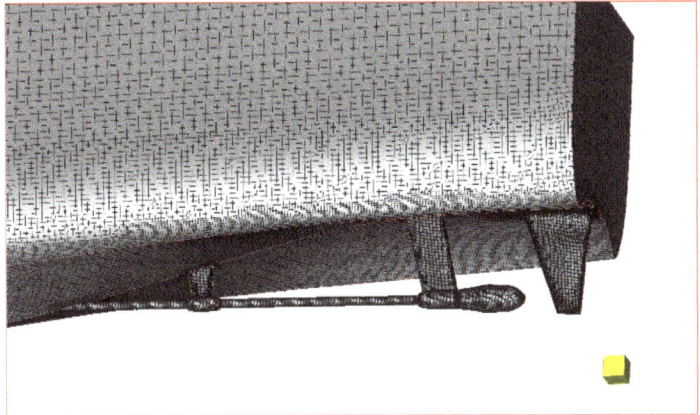

Figure 11. Numerical mesh around the stern tunnel area for the local optimization of the Stavanger demonstrator.

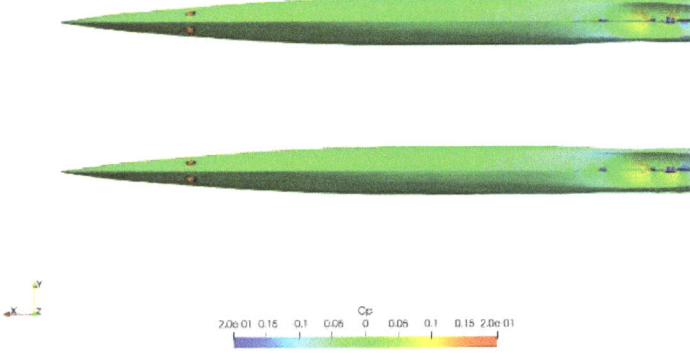

Figure 12. Cp distribution on the demihulls at 23 knots.

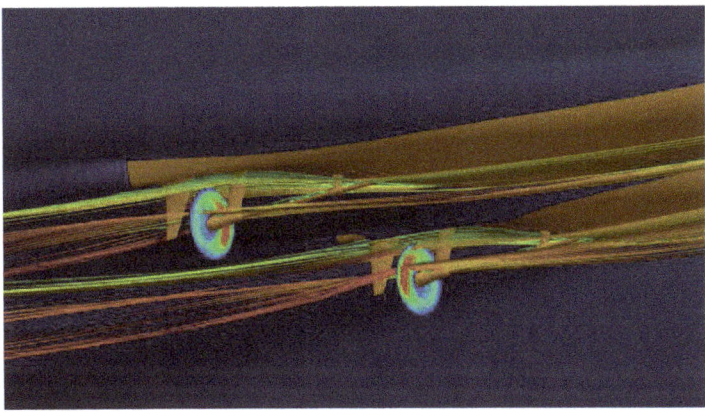

Figure 13. Streamlines through propeller discs and propeller body force distribution at 23 knots.

8. Experimental Verification

As an essential part of the design process of a highly complex and innovative ship, physical model tests play a crucial role considering the verification of the anticipated full-scale speed-power performance.

8.1. Tested Model

The determination of a suitable model scale ratio is one of the first and most important steps in the process of planning a model test campaign. The principle goal of minimizing scale effects by building a large model needs to be balanced with limiting factors such as basin constraints, carriage speed, estimated loads, measurement equipment and certainly building costs. For the TrAM model, a very good trade-off between these factors resulted in a scale ratio of 1/5.6, namely a 5.34 m long catamaran model. This allowed very precise measurements and minimized scale effects. The two separate demihull models were manufactured from wood and were coupled by high-strength metal beams. Proper alignment and positioning of the demihulls were ensured by special high precise measurement gauges individually designed for this test setup. A 3D wireframe and faired surface view of the numerically optimized hull is shown in Figure 14, while a close-up of the stern area of the tested model with fitted CP propellers and (twisted) rudders is shown in Figure 15.

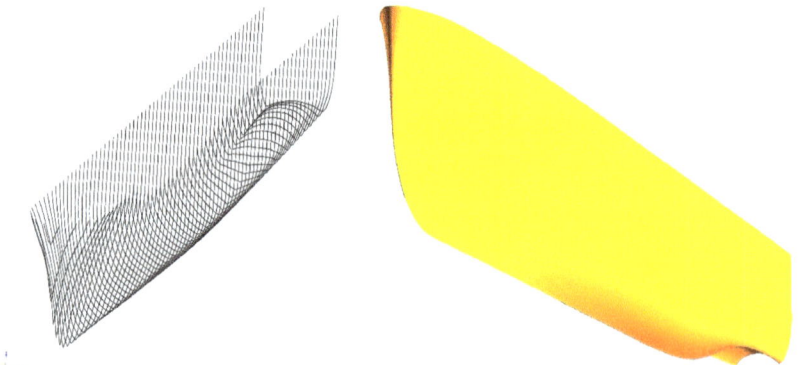

Figure 14. 3D wireframe and faired surface view of the numerically optimized Stavanger hull form.

Figure 15. Close-up of the stern area of the tested Stavanger model, with fitted propellers, shafts, brackets and rudders.

8.2. Test Scope

The calm water model tests were carried out in HSVA's large basin, which is 300 m long, 18 m wide and 6 m deep. The speed of the model ranged from 1.5 to 6.5 m/s which corresponds to a ship speed of 8 to 29 knots (Figure 16). During the test runs, all relevant forces and movements of the model were recorded, while also including wave profile measurements for the generated wave wash downstream. The test program included both towing resistance and self-propulsion tests for three different displacements—Δ1, Δ2, Δ3—and a range of trims. Besides the variation of the calm water resistance for tested conditions, special attention was paid to the propulsive efficiency of the fitted propulsion plant and the hull–propeller–rudder interaction (wake and thrust deduction fractions). It should be also noted that a first test campaign with the originally optimized hull form (battery racks placed in the demihulls) was conducted in December 2019, and a second campaign repeated the test series for the finally selected hull form in May 2020.

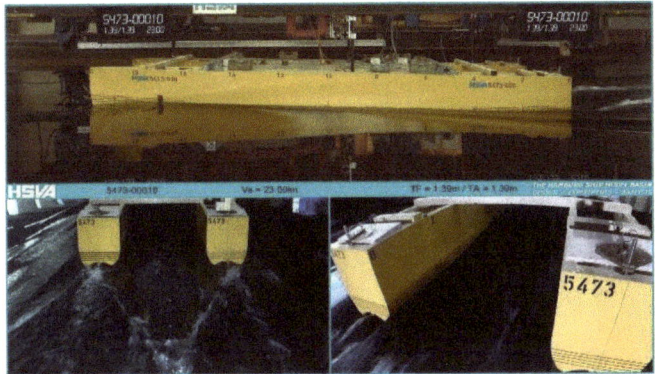

Figure 16. Self-propulsion model of the Stavanger demonstrator at 23 knots full scale speed.

8.3. Test Results

The numerically predicted model and full-scale values obtained by CFD simulations could be very well confirmed by the test campaigns. In the conducted second test series with the revised hull form (May 2020), the resistance and propulsion power could even be reduced significantly for the relevant speed range above 14 knots (see Figures 17 and 18). A remarkable result of the model tests was the extraordinarily high propulsive efficiency that could be achieved by the refined local optimization of the transom stern and of the hull–propeller–appendices interaction (see Table 6). The very low thrust deduction and wake fraction, on the one hand, and the achievement of a hardly disturbed propeller inflow condition, on the other hand (note that ETAR is even larger than 1.00 for speeds over 19.00 knots), resulted in a propulsion efficiency of up to 80% at higher speeds. A systematic variation of the static pretrim of the vessel delivered valuable information for a beneficial arrangement of the ship's weight distribution in terms of power reduction. The entire test series was live-broadcasted ("live-stream") and recorded by several cameras showing the model and the flow around it from different perspectives. This allowed a detailed observation of the vessel's hydrodynamic behavior remotely and even after the tests, as necessary.

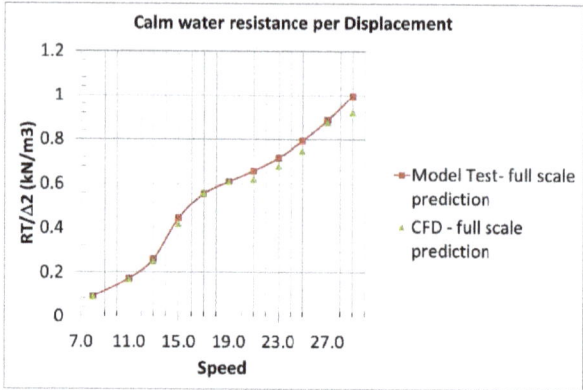

Figure 17. Prediction of the rated full-scale calm water resistance for the Stavanger demonstrator on the basis of model experiments and CFD calculations by Hamburgische Schiffbau Versuchsanstalt (HSVA) (revised hull form, battery racks on deck).

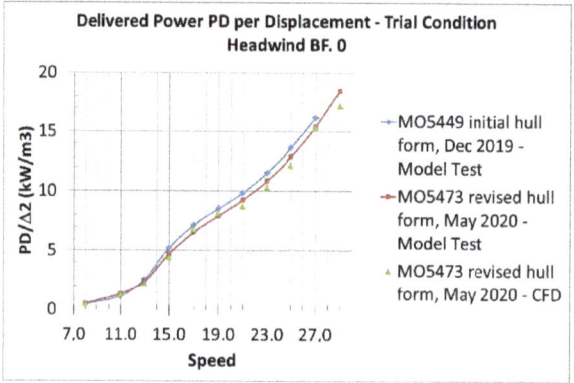

Figure 18. Prediction of rated delivered horsepower under trail condition for the Stavanger demonstrator on the basis of model experiments and CFD calculations by HSVA (originally optimized and revised hull form).

Table 6. Experimentally measured hull and propulsive efficiencies for the design displacement of the Stavanger demonstrator (full scale)—V: speed in knots; Fn: Froude number; t: thrust deduction fraction; w: Taylor wake fraction; ETAH: hull efficiency; ETAO: open water propeller efficiency; ETAR: relative rotative efficiency; ETAD: propulsive efficiency.

V	Fn	t	w	ETAH	ETAO	ETAR	ETAD	Notes
kn	-	-	-	-	-	-	-	
8.00	0.243	0.003	0.002	0.999	0.782	0.982	0.762	
10.00	0.334	0.021	0.006	0.985	0.785	0.974	0.753	
13.00	0.394	0.030	0.032	1.002	0.764	0.989	0.757	
15.00	0.455	0.015	0.016	1.001	0.744	0.994	0.741	
17.00	0.516	0.011	0.000	0.989	0.753	1.000	0.745	
19.00	0.576	0.017	−0.003	0.980	0.765	1.006	0.755	ETAR > 1.000
21.00	0.637	0.025	0.007	0.982	0.773	1.013	0.770	ETAR > 1.000
23.00	0.698	0.036	0.020	0.983	0.777	1.024	0.782	ETAR > 1.000 ETAD > ETAO!
25.00	0.758	0.045	0.029	0.983	0.779	1.031	0.789	ETAR > 1.000 ETAD > ETAO!
27.00	0.819	0.053	0.038	0.985	0.780	1.039	0.798	ETAR > 1.000 ETAD > ETAO!

9. Summary and Conclusions

A two-stage parametric optimization of the hull form and propulsion plant of twin-hull vessels was developed and applied to the design of a fast catamaran vessel aiming to operate a multistop commuter route in the Stavanger area, the so called "Stavanger demonstrator". The parametric model of the hull form was developed in CAESES® and offers the possibility to automatically generate smooth hull forms in the specified range of the main particulars of the demihulls along with the possibility to control and modify a series of important hull form details.

A large number of about 1000 alternative hull forms was elaborated and assessed with the potential theory 3D panel code v-SHALLO of HSVA to form the basis for the development of surrogate models (response surfaces) for the estimation of calm water resistance during the first-stage, global optimization studies. Global optimization studies were carried out using the NSGA-II optimization algorithm, and two of the most promising designs were selected for more refined optimization. These hull forms were further optimized using the same parametric model in CAESES® and HSVA's CFD tool FreSCo$^+$, while focusing on the unique stern-transom area and the propulsive efficiency, which reached a value of a remarkable 78.2% at a design speed of 23 knots (and even 80% at 27 knots).

The resulting hull form was model-tested at HSVA's towing tank in December 2019 and obtained model test results verified the numerical predictions. A second design alternative was also considered, in which the battery racks were fitted on the deck area, rather than in the demihulls. This design option proved better than the first one, as could be expected, due to the relaxed constraints on the demihull's minimum width, and will form the basis for the final selection of battery capacity and electric motors power for the desired speed profile of the Stavanger demonstrator.

The obtained results prove the feasibility of the TrAM zero emission, fast catamaran concept and will decisively support decisions in the final design of the Stavanger demonstrator, planned to enter service in the Stavanger area before the end of the TrAM project in the first half of 2022.

Author Contributions: Conceptualization, A.P.; methodology, A.P., G.Z. and Y.X.-K.; software, Y.X.-K. and A.K.; validation, Y.X.-K., A.K. and Johannes Strobel; formal analysis, A.P., G.Z. and Y.X.-K.; investigation, Y.X.-K., A.K. and J.S.; resources, Y.X.-K., A.K. and J.S.; data curation, Y.X.-K., A.K., J.S. and E.T.; writing—original draft preparation, A.P.; writing—review and editing, A.P. and E.T.; visualization, Y.X.-K., A.K. and J.S.; supervision, A.P., G.Z. and E.T.; project administration, Y.X.-K., G.Z. and E.T.; funding acquisition, A.P., G.Z. and E.T. All authors have read and agreed to the published version of the manuscript.

Funding: This research was funded by the European Union's Horizon 2020 research and innovation program under grant agreement No 769303.

Acknowledgments: The authors acknowledge the support of a several of TrAM project partners in the presented research, particularly of Servogear (propeller designer) and Kolumbus (ship operator).

Conflicts of Interest: The authors declare no conflict of interest.

References

1. TrAM—Transport: Advanced and Modular, H2020 EU Research Project, Grant Agreement No 769303, H2020. Available online: https://tramproject.eu/ (accessed on 25 August 2020).
2. Papanikolaou, A.; Androulakakis, M. Hydrodynamic Optimization of High-Speed SWATH. In Proceedings of the 1st International Conference on Fast Sea Transportation (FAST'91), Trondheim, Norway, 17–21 June 1991.
3. Papanikolaou, A.; Kaklis, P.; Koskinas, C.; Spanos, D. Hydrodynamic Optimization of Fast Displacement Catamarans. In Proceedings of the 21st International Symposium on Naval Hydrodynamics, ONR'96, Trondheim, Norway, 1 June 1996.
4. Zaraphonitis, G.; Papanikolaou, A.; Mourkoyiannis, D. Hull Form Optimization of High-Speed Vessels with Respect to wash and Powering. In Proceedings of the 8th International Marine Design Conference (IMDC), Athens, Greece, 5–8 May 2003.
5. Skoupas, S.; Zaraphonitis, G.; Papanikolaou, A. Parametric Design and Optimisation of High-Speed Ro-Ro Passenger Ships. *Ocean Eng.* **2019**, *189*, 106346. [CrossRef]

6. Bertram, V.; Krebber, B.; Hochkirch, K.; Frers, G.; Flynn, M. Advanced simulation-based design for an advanced power trimaran. In Proceedings of the 11th Symposium on High Speed Marine Vehicles (HSMV), Naples, Italy, 25–26 October 2017.
7. Molland, A.; Wellicome, J.; Couser, P. *Resistance Experiments on a Systematic Series of High-Speed Displacement Catamaran Forms: Variation of Length-Displacement Ratio and Breadth-Draught Ratio*; Ship Science Report No. 71; Department of Ship Science, University of Southampton: Southampton, UK, 1994.
8. Insel, M. An Investigation into Resistance Components of High-Speed Displacement Catamarans. Ph.D. Thesis, Department of Ship Science, University of Southampton, Southampton, UK, 1990.
9. Brizzolara, S. Parametric Optimization of SWATH Hull Forms by a Viscous-Inviscid Free Surface Method Driven by a Differential Evolution Algorithms. In Proceedings of the 25th Symposium on Naval Hydrodynamics, St. John's, NL, Canada, 8–13 August 2004.
10. CAESES® Software Platform. Friendship Systems. Available online: https://www.caeses.com/ (accessed on 25 August 2020).
11. Abt, C.; Harries, S.; Wunderlich, S.; Zeitz, B. Flexible Tool Integration for Simulation-driven Design using XML, Generic and COM Interfaces. In Proceedings of the International Conference on Computer Applications and Information Technology in the Maritime Industries (COMPIT 2009), Budapest, Hungary, 10–12 May 2009.
12. Papanikolaou, A. Holistic Ship Design Optimization. *J. Comput. Aided Des.* **2010**, *42*, 1028–1044. [CrossRef]
13. Jensen, G.H.; Mi, Z.; Söding, X. Rankine source methods for numerical solutions of the steady wave resistance problem. In Proceedings of the 16th Symposium on Naval Hydrodynamics, Berkeley, CA, USA, 13–16 July 1986.
14. Hafermann, D. Proceedings of the New RANSE Code FreSCo for Ship Application, Jahrbuch der Schiffbautechnischen Gesellschaft (STG), Hamburg, Germany, 9 October 2007; Volume 101.
15. Gatchell, S.; Hafermann, D.; Jensen, G.; Marzi, J.; Vogt, M. Wave resistance computations—A comparison of different approaches. In Proceedings of the 23rd Symposium Naval Hydrodynamics (ONR), Val de Reuil, France, 17–22 September 2000.
16. HOLISHIP: Holistic Optimisation of Ship Design and Operation for Life Cycle. Project Funded by the European Commission, H2020-DG Research, Grant Agreement 689074. Available online: http://www.holiship.eu (accessed on 25 August 2020).
17. Friedman, J.H. Multivariate Adaptive Regression Splines. *Ann. Stat.* **1991**, *19*, 1–141. [CrossRef]
18. Hagan, M.T.; Demuth, H.B.; Beale, M.H.; De Jesus, O. *Neural Network Design*, 2nd ed. 2014. Available online: https://www.google.com.hk/url?sa=t&rct=j&q=&esrc=s&source=web&cd=&ved=2ahUKEwjhxdru7rfrAhUXH3AKHTmQByoQFjABegQIAxAB&url=https%3A%2F%2Fhagan.okstate.edu%2FNNDesign.pdf&usg=AOvVaw0Bfkk6GUwTvySiX6_rUNJ0 (accessed on 25 August 2020).
19. McCulloch, W.; Pitts, W. A logical calculus of the ideas immanent in nervous activity. *Bull. Math. Biophys.* **1943**, *5*, 115–133. [CrossRef]
20. Srinivas, N.; Deb, K. Multiobjective function optimization using nondominated sorting genetic algorithms. *Evol. Comput. J.* **1994**, *2*, 221–248. [CrossRef]
21. Kanellopoulou, A.; Xing-Kaeding, Y.; Papanikolaou, A.; Zaraphonitis, G. Parametric Design and Hydrodynamic Optimisation of a battery-driven fast catamaran vessel. In Proceedings of the RINA Conference on Sustainable and Safe Passenger Ships, Athens, Greece, 4 March 2020.
22. Kerwin, J.E.; Lee, C.S. Prediction of Steady and Unsteady Marine Propeller Performance by Numerical Lifting-Surface Theory. *SNAME Trans.* **1978**, *86*, 1–30.
23. Greeley, D.S.; Kerwin, J.E. Numerical Methods for Propeller Design and Analysis in Steady Flows. *SNAME Trans.* **1982**, *90*, 415–453.
24. Nakamura, N. Estimation of Propeller Open-Water Characteristics Based on Quasi-Continuous Method. *J. Soc. Nav. Archit. Jpn.* **1985**, *157*, 95–107. [CrossRef]
25. Lan, C.E. A Quasi-Vertex-Lattice Method in thin Wing Theory (E). *J. Aircr.* **1974**, *11*, 518–527. [CrossRef]
26. Xing-Kaeding, Y.; Gatchell, S.; Streckwall, H. Towards Practical Design Optimization of Pre-Swirl Device and its Life Cycle Assessment. In Proceedings of the Fourth International Symposium on Marine Propulsors, SMP'15, Austin, TX, USA, 31 May–4 June 2015.

© 2020 by the authors. Licensee MDPI, Basel, Switzerland. This article is an open access article distributed under the terms and conditions of the Creative Commons Attribution (CC BY) license (http://creativecommons.org/licenses/by/4.0/).

Article

Electrical Swath Ships with Underwater Hulls Preventing the Boundary Layer Separation

Igor Nesteruk [1], Srecko Krile [2] and Zarko Koboevic [2,*]

1. Institute of Hydromechanics, National Academy of Sciences of Ukraine, 03057 Kyiv, Ukraine; inesteruk@yahoo.com
2. Maritime Department, University of Dubrovnik, Ćira Carića 4, 20000 Dubrovnik, Croatia; srecko.krile@unidu.hr
* Correspondence: zarko.koboevic@unidu.hr

Received: 20 July 2020; Accepted: 17 August 2020; Published: 25 August 2020

Abstract: The body shapes of aquatic animals can ensure a laminar flow without boundary layer separation at rather high Reynolds numbers. The commercial efficiencies (drag-to-weight ratio) of similar hulls were estimated. The examples of neutrally buoyant vehicles of high commercial efficiency were proposed. It was shown that such hulls can be effectively used both in water and air. In particular, their application for SWATH (Small Water Area Twin Hulls) vehicles is discussed. In particular, the seakeeping characteristics of such ships can be improved due to the use of underwater hulls. In addition, the special shaping of these hulls allows the reducing of total drag, as well as the energetic needs and pollution. The presented estimations show that a weight-to-drag ratio of 165 can be achieved for a yacht with such specially shaped underwater hulls. Thus, a yacht with improved underwater hulls can use electrical engines only, and solar cells to charge the batteries.

Keywords: environment protection; drag reduction; commercial efficiency; boundary layer separation; SWATH vehicles

---

1. Introduction

The urgent task of reducing negative impacts on the environment requires economical vehicles with minimal emissions of carbon dioxide and toxic substances. In particular, many ships burn dirty fuels. In particular, in was estimated in [1] that ships cause up to 40% of the air pollution in coastal towns around the Mediterranean. A report of the French environment ministry [1] states that shipping pollution causes approximately 6000 premature deaths around the Mediterranean each year.

Croatian tourism is continuously growing. Each year, protected areas (e.g., national parks, the Plitvice Lakes, Krka and Kornati) attract more visitors. Cruising and nautical tourism are the fastest-growing types of tourism, and require efforts to protect the marine environment [2]. The relatively small number of COVID-19 cases in Croatia gives hope for a rapid recovery of the tourism business.

To reduce ships' contribution to global warming, using fossil fuels must be stopped in the next few decades [1]. Norway and China are already using electric ships [3,4]. Recreation yachts and ferries are a perfect place to start. since they travel only short distances and stay for relatively long periods of time at ports, where they can be charged and use battery packs [3,4].

The delay in the electrification of maritime transport (in comparison with automobiles or railways) is probably connected with the occurrence of higher drag in water, due to its much higher density. To overcome this drag much more powerful engines are necessary. As such, the problem of drag reduction is very important for maritime transport in general, and especially as regards its electrification. The low drag of vehicles allows for the increasing of their commercial efficiency [5] and range with the use of one charge.

To have all-season ships, their seakeeping characteristics must be improved. In particular, SWATH (Small Waterplane Area Twin Hull) technology uses underwater hulls and allows the vehicle to move smoothly at rather high waves [6–9]. Improving the underwater shape of these hulls grounds the possibility of having comfortable low-drag ships with electrical or even solar propulsion.

Vehicles or animals that ensure a laminar attached flow pattern are expected to be the most effective, since separation and turbulence cause intensive vortexes in the flow, increase drag and produce noise. The high swimming velocities of dolphins and other aquatic animals continue to attract the interest of researchers [10–15]. From the point of view of biomechanics, the body shape of good swimmers ensures the associated flow patterns. To prove this fact, testing of rigid bodies similar to animal shapes was carried out, with Reynolds number values close to real ones [16]. Observations of gliding dolphins indicated that a flow pattern without boundary layer separation [17] explains the fact of the low drag of very good shapes only. Thus, from the point of view of this research, the absence of separation on the bodies of good swimmers is a reason of their low drag.

On the other hand, most researchers believe that the minimum level pressure is located near the midline of the body, so separation is inevitable downstream from the cross section of the maximum area [18,19]. Nevertheless, theoretical investigations have shown that pressure decrease is possible near the tail of some specially shaped axisymmetric bodies (e.g., [20–22]). Examples of such hulls were also manufactured and tested in wind tunnels [23,24]. Unfortunately, a negative pressure gradient downstream of the maximum thickness section is not enough to preclude separation. For example, the attached flow pattern was achieved with the specially shaped Goldschmied's body [23], but boundary layer suction was used. A short survey of the theoretical and experimental results concerning these specially shaped bodies of revolution are presented in Section 2 of this paper. The commercial efficiency of hulls preventing boundary layer separation and their possible areas of application will be discussed in other Sections.

2. Special Shaped Underwater Hulls

In the case of the attached flow pattern, slender bodies of revolution ensure low pressure drag, and can delay laminar–turbulent transitions on their surfaces [22,25,26]. Therefore, the skin friction drag and total drag can be reduced on such bodies. Rigid bodies with a laminar attached have been investigated at the Institute of Hydromechanics (IHM) of National Academy of Sciences, Kyiv, Ukraine [24]. In particular, a UA-2c shape similar to the dolphin body was calculated (see Figure 1).

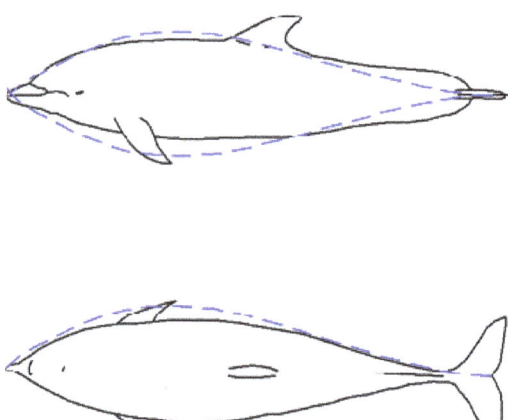

Figure 1. The body of a bottlenose dolphin compared with the closed body of revolution UA-2c [27].

Some support tubes are necessary in order to fix models in wind or water tunnels. An example of such an unclosed shape—UA-2—was tested at IHM and at the Institut für Strömungsmechanik (ISM) at Technische Universität Braunschweig, Germany [28]. A good agreement between theoretical and experimental pressure distributions and unseparated and laminar flow patterns was obtained [28]. In comparison, on the Goldschmied's body the separation was precluded only with the use of boundary layer suction [23].

The method of calculation of shapes UA-2 and UA-2c was applied to obtain other bodies of revolution (with different thickness ratios, D/L, and positions of their maximum thickness point; D is the maximum diameter of the body, L is its length) and 2D profiles as well [22,24,27]. Some examples are shown in Figure 2. The separation behavior of these shapes needs further experimental investigation, but their similarities to the bodies of aquatic animals allows us to expect an attached boundary layer, as was shown in the experiments with the rigid copies of different fish [16]. It must be noted that all the bodies have concave tails. The shape corresponding to the smallest thickness ratio $D/L = 0.1$ also has a concave front body part (see Figure 2) similar to the shapes of some fast-swimming fish (e.g., the Mediterranean spearfish, Indo-Pacific sailfish, black marlin or swordfish).

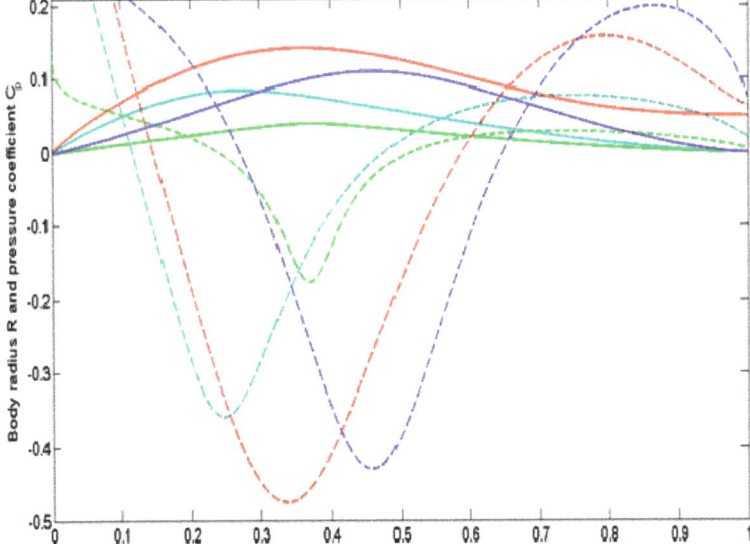

Figure 2. Specially shaped bodies of revolution [22]. Radius R (solid lines) and pressure coefficient C_P (dashed lines) versus dimensionless axis coordinate x/L. Unclosed body UA-2 without boundary layer separation ($L/D = 3.52$; red lines). Closed bodies UA-4.5c "Albacore" ($L/D = 4.5$; dark blue lines); UA-5.9c "Blue shark" ($L/D = 5.9$; blue lines); UA-12.4c "Sailfish" ($L/D = 12.4$; green lines).

The attached flow pattern allows us to delay the boundary layer's turbulent motion (see [26,27]) and reduce the friction drag (the pressure drag is close to zero due to the D'Alembert paradox). Then the total drag on such a body of revolution can be estimated via the following formula [27,29]:

$$C_V \approx \frac{4.7}{\sqrt{Re_V}}, Re_V = UV^{\frac{1}{3}}/\nu \qquad (1)$$

where ν is the kinematic viscosity coefficient of water. Equation (1) is in good agreement with the Hoerner formula [30] for the laminar drag on the standard elongated bodies of revolution:

$$C_S = \frac{2X}{\rho U^2 S} = C_{fl}\left[1 + 1.5(D/L)^{1.5}\right] + 0.11(D/L)^2, C_{fl} = \frac{1.328}{\sqrt{Re_L}}, Re_L = \frac{UL}{\nu}$$

where the term 0.11 $(D/L)^2$ (connected with separation) can be neglected. The body surface area S and the Reynolds number Re_L can be connected with the body volume V and the volumetric Reynolds number Re_V by the empirical Hoerner equations [30],

$$V \approx 0.65 L \pi D^2 / 4, \quad S \approx 0.75 L \pi D$$

(see also [22], Figure 2).

In order to have a laminar boundary layer on the entire hull surface, the speed, length and volume of such hulls are related by the following inequality [25,26]:

$$V < \frac{59558\pi L^3}{Re_L} = \frac{59558\pi \nu L^2}{U} \tag{2}$$

To estimate possible drag reduction, an axisymmetric shape similar to the body of Dolphinus delphis ponticus Barab. (L/D = 4.76) [16] was taken to calculate the critical Reynolds number according to relationship (2). If $Re_L < 1.4 \times 10^7$, the boundary layer is laminar on the entire body surface, and its drag can be estimated by (1) (see laminar curve in Figure 3). With increasing the Reynolds number, the turbulent boundary layer zone near the tail expands and leads to the drag increasing. Simple estimations of the turbulent drag in this zone can be done with the flat plate concept [30], and are also shown in Figure 3. A comparison with the experimental drag measurements on the Hansen and Hoyt body [31] shows that specially shaped hulls can ensure almost twofold lower drag.

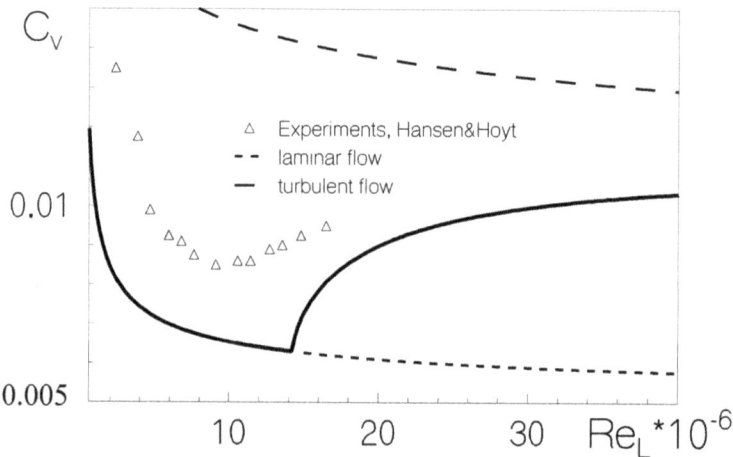

Figure 3. Drag coefficient estimations for a body of revolution similar to the shape of Dolphinus delphis ponticus Barab. (L/D = 4.76) and comparison with the experimental measurements of drag on the Hansen and Hoyt model [31].

3. Estimations of Commercial Efficiency of Neutrally Buoyant Vehicles

An estimation of the commercial efficiency of vehicles is the drag-to-weight ratio $1/k$. The minimal value of this parameter yields the maximum of tons × kilometers that can be transported by the vehicle per unit of time [5]. With the fixed fuel (or another energy) capacity onboard, a vehicle with the maximum value of k has the maximum range. The drag-to-weight ratio can also be treated as the cost of motion, i.e., how much energy is used to move 1N of weight the distance of 1m. Usually in the literature this characteristic is related to 1 kg of mass or weight—$Jkg^{-1}m^{-1}$ (see e.g., [32]). By dividing the values in $Jkg^{-1}m^{-1}$ by 9.8 (the value of gravity acceleration g), we obtain the dimensionless criterion, coinciding with $1/k$.

For example, the mass m of a SWATH yacht is related to the volume of an underwater hull V by the simple formula $m \approx 2\rho V$, where ρ is the water density. Taking into account that the total drag of such a vehicle is approximately two times higher than the drag X of each underwater hull and the volumetric drag coefficient $C_V = 2X/(\rho U^2 V^{\frac{2}{3}})$, we can obtain the formula

$$\frac{1}{k} = \frac{C_V \rho U^2 V^{\frac{2}{3}}}{mg} = 0.5 C_V Fr_V^2 \qquad (3)$$

where the volumetric Froude number is related to the standard one, $Fr_L = U/\sqrt{gL}$ (based on the vehicle length L and speed U), by the following equation:

$$Fr_V^2 = \frac{U^2}{gV^{\frac{1}{3}}} = \frac{Fr_L^2 L}{V^{\frac{1}{3}}} \qquad (4)$$

Putting (1) and (2) into (3) allows us to obtain [16]:

$$1/k = 2.35 v^{\frac{1}{2}} U^{\frac{3}{2}} V^{-\frac{1}{2}} g^{-1} \qquad (5)$$

Estimation (5) can be treated as the lowest possible value of the drag–weight ratio not only for the vehicles with underwater hulls, since the drag of those with floating hulls is greater, due to the wave resistance. Equation (5) shows that the easiest means of reducing the energetic needs and pollution is to reduce the speed, to increase the volume and to use specially shaped hulls with the attached laminar flow pattern.

Equation (2) yields the limitations for the maximal mass of the neutrally buoyant vehicles with such hulls. The results of corresponding estimations are shown in Figure 4 for water and air. According to the results shown in Figure 4, it is possible to have a variety of fully laminar airships with very high commercial efficiency. For example, a stratospheric airship with $L/D = 20$ operating at an altitude of 20 km can achieve the velocity $U \approx 100$ m/s, and its mass can be approximately 6 t. At the attitude of 10 km, the commercial effective laminar airships are not so fast, but can be much larger. For example, an airship with $L/D = 20$ can have a velocity $U \approx 20$ m/s, and a mass of approximately 40 t. To increase the velocity of the airship, its L/D must be higher. For example, for $L/D \approx 50$, it is possible to have an effective laminar airship at $U \approx 100$ m/s with a mass of approximately 15 t.

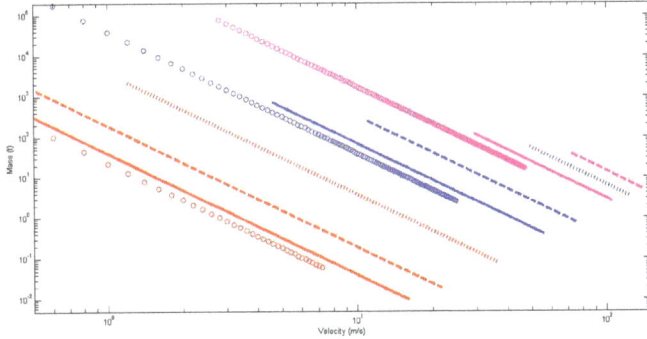

Figure 4. Maximum mass of the laminar neutrally buoyant hulls in water ("circles") and in air at attitudes 0, 10 and 20 km (solid, dashed and dotted lines respectively). The color of lines and "circles" corresponds to the body thickness $D/L = 0.02, 0.05$ and 0.278 (magenta, blue and red respectively).

Airships operating at small attitudes can also be rather fast and large. For example, an airship with $L/D = 20$ can have the velocity $U \approx 20$ m/s and a mass of approximately 9 t. To increase the mass of the commercial effective airship, we need to decrease its velocity. For example, at L/D = 20 and velocity 5 m/s, the mass is approximately 500 t.

4. Wave Drag and Critical Froude Number

The neutrally buoyant vehicles are efficient at a small Froude number only. The estimation of the critical value of the Froude number (based on the hull length) can be found in [25]. In particular, for a hull with a laminar attached boundary layer,

$$Fr^*_{L,\,lam} = 13.6 k_W^{-\frac{1}{2}}$$

where k_W is the aerodynamic efficiency for airplanes or planning ships (lift-to-drag ratio). Usually, k_W cannot exceed the value 60 (and is much smaller for planning boats). Thus, we can use the estimation

$$Fr^*_{L,\,lam} \approx 2 \qquad (6)$$

At smaller values of the Froude number, the commercial efficiency of the neutrally buoyant vehicles (in particular, SWATH yachts) is higher than that of vehicles with dynamical weight support (like high speed planing boats).

Estimation (6) exceeds another critical Froude number, $Fr^*_L \approx 0.4$, which corresponds to the drastic increase in the wave drag on floating ships at supercritical Froude numbers. SWATH technology uses underwater hulls, and therefore this increase can be neglected, and faster yachts can be made more efficient in comparison with the standard floating boats.

5. Some Suggestions of Low Drag Swath Vehicles

The specially shaped bodies of revolution can be applied both for the underwater hulls of SWATH vehicles and for the hulls located above water. It is also possible to have a vehicle with high commercial efficiency at supercritical Froude numbers $Fr_L > 0.4$. We will give two schematic examples of SWATH ships without discussion of strength, stability and immersion. We will illustrate only the relationship between the speed and the size of vehicles with laminar underwater hulls, which provide a high commercial efficiency and the possibility of electrification.

As an example, we propose to use this technology for a fast SWATH ship (speed up to 50 m/s, weight up to 30 t). Its speed is almost two times higher in comparison with the existing SWATH vehicles (e.g., Sea Fighter FSF-1 and Francisco High-Speed Ferry). In addition, its weight-to-drag ratio is expected to be around 20. The sketch of a 1:4 model is shown in Figure 5. Air propulsion and specially shaped hulls with the laminar-attached boundary-layer both in water and in air can be used. This concept can be employed for both small and middle-sized fast economy ferries and special ships for all seasonal operations (in particular, for high speed and seakeeping ferries and patrol ships). The use of shapes with minimal possible drag allows the reduction of the capacity of engines and their negative impact on the environment.

In the case of recreation yachts, the velocity demand is not very high, and it is possible to achieve very small values of $1/k$, provided separation could be precluded for the underwater hulls. For example, assuming the laminar flow of the entire underwater hulls and neglecting the wave drag, the value of k could be estimated as 165 at U= 10 m/s and V = 2 m^3. This figure is approximately three times higher than the lift-to-drag ratio of the Solar Impulse 2 plane [19], which the rounded globe with the use of solar energy only. Such yachts can be electric, using solar cells to charge the batteries, and therefore reduce pollution. The main characteristics of the yacht are as follows: U= 10 m/s; V = 2 m^3; weight 4 t; length 9.2 m; two underwater hulls with maximal diameter 0.92 m (L/D = 10), with two electrical engines, propellers located on their tails and batteries; and one overwater hull of the same length, maximal diameter 2.6 m and volume 30 m^3 (see Figure 6). It must be noted that there is no need to use very slender hulls, and there are no problems with their strength and stability.

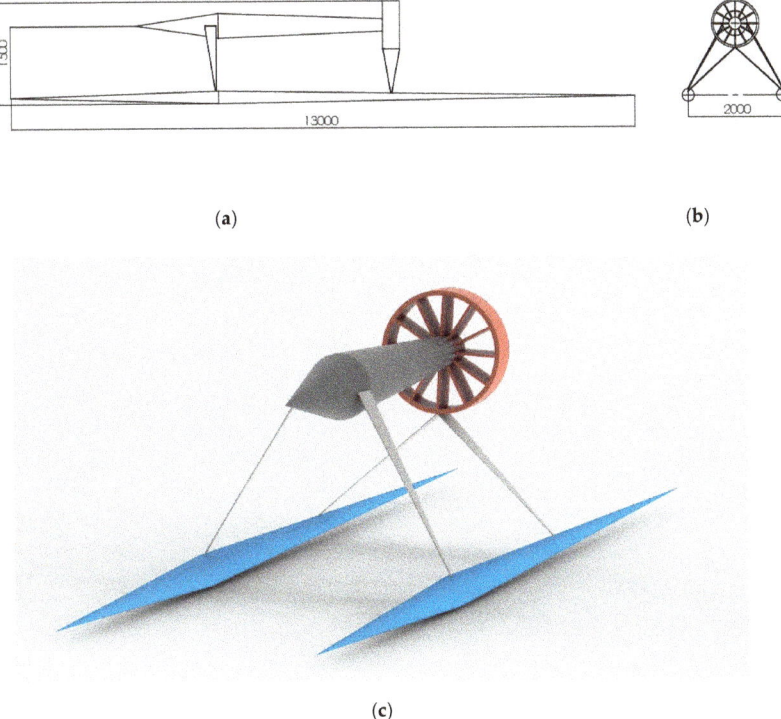

Figure 5. Sketch of a 1:4 model of the high-speed laminar SWATH ferry. (**a**) Side view; (**b**) front view; (**c**) 3D view.

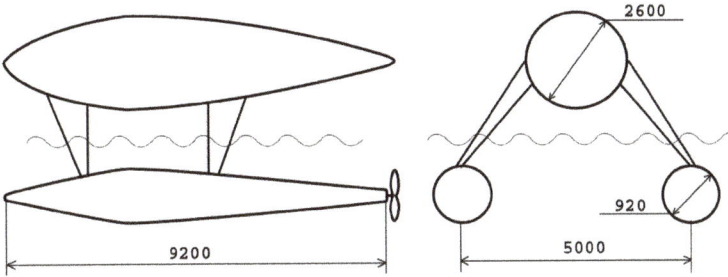

Figure 6. Sketch of SWATH low drag yacht.

The proposed technology could have a huge area of application, since it is economically efficient, green and comfortable (due to the high seakeeping). The Froude number is approximately 1.05, and is much higher than the critical value for the conventional ships. This means that conventional yachts of the same speed must be at least 4 times longer in order to avoid a huge increase in wave drag.

6. Conclusions

The specially shaped hulls without boundary layer separation can be used both in air and water in order to reduce the drag and pollution. In particular, SWATH electrical yachts with specially shaped underwater hulls could have a huge area of application, since they are economically efficient, green and comfortable (due to the high seakeeping). The Froude number of such yachts can be much higher than the critical value of conventional ships. The realized yacht (or its self-propulsion model) could also be a prototype for faster and larger SWATH ferries.

Author Contributions: Conceptualization, Z.K. and S.K.; methodology, I.N.; software, I.N.; validation, Z.K. and S.K.; investigation, I.N.; resources, Z.K. and S.K.; data curation, Z.K. and S.K.; writing—original draft preparation, I.N.; writing—Z.K.; visualization, I.N.; supervision, S.K.; project administration, Z.K. All authors have read and agreed to the published version of the manuscript.

Funding: This research received no external funding.

Conflicts of Interest: The authors declare no conflict of interest.

References

1. Michael Le Page. France and Others Plan to Tackle Air Pollution in Mediterranean Sea. Available online: https://www.newscientist.com/article/2191427-france-and-others-plan-to-tackle-air-pollution-in-mediterranean-sea/ (accessed on 20 April 2020).
2. Croatia Country Briefing—The European Environment—State and Outlook. 2015. Available online: https://www.eea.europa.eu/soer-2015/countries/croatia (accessed on 20 April 2020).
3. Fred Lambert. All-Electric Ferry Cuts Emission by 95% AND Costs By 80%, Brings IN 53 Additional Orders. Available online: https://electrek.co/2018/02/03/all-electric-ferry-cuts-emission-cost/ (accessed on 3 February 2018).
4. Fred Lambert. A New All-Electric Cargo Ship with a Massive 2.4 MWh Battery Pack Launches in China. Available online: https://electrek.co/2017/12/04/all-electric-cargo-ship-battery-china/ (accessed on 4 December 2017).
5. Gabrielly, Y.; von Karman, T. What price speed. In *Mechanical Engineering*; ASME: New York, NY, USA, 1950; Volume 72, pp. 775–781.
6. Lamb, G.R. Relationship between seakeeping requirements and swath ship geometry. In *OCEANS 88: A Partnership of Marine Interests. Proceedings*; IEEE: Miami, FL, USA, 1988; pp. 1131–1143.
7. Dubrovsky, V.; Matveev, K.; Sutulo, S. *Ships with Small Water-Plane Area*; Backbone Publishing Co.: Hoboken, NJ, USA, 2007; 256p.
8. Pérez-Arribas, F.; Calderon-Sanchez, J. A Parametric Methodology for The Preliminary Design of Swath Hulls. *Ocean Eng.* **2020**, *197*, 106823. [CrossRef]
9. Available online: https://www.atlanticyachtandship.ru/catalog/file/227371/ghost-juliet_marine-english.pdf (accessed on 20 April 2020).
10. Gray, J. Studies in animal locomotion VI. The propulsive powers of the dolphin. *J. Exp. Biol.* **1936**, *13*, 192–199.
11. Greiner, L. (Ed.) *Underwater Missile Propulsion*; Compass Publications: Arlington, VA, USA, 1967; Chapter 4.
12. Fish, F.E.; Rohr, J. *Review of Dolphin Hydrodynamics and Swimming Performance*; Space and Naval Warfare Systems Command: San Diego, CA, USA, 1999.
13. Fish, F.E. The myth and reality of Gray's paradox: Implication of dolphin drag reduction for technology. *Bioinspiration Biomim.* **2006**, *1*, R17–R25. [CrossRef] [PubMed]
14. Fish, F.E.; Legac, P.; Williams, T.M.; Wei, T. Measurement of hydrodynamic force generation by swimming dolphins using bubble dpiv. *J. Exp. Biol.* **2014**, *217*, 252–260. [CrossRef] [PubMed]
15. Bale, R.; Hao, M.; Bhalla, A.P.S.; Patel, N.; Patankar, N.A. Gray's paradox: A fluid mechanical perspective. *Nat. Sci. Rep.* **2014**, *4*, 1–5. [CrossRef] [PubMed]
16. Aleyev, Y.G. *Nekton*; Junk, W., Ed.; Dr W. Junk Publishers: The Hague, The Netherlands, 1977; Chapter 5.
17. Rohr, J.; Latz, M.I.; Fallon, S.; Nauen, J.C. Experimental approaches towards interpreting dolphin stimulated bioluminescence. *J. Exp. Biol.* **1998**, *201*, 1447–1460. [PubMed]
18. Loitsyanskiy, L.G. *Mechanics of Liquids and Gases*, 6th ed.; Begell House: New York, NY, USA; Wallingford, UK, 1995; Chapter 10.
19. Landau, L.D.; Lifshits, E.M. Volume 6: Course of Theoretical Physics. In *Fluid Mechanics*, 2nd ed.; Butterworth-Heinemann: Oxford, UK, 1987; Chapter 4.
20. Zedan, M.F.; Seif, A.A.; Al-Moufadi, S. Drag Reduction of Fuselages Through Shaping by the Inverse Method. *J. Aircr.* **1994**, *31*, 279–287. [CrossRef]

21. Lutz, T.; Wagner, S. Drag Reduction and Shape Optimization of Airship Bodies. *J. Aircr.* **1998**, *35*, 345–351. [CrossRef]
22. Nesteruk, I. Maximal speed of underwater locomotion. *Innov. Biosyst. Bioeng.* **2019**, *3*, 152–167. [CrossRef]
23. Goldschmied, F.R. Integrated hull design, boundary layer control and propulsion of submerged bodies: Wind tunnel verification. In *AIAA (82-1204), Proceedings of the AIAA/SAE/ASME 18th Joint Propulsion Conference*; AIAA: Cleveland, OH, USA, 1982; pp. 3–18.
24. Nesteruk, I. Rigid Bodies without Boundary-Layer Separation. *Int. J. Fluid Mechan. Res.* **2014**, *41*, 260–281. [CrossRef]
25. Nesteruk, I. Efficiency of Steady Motion and its Improvement with the Use of Unseparated and Supercavitating Flow Patterns. *Naukovi Visti NTUU Kyiv Polytech. Inst.* **2016**, *6*, 51–67. [CrossRef]
26. Nesteruk, I. Peculiarities of Turbulization and Separation of Boundary-Layer on Slender Axisymmetric Subsonic Bodies. *Naukovi Visti NTUU Kyiv Polytech. Inst.* **2002**, *3*, 70–76. (In Ukrainian)
27. Nesteruk, I.; Passoni, G.; Redaelli, A. Shape of Aquatic Animals and Their Swimming Efficiency. *J. Mar. Biol.* **2014**. [CrossRef]
28. Nesteruk, I.; Brühl, M.; Möller, T. Testing a special shaped body of revolution similar to dolphins trunk. *KPI Sci. News* **2018**, 44–53. [CrossRef]
29. Nesteruk, I. Reserves of the hydrodynamical drag reduction for axisymmetric bodies. *Bull. Kiev Univ. Ser. Physic. Math.* **2002**, *4*, 112–118.
30. Hoerner, S.F. *Fluid-Dynamic Drag*; S.F.Hoerner: Midland Park, NJ, USA, 1965; Section 6; 416p.
31. Hansen, R.J.; Hoyt, J.G. Laminar-To-Turbulent Transition on a Body of Revolution with an Extended Favorable Pressure Gradient Forebody. *J. Fluids Eng.* **1984**, *106*, 202–210. [CrossRef]
32. Saibene, F.; Minetti, A.E. Biomechanical and physiological aspects of legged locomotion in humans. *Eur. J. Appl. Physiol.* **2003**, *88*, 297–316. Available online: https://en.wikipedia.org/wiki/Solar_Impulse (accessed on 20 April 2020). [CrossRef] [PubMed]

 © 2020 by the authors. Licensee MDPI, Basel, Switzerland. This article is an open access article distributed under the terms and conditions of the Creative Commons Attribution (CC BY) license (http://creativecommons.org/licenses/by/4.0/).

Article

Application of Radial Basis Functions for Partially-Parametric Modeling and Principal Component Analysis for Faster Hydrodynamic Optimization of a Catamaran

Stefan Harries [*,†] **and Sebastian Uharek** [*,†]

FRIENDSHIP SYSTEMS AG, 14482 Potsdam, Germany
* Correspondence: harries@friendship-systems.com (S.H.); uharek@friendship-systems.com (S.U.)
† These authors contributed equally to this work.

Citation: Harries, S.; Uharek, S. Application of Radial Basis Functions for Partially-Parametric Modeling and Principal Component Analysis for Faster Hydrodynamic Optimization of a Catamaran. *J. Mar. Sci. Eng.* **2021**, *9*, 1069. https://doi.org/10.3390/jmse9101069

Academic Editor: Gregory Grigoropoulos

Received: 30 August 2021
Accepted: 9 September 2021
Published: 29 September 2021

Publisher's Note: MDPI stays neutral with regard to jurisdictional claims in published maps and institutional affiliations.

Copyright: © 2021 by the authors. Licensee MDPI, Basel, Switzerland. This article is an open access article distributed under the terms and conditions of the Creative Commons Attribution (CC BY) license (https://creativecommons.org/licenses/by/4.0/).

Abstract: The paper shows the application of a flexible approach of partially-parametric modelling on the basis of radial basis functions (RBF) for the modification of an existing hull form (baseline). Different to other partially-parametric modelling approaches, RBF functions allow defining sources which lie on the baseline and targets which define the intended new shape. Sources and targets can be corresponding sets of points, curves and surfaces. They are used to derive a transformation field that subsequently modifies those parts of the geometry which shall be subjected to variation, making the approach intuitive and quick to set up. Since the RBF approach may potentially introduce quite a few degrees-of-freedom (DoF) a principal component analysis (PCA) is utilized to reduce the dimensionality of the design space. PCA allows the deliberate sacrifice of variability in order to define variations of interest with fewer variables, then being called principal parameters (prinPar). The aim of combining RBFs and PCA is to make simulation-driven design (SDD) easier and faster to use. Ideally, the turn-around time within which to achieve noticeable improvements should be 24 h, including the time needed to set up both the CAD model and the CFD simulation as well as to run a first optimisation campaign. An electric catamaran was chosen to illustrate the combined approach for a meaningful application case. Both a potential and a viscous solver were utilized, namely, SHIPFLOW XPAN (SHF) and Neptuno (NEP), respectively. Rather than to compare the two codes in any detail the purpose of this was to study the efficacy of the proposed approach of combining RBF and PCA for solvers of different fidelity. All investigations were realized within CAESES, a versatile process integration and design optimisation environment (CAESES). It is shown that meaningful reductions of total resistance and, hence, improvements of energy efficiency can be realized within very few simulation runs. If a one-stop steepest descent is applied as a deterministic search strategy, for instance, some 10 to 12 CFD runs are needed to already identify better hulls, rendering turn-around times of a day of work and a night of number crunching a realistic option.

Keywords: computer aided design (CAD); partially-parametric modeling; radial basis functions (RBF); principal component analysis (PCA); simulation-driven design (SDD); hull form optimisation; computational fluid dynamics (CFD); potential flow code; viscous flow code; computer aided engineering (CAE); electric catamaran

1. Introduction

In the maritime industry, simulation-driven design (SDD) has become a widely accepted approach of developing functional surfaces such as ship hulls and appendages as well as turbochargers and engine component [1]. In simulation-driven design variants of a functional surface are analysed by means of simulations, often resource-intensive simulations of computational fluid dynamics (CFD) and finite element analysis (FEA). The simulation results are also used to come up with new variants which are then analysed. Instead of running the process manually, formal explorations and exploitations such as Designs-of-Experiment (DoE) and local or global search strategies, respectively, are put to use. As a consequence, the simulations—more precisely, the results from the simulations for each of the variants investigated—drive the process, the aim being to improve selected objectives while complying to a set of constraints, see [1].

In SDD, the role of the design team is that of formulating the design task, setting up an efficient variation approach via a suitable computer aided design (CAD) model, selecting and configuring a meaningful simulation approach as well as monitoring and adjusting the process. Ideally, the busy work of changing geometry, pre-processing the simulations, transferring, extracting, post-processing and aggregating data, as well as of managing variants and results, is done by a process integration and design optimisation environment (PIDO). This frees the design team from cumbersome, repetitive and error-prone work and, naturally, allows and encourages the generation and analysis of very many variants. Simulation-driven design generally lives off the abundance and often the number of variants considered is one to two and sometimes even three orders of magnitude higher than in the more laborious traditional approach of manually generating a new variant and afterwards analysing its performance in a separate step.

Not surprisingly, many variants can be afforded the chance of identifying an exceptionally good design increase. Still, there are several good reasons why SDD approaches are being investigated and proposed that need as little time and as few simulations as possible:

- More and more high-fidelity simulations are used which require considerable computational resources, sometimes several hours or even days per variant.
- A number of operational points are considered concurrently which calls for more simulations per variant.
- A wider range of objectives from different disciplines are taken into account in parallel, again intensifying the computational burden.
- Time pressure to improve a product is continuously rising, rendering faster optimisation campaigns more attractive.
- Less expensive products developed with smaller budgets could and should also benefit from SDD.

Hence, one of the challenges is to reduce the number of simulation runs meaningfully, possibly by sacrificing some of the improvement potential for the sake of conducting a faster campaign. The actual number of variants to be analysed quickly scales up with the system's degrees-of-freedom (DoF), i.e., the number of free variables defining the functional surface or its modifications. Two recent approaches of counteracting this and of reducing turn-around times in SDD were discussed in [2], namely, parametric-adjoint simulation and dimensionality reduction. Both approaches are meant to improve the fluid-dynamic performance of a functional surface with not too many simulation runs, i.e., within less than 100 variations.

Mathematically speaking, parametric-adjoint simulation is a very elegant approach but requires the solution of the adjoint equations which, so far, only a few CFD solvers provide. The gradient of the objective with respect to the free variables is numerically approximated by concatenating the CAD model's design velocities and the CFD code's adjoint sensitivities, the latter resulting from the solution of the so-called adjoint equations. The complementing effort of determining the adjoint sensitivities is similar to solving the primal equation system, for instance the Reynolds-averaged Navier–Stokes equations (RANS), and independent of the number of free variables. Intrisically, the approach is confined to a local search unless systematically repeated in various regions of the design space.

Dimensionality reduction is also mathematically elegant. It builds on a principal component analysis (PCA) of the design space spanned by the CAD variables, i.e., the geometric parameters that define the functional surface itself or its modifications. The original set of CAD variables is replaced by a (considerably) smaller set of principal parameters (or modes) which then capture the variability of the possible shape variations up to a user specified level. It is independent of the chosen simulation code(s) but room for improvement is literally speaking a bit smaller due to deliberately forgoing variability.

Another option of reducing the number of free variables and, hence, the number of necessary simulations, is to conduct an initial DoE from which the most important CAD variables are identified and kept for subsequent optimisation runs while the less important CAD variables are held constant. Quite a few variants need to be investigated before being able to select the subset of variables with which to continue. The potential of tangibly reducing simulations is therefore somewhat limited. Importantly, this approach should not to be confused with the dimensionality reduction based on PCA. The principal parameters of the PCA form a new and orthonormal coordinate system, avoiding many inherent geometric dependencies between CAD variables.

Another challenge that design teams encounter when commencing with an SDD campaign is the time and expertise needed to produce a suitable model of variable geometry. In general, two approaches can be distinguished as elaborated in [3]: partially-parametric and fully-parametric modelling. In short, partially-parametric modelling builds on an existing geometry, from wherever it may originate, and only defines the modifications parametrically while fully-parametric modelling brings about a geometry from scratch by means of a self-sufficient hierachical CAD model. A fully-parametric model usually is dedicated to a specific application, say a specific ship type, and supports different levels of modifications, from changing main dimensions down to fine-tuning specific regions, but it needs time and expertise to be developed. A partially-parametric model is often quicker to set up but rarely supports a wide range of modifications or the same level of sophistication as a fully-parametric model does.

A promising combination for an SDD campaign which brings about good optimisation results within reasonable effort, i.e., a campaign of high efficacy, appears to be a combination of partially-parametric modelling and dimensionality reduction. Ideally, the design team would see some improvements of their design within 24 h. In order to shed light on this, a comprehensive investigation of improving the energy efficiency of a ship hull by means of modifying its geometry via radial basis functions (RBF)—an intuitive and flexible approach of partially-parametric—and by applying a range of optimisation strategies, including dimensionality reduction via PCA, was undertaken. A fast catamaran was chosen as an illustrating application case. A thorough comparison is given between number of simulation runs and improvements achieved.

The paper first describes the design task and its deliberately chosen simplifications. It then explains the RBF approach along with the partially-parametric model realized with it. This is followed by a discussion of dimensionality reduction based on PCA. Both the RBF approach and the PCA were implemented in CAESES which was also utilized as the PIDO environment. Two different CFD codes were used for the simulations, a potential flow code (SHIPFLOW XPAN from FLOWTECH) and a RANS code (Neptuno from the Technical University Berlin), both of which were coupled to CAESES. The potential flow code is a non-linear free surface code with free sinkage and trim. It only needs a few minutes of CPU time per variant and was employed to cover the bulk of the investigations, i.e., the systematic testing and comparison of different optimisation approaches. The viscous code is a high-fidelity code that solves the RANS equations, taking into account the free surface and the cat's sinkage and trim for given weight and centre of gravity. It needs several hours per variant and was therefore employed to test and showcase the initial findings for substantially more expensive simulations. An overview of the two CFD codes and for the selected optimisation strategies is given as needed to appreciate the campaigns. This is followed by an elaboration of the results. The paper ends with conclusions and recommendations for future work.

2. Design Task And Simplifications

2.1. Design Task

A small catamaran ferry of 20 m length between perpendiculars at a design speed of 13 kn and a displacement mass of 25 t at 0.85 m draft shall be optimised for energy efficiency, see Table 1. The cat is meant to be solely battery powered and should serve up to 50 passengers. The energy density of available batteries being substantially lower in comparison to the energy density of fossil fuels, the battery mass becomes critical for a given operational range. Consequently, highest energy efficiency turns out to be even more important for an electric ferry as it would be for a conventionally powered vessel.

Table 1. Main particulars.

length between perpendiculars	L_{PP}	20 m
beam demi-hull	B_{dH}	1.58 m
beam over all	B_{OA}	5.38 m
clearance between demi-hull centre planes	c_{dH}	3.8 m
design draft	T_0	0.85 m
displacement	$\forall$	24.39 m^3
design speed	v_0	13 kn
Froude number	Fn	0.477

An existing hull shape, from hereon referred to as baseline, see Figure 1, was provided which featured a slender round-bilge demi-hull with a slightly submerged transom stern. The term baseline is used here to refer to the initial design, i.e., the parent hull from wich variations are derived. Figure 2 gives an impression for various views, omitting the deck and superstructure. The blue demi-hull illustrates the cat's port side up to the deck while the green demi-hull shows the underwater portion of the cat's starboard side cut at design draft. The baseline's demi-hulls were symmetric with regard to both their centre planes and, naturally, the cat's mid plane. They neither featured any knuckle lines nor spray rails. No topological changes ought to be introduced during the optimisation campaign, but an asymmetry of the demi-hulls would be acceptable.

Each side of the demi-hull was modelled with just one single B-spline surface that featured a polyhedron of ten vertices in longitudinal direction (u-direction) and six vertices in vertical direction (v-direction). It should be noted that the B-spline surfaces were set up within CAESES which can also be used as system for free-form surface modelling. However, any other CAD system could have been used to establish the baseline. For the B-spline surfaces the degrees in u- and v-direction were five and three, respectively. Both the transom and the deck were modelled as ruled surfaces, following the B-spline surface's aft edge and upper edge, respectively.

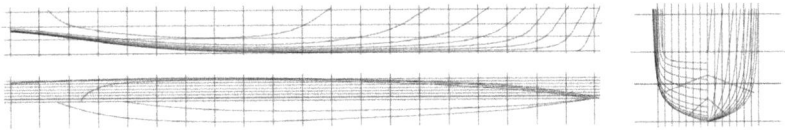

Figure 1. Linesplan baseline.

The Froude number at 20 m length and 13 kn is 0.4774 which, from a general point of view, falls into a rather unfavourable speed range. A resistance curve was computed for the baseline with both the potential flow code and with the RANS code checked for one speed, see Figure 3. It showed, not surprisingly, that the cat's total resistance steadily increases with speed. From the wave resistance coefficient it can be seen that the cat would indeed sail close to a resistance hump, the maximum wave resistance coefficient occurring at 13.5 kn. The bare hull's total resistance R_{T0} (both demi-hulls taken into account) as computed from SHIPFLOW amounted to 10.3 kN. All potential flow results were normalized with this value since any absolute values ought to be looked at with some caution. First of all, appendages and openings were not taken into account. Secondly, even though both potential flow and boundary layer theory was utilized, the form factor was simply set to zero. Thirdly, by definition the interactions fo the free surface and viscous flow are not accounted for the potential flow analyses.

Figure 2. Baseline.

A pure variation of the distance between the demi-hulls revealed that the cat's resistance would steadily fall with increasing clearance up to its upper bound, corresponding to a beam over all of 5.376 m. It was therefore decided to set the clearance to its maximum value and exclude it from the optimisation campaign afterwards. Apart from thus reducing the number of free CAD variables—which is always beneficial—the maximum clearance would also yield the highest stability.

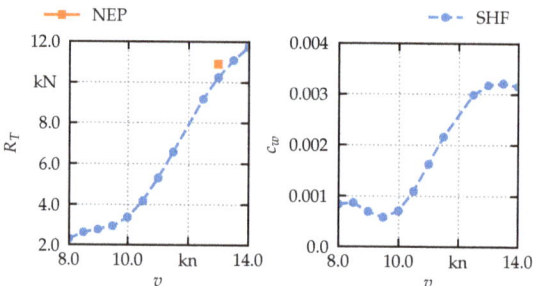

Figure 3. Resistance curve computed with SHIPFLOW XPAN (SHF) for the baseline and cross-checked with RANS solver Neptuno (NEP).

2.2. Simplifications

The distance from the keel K to the metacentre M, KM, was 7.58 m for the baseline and varied between 7.1 m and 7.8 m for the shapes investigated, keeping the clearance constant. This points towards a sufficiently stable vessel so that no constraints had to be monitored with regard to stability.

For a propulsion system a standard arrangement with shaft, bracket and propeller plus a spade rudder for each of the demi-hulls was chosen. As this was not subject to any changes during the optimisation campaign, it was assumed—for the sake of simplicity—that no unfavourable effects would be encountered for any of the anticipated variants. Consequently, the propulsive efficiency was assumed to stay constant so that the cat with least total resistance would also yield the design of highest energy efficiency. It needs to be pointed out that this, naturally, neglects an important component of calm-water hydrodynamics which can be justified solely on the grounds of focusing on methodology and not on proposing a final design.

Additional hydrodynamic factors would also actually have to be considered, such as seakeeping and manoeuvring performance as well as possible resistance increase due to canal and shallow water effects, depending on the cat's final operational profile. It could be argued that the anticipated shape modifications would primarily influence calm-water performance and should only lead to minor effects in seakeeping etc. as main dimensions, including length, draft, beam and displacement, were all kept constant.

An upfront design-of-experiment (DoE) for two speeds (applying a Sobol sequence [4]), using the potential flow code, showed that an improvement of the hull at its design speed of 13 kn would not necessarily give an improvement also at a lower speed of interest, here 9 kn. The normalized resistance values for the different variants at both speeds are shown in Figure 4. The general tendency is that variants which perform well at 13 kn may come with slightly higher resistance values at 9 kn. For certain designs, however, this is not necessarily the case, indicating that there would be favourable compromises.

In addition, a similar DoE study (also applying a Sobol sequence) was run at 13 kn for the design draft of 0.85 m and at a higher draft of 0.95 m, see Figure 5. It shows that performance is correlated but that variants do not always yield improvements at both drafts. Both DoEs clearly indicate that a thorough optimisation for calm-water hydrodynamics would have to be multi-objective, including (at least) two speeds, possibly two drafts and the influence of the propulsion system. If a design team has to consider all these aspects, further complemented by non-hydrodynamic aspects, it becomes all the more clear that the number of variants to be investigated needs to be as small as possible when running expensive simulations.

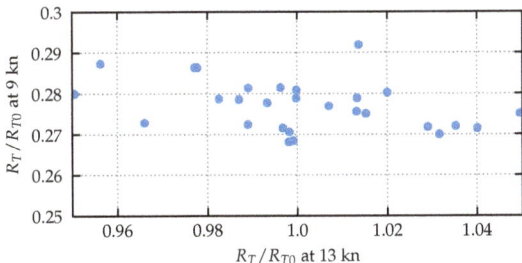

Figure 4. Non-dimensional resistance at design speed vs. non-dimensional resistance at lower speed.

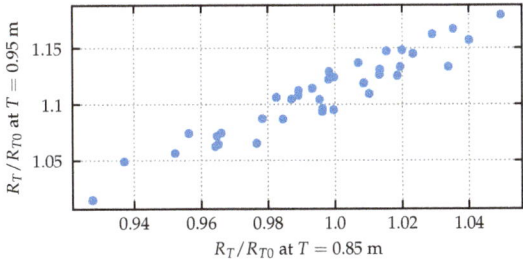

Figure 5. Non-dimensional resistance at design draft vs. non-dimensional resistance at higher draft.

3. Parametric Model

3.1. Partially-Parametric Model

With the aim of realizing first and meaningful improvements of a given hull form within a short turn-around time, ideally 24 h, a partially-parametric modelling approach is proposed. While fully-parametric modelling is more powerful and more precise, a partially-parametric model is faster and, generally speaking, a bit more intuitive to build, requiring less mathematical know-how, see [3].

A flexible partially-parametric modelling approach is based on radial basis functions (RBFs). The basic idea of the approach is to obtain a smooth transformation between a source and target geometry by using the RBFs to translate the control polygon of the surface, see Figure 6. While the present application features a rather simple hull shape, RBFs as implemented in CAESES are capable of handling complex geometries as well, e.g., ships with bulbous bows but also engine and turbo machinery components [5]. The implementation in CAESES is based on the work by [6] and explained in some detail in [7]. In general, radial basis functions (RBFs) are used in the context of scattered data interpolation. In particular, when utilizing RBFs for partially-parametric modeling a vector-valued space deformation

$$\vec{d}(\vec{x}) = \sum_j \vec{w}_j \cdot \phi_j(\vec{x}) + \vec{p}(\vec{x}) \tag{1}$$

is to be computed. It makes use of influence functions, here the triharmonic RBF

$$\phi_j(\vec{x}) = \left| (\vec{x} - \vec{c}_j)^3 \right| \tag{2}$$

featuring control points vector $\vec{c}(\vec{x})$ along with a triharmonic quadratic polynomial vector $\vec{p}(\vec{x})$. A set of weights vector $\vec{w}(\vec{x})$ has to be found to smoothly interpolate a number of user defined and, hence, known displacements. To this end, fixed points, so-called clamped supports and sources, as well as displaced points, so-called targets, are used, see Figure 7. The targets are made subject to variation, establishing a flexible parametric model. For each fixed point, lying either on a clamped support or on a source, a new position in space is known from the associated point on the topologically identical yet geometrically different target, determining the vector-valued space deformation.

Figure 6. Polyhedron of the Brep and RBF source curves.

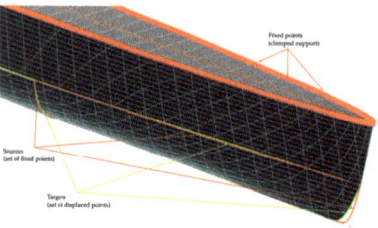

Figure 7. Fixed point (clamped support and sources) and displaced points (targets) for the forebody of the cat.

The advantage of this procedure is that it can be established within just a few hours of interactive work. Points, curves and, if of interest, surfaces are identified on the baseline which capture prominent features of the shape to be varied, keeping in mind a design team's background knowledge and intuition about which parts of a shape may positively influence the objective functions when modified. These entities, here points and curves, form the sources of the RBF model, see Figure 8.

Matching entities are then introduced, i.e., a point for a point and a curve for a curve etc., which form the targets. The RBFs are computed, here within CAESES, such that a vector-valued space deformation is established that gives the desired displacements for all points in space. For each point of interest it can thus be determined to which new position it shall be transferred, see Figure 9.

The space deformation is calculated from a set of control points, typically a few thousand, which comprise fixed points, i.e., boundary points not to be moved, and points for which the initial and the new positions are known from the sources and targets, respectively. For the cat a total of 14 CAD variables were used to control the target functions, 13 of which are related to RBF morphing and one from a subsequent Lackenby transformation (namely, deltaXCB), see Table 2. Their corresponding bounds are listed in Table 3. The Lackenby transformation, see [8], was utilized to maintain the displacement and actively control the longitudinal centre of buoyancy.

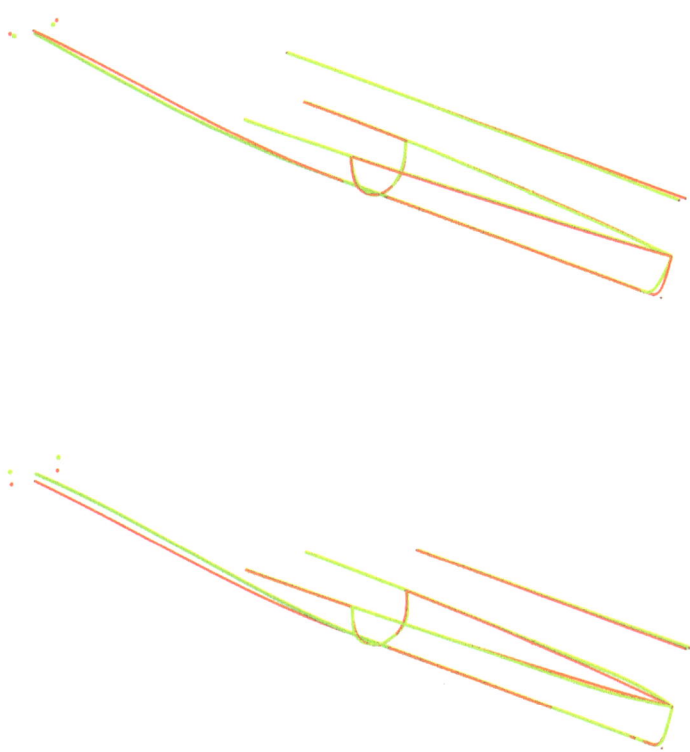

Figure 8. Matching source (red) and target (green) curves corresponding to minimum and maximum values of free variables in CAD space.

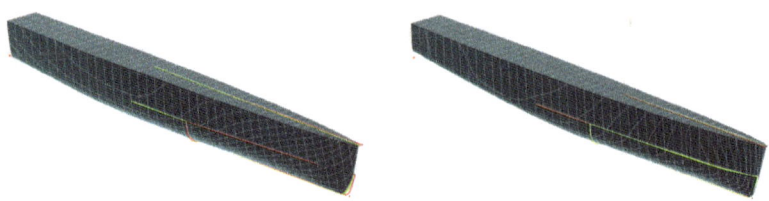

Figure 9. Resulting geometry of the cat's demi-hull for target curves with minimum and maximum values of free variables.

Table 2. Design variables in CAD space used for optimisation.

Name	Explanation
iE	half-angle of the design waterline at forward perpendicular (FP)
touchFor	change of the transverse position of the design waterline (DWL) midway between the FP and the maximum section
bowAsymmetry	change of the transverse position of the stem contour at the deck forward of the FP
stemContour	weight for one of the vertices of the NURBS curve defining the stem contour, influencing the stem's sharpness
keepCPCfor	position along the forward part of the center plane curve (CPC) at which an asymmetry of the demi-hull, if present, starts
modTranAngleCPC	modification of the angle of the CPC at the transom stern (within the demi-hull's center plane)
moveParallel	change of the aft position of the parallel part of CPC
startAsymmetry	longitudinal position at which an asymmetry of the demi-hull, if present, starts
deltaAreaMid	change of the sectional area at midship
deltaDeadrise	change of the deadrise angle of the section at midship
deltaY	transversal shift of the rounded part of the transom, causing changes in the transom's beam
deltaZ	vertical shift of the rounded part of the transom, causing changes of the deadrise
deltaTransom	change of the submergence of the transom stern (in the demi-hull's center plane)
deltaXCB	change of the longitudinal center of buoyancy

3.2. Dimensionality Reduction

A flexible parametric model may easily feature quite a few free variables, i.e., parameters that a design team feels could be advantageous to change. Since the effort of evaluating the objective function scales up quickly with the number of free variables, i.e., the systems' degree-of-freedom (DoF), it readily appears to be mandatory to keep the DoF as low as possible. Parametric modelling already is key to small numbers of free variables, often between 10 and 50 for hull shapes, propellers and appendages [1]. If a representative evaluation of the objective function takes just one hour per variant than a turn-around time of 24 h can only be achieved if very few variants need to be investigated, say 20 when also taking into account the set-up time for modelling and simulation. This is naturally based on the assumption that the people involved are familiar with the design task in general and with all tools in particular. N.B. One can always argue that more computational power can be brought in so as to compute the objective function more quickly. While this is true, it should be kept in mind that if computational resources are high more objective functions are usually taken into consideration, see Section 1, and that, more often than not, a human desire quickly grows to utilize higher-fidelity tools, for instance, to increase accuracy as will also be addressed below.

Table 3. Bounds of design variables used for optimisation (CAD space).

	Name	Lower Bound	Baseline Value	Upper Bound
1	iE	5	7.5	10
2	touchFor	−0.05	0	0.05
3	bowAsymmetry	−0.15	0	0.1
4	stemContour	−0.05	1.75	2.5
5	keepCPCfor	0.55	0.7	0.85
6	modTranAngleCPC	−3	0	2
7	moveParallel	−0.75	0	1
8	startAsymmetry	8	10	12
9	deltaAreaMid	−0.008	0	0.008
10	deltaDeadrise	−10	0	10
11	deltaY	−0.1	0	0.03
12	deltaZ	−0.02	0	0.1
13	deltaTransom	−0.05	0	0.15
14	deltaXCB	−0.0025	0	0.00125

A powerful and rather new approach of reducing the dimensionality of the design space is to undertake a principal component analysis (PCA) as first introduced in naval architecture by Diez et al. [9]. The idea is to identify how the parameters of a CAD model selected as free variables, here the partially-parametric RBF model, are actually coupled and to replace these CAD variables by a new and smaller set of free variables that are independent of each other albeit less easy to interpret. These new and independent free variables are called principal parameters (or super parameters). They result from the eigendecomposition of the covariance matrix for point data derived by a large sample of geometric variants, as described in [2,9–11].

To this end a DoE, here a Sobol sequence, of purely geometric variants is produced from the CAD variables, say 1000 variants for a set of 10 to 15 free variables. On each of the variants, an equal number of evenly distributed points are generated whose geometric positions in cartesian space necessarily vary but which are topologically identical. Here some 6000 points per cat's demi-hull were used. For a sample of 1000 variants this forms a collection of 60,000 points that serves as input to the PCA. The main benefit of employing a PCA lies in deliberately sacrificing some of the variability of the CAD model for the sake of subsequently working with a (considerably) smaller set of principal parameters [10]. Typically, only the first and most important principal parameters are put to use while the less contributing modes modes are cut off. For the cat's demi-hull the variability achieved by accumulating the first principal parameters is given in Table 4

Table 4. Influence of the first seven principal parameters (prinPar) and the PCA model's variability in comparison to the variability of the CAD model.

	Principal Parameter	Variability	Accumulation
1	prinPar1	37.93%	37.93%
2	prinPar2	27.66%	65.59%
3	prinPar3	14.20%	79.79%
4	prinPar4	6.06%	85.85%
5	prinPar5	4.80%	90.65%
6	prinPar6	3.34%	93.99%
7	prinPar7	2.25%	96.24%

It needs to be pointed out that the principal parameters from the PCA are less easy to be interpreted then any of the CAD variables. Mathematically speaking, they represent the Eigenvalues associated to the Eigenvectors that span the orthonormal PCA space. Practially, they bring together the free variables defined in CAD space in a new form.

For the investigation it was thus decided to run the optimisation campaigns with the first seven principal parameters which together yield a pretty high variability while cutting the number of free variables to half the number of CAD variables. Figure 10 illustrates the influence that the first six principal parameters have on the hull shape. From the colour distributions it can be nicely seen that indeed different regions are addressed by each of the principal parameters. For further details and elaborations of the PCA, see [2,11].

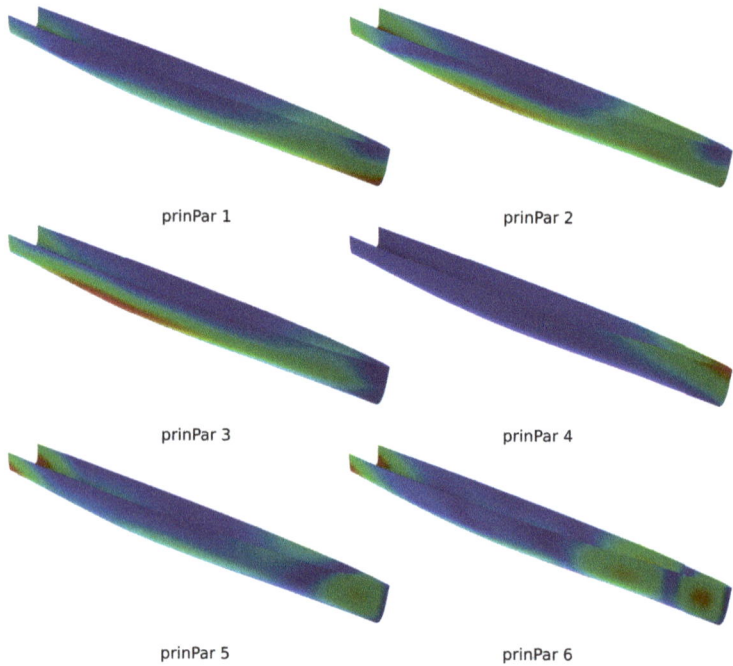

Figure 10. Influence of first six principal parameters (prinPar) on the hull shape.

Figure 11 shows the correlations between selected design variables in CAD space (bowAsymmetry, keepCPCfor, iE and deltaTransom) and the first four principal parameters obtained from the dimensionality reduction (prinPar1, prinPar2, prinPar3 and prinPar4). It is obvious that in some cases a strong coupling exist, such as, for example between the parameter bowAsymmetry and prinPar4. In other cases, the coupling between principal parameters and CAD variables is a bit less pronounced, e.g., keepCPCfor and prinPar1 or even less obvious, e.g., bowAsymmetry and prinPar2. In addition, one principal parameter usually influences several CAD parameters, such as, for example keepCPCfor and stemContour in relationship to prinPar1. This behaviour is expected and exactly this correlation between CAD variables is used by the dimensionality reduction to reduce the degrees-of-freedom of the system.

In order to check by how much the design space is reduced hydrodynamically when employing a dimensionality reduction in the process, a design space exploration using the Sobol algorithm is performed. The results are presented and discussed in Section 5.1.

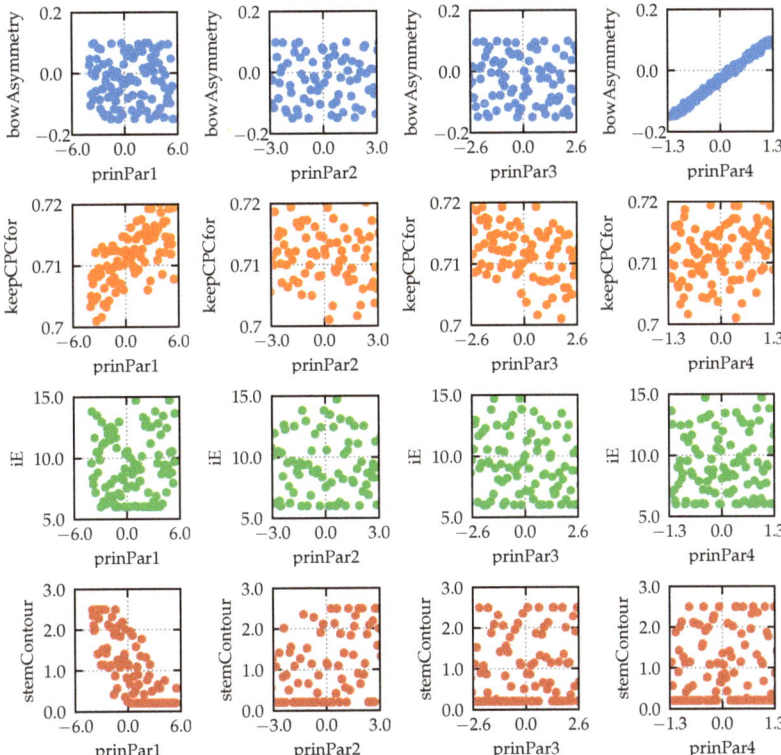

Figure 11. Correlations between several selected CAD variables and the first four principal parameters resulting from the PCA.

4. CFD Codes and Numerical Set-Ups

For the evaluation of the towing resistance of the cat, which serves as an objective function for the optimisation study, two numerical codes are used. The first code is SHIPFLOW XPAN/XBOUND, a non-linear potential flow code including sinkage and trim, complemented with a boundary layer simulation, developed by FLOWTECH. The advantage of using a potential flow code in numerical optimisation studies is the rather short computation time, which allows the evaluation of the objective function in just a few minutes per variant. Especially for studies where the geometry variation is limited to the bow area, potential flow results provide at least a good ranking of the designs.

For a few years, however, it has been rather common to use more sophisticated methods such as a RANS solver to compute the viscous flow and the free surface around the hull so as to predict the absolute values with higher accuracy for the price of a significantly higher computation time. In the present study, selected computations have been performed with the RANS code Neptuno, which has proven to yield very accurate results in diverse marine applications, see for example [12].

4.1. Potential Flow Computation

In the SHIPFLOW package (SHF), the non-linear potential flow code XPAN is shipped together with the panel generator XMESH and a module for thin turbulent boundary layer analysis, XBOUND, which solves the momentum equation along the streamlines traced from the potential flow solution. All these programs are tightly integrated into CAESES and their set-up is rather straightforward. The geometry is exported as offsets at each station. The free surface mesh extends $1L_{PP}$ upstream of the bow, $2L_{PP}$ downstream of the stern and $1.5L_{PP}$ to the side.

For the present application case, some fine-tuning was required in order to obtain converging solutions for all variants. For example, the panel mesh size between the two demi-hulls was adjusted, since some wave breaking between the demi-hulls would occur, leading to divergence unless numerically suppressed. Figure 12 shows the final panel mesh on the free surface and on the hull. During the course of the computation, sinkage and trim is taken into account and at the end, the results are imported back into CAESES for visualization.

It should be noted that the cat's hydrodynamics approaches the limits of a potential flow analysis, with possible wave breaking and modifications of the transom. However, XPAN provides a very fast way to evaluate the objective function and since the main goal of the present work is to study the capabilities of the RBF and PCA approach, these restrictions were deliberately accepted.

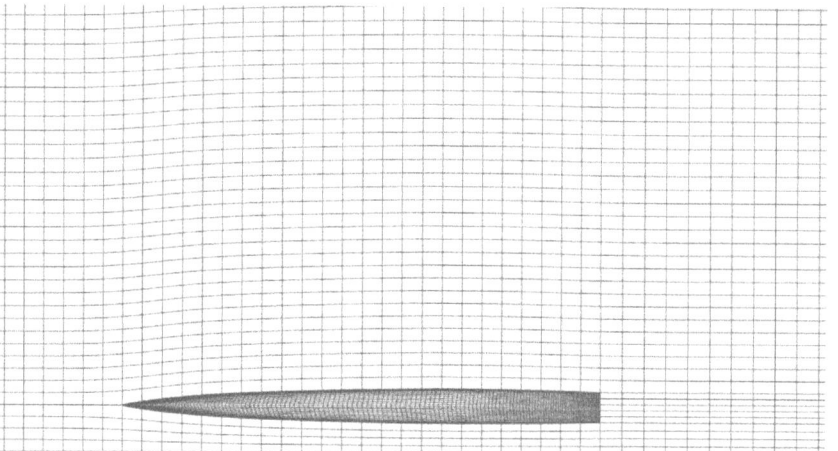

Figure 12. Close-up of the panel mesh on the hull and free surface in the vicinity of the hull.

4.2. Viscous Flow Computation

The RANS code used as a high-fidelity solver is Neptuno (NEP). Neptuno is an in-house CFD code developed by Prof. Cura [13,14]. It solves the RANS equation on a multi-block structured grid using the finite volume method. The pressure–velocity coupling is done using the SIMPLE algorithm from Patankar [15]. The turbulence is modelled using the standard two equation k-ω model from Wilcox [16]. The free surface is captured using a two phase level set method [17,18]. The code has been validated in several workshops, see [19–21].

The numerical grid required for the computations is built using the commercial meshing software GridPro. The main advantage of GridPro is the ability to separate topology and geometry. This allows setting up the initial topology (block structure) of the grid prior to the optimisation study and simply re-run the meshing process for each variant. The resulting grids are all of high quality and due to the identical block structure, the dependency of the objective function on the numerical grid is reduced. By using a symmetry boundary condition at the centre plane between the two demi-hulls, only one side of the domain needed to be meshed. The boundaries are located $1L_{PP}$ upstream, $2L_{PP}$ downstream, $1L_{PP}$ beside and $1L_{PP}$ below the ship. The selected boundary conditions were slip-wall (symmetry) at the centre plane and starboard boundary while at the inlet, top and bottom side of the domain the velocity is prescribed. At the outlet the hydrostatic pressure is given.

Since the code has been widely and successfully validated at model scale, it was decided to carry out all computations at a fictive model scale of $\lambda = 4$, which would correspond to a model length of 5 m and a Reynolds number of $Re = 1.67 \cdot 10^7$. For the optimisation and comparison with XPAN, the values are extrapolated to full scale using the standard ITTC procedure, see [22].

Only for some geometric variants, e.g., the ones with an asymmetric bow, a slight modification of the topology is required. Figure 13 shows a slice of the resulting mesh in the bow region for two rather different geometries cut at the same longitudinal postition. As can be seen, GridPro allows to change the grid in the vicinity of the hull, leaving the far-field grid untouched. In addition, the topology of the grid is practically identical, so it can be assumed, that changes in the grid have a rather small influence when evaluating the resistance.

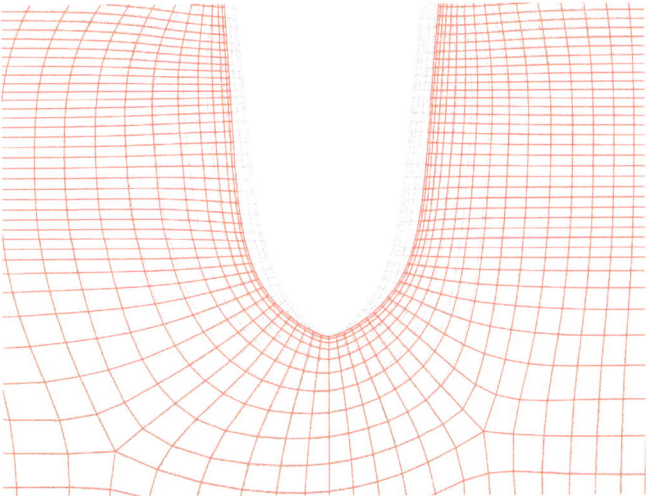

Figure 13. Mesh in the bow region for the baseline (grey) and a modified hull shape (red).

For the cat simulation with free surface, a constant number of 8000 time steps are computed. The non-dimensional time step is chosen at 0.008 and one SIMPLE iteration per timestep is performed. A numerical beach for the damping of waves in the far field is used, for details see [23]. Sinkage and trim is taken into account by a stepwise movement of the free surface relative to the ship. This is performed at the time steps 1500, 2500, 3500 and 4500. As an example, Figure 14 shows the time traces of the earth fixed longitudinal component of the hydrodynamic force (F_ξ) acting on the hull, the iterative history of sinkage (ζ_O) and trim angle (θ) along with the residuals. As can be seen, even after the force stabilized, the residuals keep dropping and thus the solution converges. Only very minor changes occur after 8000 steps, so in order to save time, only 8000 iterations are performed during the optimisation study.

In order to assess the dependence of the computed hydrodynamic forces on the grid resolution, a systematic variation has been performed with cell counts ranging from just 460,000 to 3.7 million cells. Figure 15 shows the force components resulting from the integration of pressure (F_P), and shear stresses (F_F) as well as the total force (F_T). In addition, Table 5 shows the results of the convergence study, with r_g being the convergence radius and u the uncertainty of the solution, as proposed by the ITTC [24]. As can be seen a monotonic convergence is achieved for each individual component and the quality of the prediction is quite satisfactory. It should be noted that the grid convergence study is carried out without considering sinkage and trim, since for each grid different trim angles would have resulted and thus it would have been impossible to run an uncertainty study for the resistance.

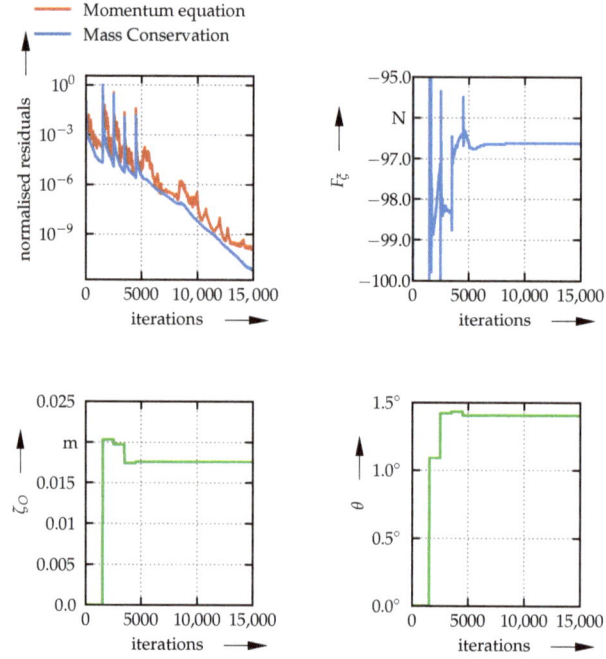

Figure 14. Convergence history, time history of earth fixed longitudinal force for one demi-hull, sinkage and trim for straight-ahead motion with stepwise adjustment of sinkage and trim.

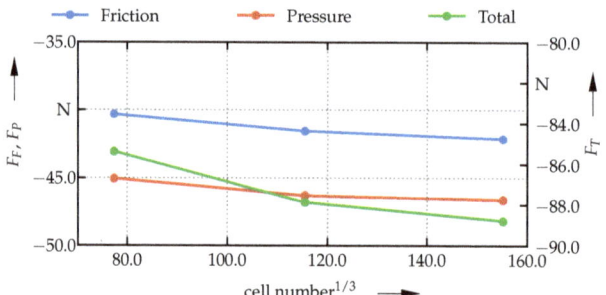

Figure 15. Grid dependency for RANS simulations.

Table 5. Results of grid convergence study for three grids with 3.7 million (S_1), 1.5 million (S_2) and 460,000 cells (S_3) for one demi-hull.

	S_3 Coarse	S_2 Medium	S_1 Fine	r_g	$u\%$
F_F	−40.331 N	−41.560 N	−42.147 N	0.477	1.57
F_P	−45.054 N	−46.312 N	−46.647 N	0.267	0.33
F_T	−85.385 N	−87.872 N	−88.794 N	0.371	0.76

Since the change in total resistance between the fine and coarse grid is only 3.85%, but the computation time differs by one order of magnitude, namely, 8 h (480 min) for

S_1 in comparison to 48 min for S_3 on a state-of-the-art workstation using eight cores, all computations for the optimisation study are carried out on the coarse grid. Figure 16 shows a comparison of the free surface elevations for the coarse and fine grid. One should keep in mind that parallel computations could be carried out on a cluster for the finer grid. However, that would accelerate the computations for all grids, naturally.

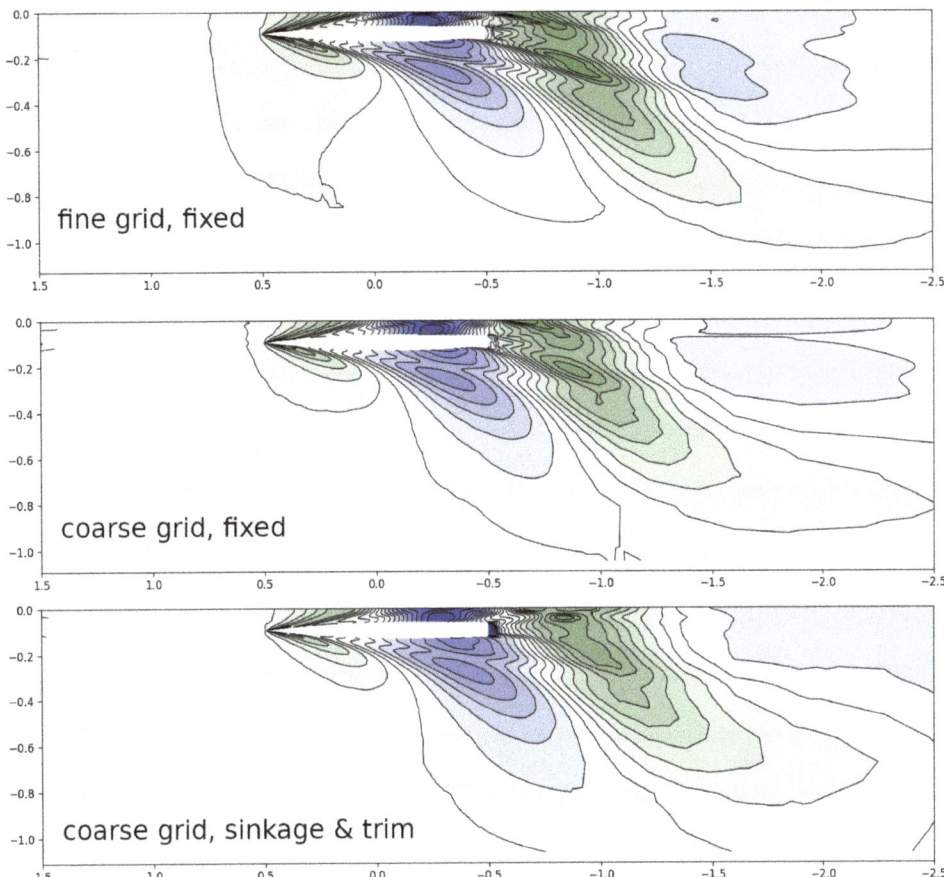

Figure 16. Contour plot of the free surface on the fine and coarse grid without sinkage and trim as well as on the coarse grid with sinkage and trim.

4.3. Exploration and Optimisation Algorithms

For the exploration of the design space, a Sobol algorithm is used which is based on a quasi-random yet deterministic sequence [4]. The values of the free variables are set such that the sampling points iteratively fill up the design space as evenly as possible, an additional sampling point always reducing the region of the design space that has been least populated so far. There are other exploration algorithms available in CAESES, for example the Latin-Hypercube sampling (LHS), which is made available through the DAKOTA optimisation package. However, an advantage of using a Sobol sequence is, that contrary to the LHS, the number of samples in the design space can be readily extended, whereas with the LHS a complete new design set would have to be generated. Furthermore, a Sobol sequence can be easily repeated with the same sampling points, provided that the bounds of the free variables have not been modified.

The exploitations, i.e., various optimisation runs, have been undertaken on the basis of three purely deterministic strategies, namely, a T-Search, a Nelder–Mead simplex and a one-stop steepest descent. This was done with the aim of fast turn-around times, i.e., improvements within 24 h. All three strategies perform local searches and are typically employed when quickly trying to understand the potential for improvement of a baseline without introducing major modifications and/or when fine-tuning a design.

The T-Search, i.e., tangent search by Hillary [25], first evaluates the baseline and then starts with a small positive perturbation of the first design variable. If this readily gives an improvement the second design variable undergoes a small perturbation. Otherwise a small negative perturbation is introduced to the first design variable. Setting out from the best design so far the second design variable is slightly changed, first in positive direction than, if no improvement was found, in negative direction. This is repeated until all free variables have been modified at least once, establishing a local search pattern. Using both the best design found during such a local search and its current starting point (e.g., the baseline at the onset) a global search direction can be determined and a global step is then undertaken, hoping to bring home a more substantial improvement. If that was successful further global steps can follow until a suitable starting point for a new local search pattern has been found. Without going into the strategy's details any further, it can be appreciated that the algorithm scales up linearly with the number of free variables. In the best case, all positive increments would readily give improvements during the local search. In the worst case, all free variables would have to be perturbed twice, once into positive and once into negative direction. Therefore, in practical situations an objective function needs to be computed once for the baseline and then anything between the number of free variables (best case) and twice that number (worst case) before a global step can be taken.

The Nelder–Mead simplex [26] also is a local search method which flips through the design space with a simplex, i.e., the simplest possible polytope in any given space. In one-dimensional space a simplex is a line segment while in two-dimensional space it is a triangle. In three-dimensional space it is a tetrahedron while, by abstraction, a simplex represents a cell with n+1 corners in an n-dimensional space. The search, in short, always takes the corner with the least favourable objective value and flips it through the centroid of the n remaining corners that offered better objective values. This step is called a reflection. Depending on the changes found for the objective an extension may take place, trying to benefit further. In general, these flips are repeated (with some additional tweaks such as contracting and shrinking the simplex), slowly advancing the simplex through the design space. As can be readily appreciated the first simplex requires n+1 evaluations of the objective, one evaluation for the baseline and n evaluations for a small perturbation of each free variable (or any independent set of n combinations). Each additional evaluation than brings about an improvement (or information how to contract or shrink the current simplex).

The one-stop steepest descent takes a similar approach as the Nelder–Mead simplex for the first n+1 evaluations. Basically, a gradient of the objective function is numerically determined by adding a small delta to each free variable while keeping all other free variables constant, establishing an approximation through forward differencing. As soon as the gradient is known, a line search is undertaken into the direction of anticipated improvement. In case the objective of the new variant is better than the baseline's a second step is taken into gradient direction, e.g., by doubling the step size. Otherwise, the first step apparently has been too far and a bisection (or any other subdivision) of the distance between the baseline and the first new variant takes place. This is repeated a few times, say three to five times, until the local search is stopped, establishing a "one-stop" strategy. Alternatively, a conjugate gradient can be utilized to advance further through the design space. In principle, the one-stop steepest descent allows setting the maximum number of evaluations of the objective function upfront, namely, one evaluation for the baseline, n evaluations for the gradient approximation plus three to five evaluations during the one-dimensional line search. This means that as soon as the time needed to simulate a single variant, say the baseline, and the time available before improvements have to be realized are known, it is possible to select the maximum number of free variables—either CAD variables or principal parameters—to take into consideration. For instance, if a single variant needs one hour to simulate for hydrodynamics, a one-stop steepest descent with seven principal parameters would take eight hours before the gradient is known (one hour for the baseline and seven hours for each perturbation). With another four variants evaluated during the line search that sums up to 12 h. That would then leave another 12 h to do both the pre- and post-processing. With some eight to ten hours of pre-processing, including parametric modelling (three to four hours), undertaking the PCA (five to six hours) and coming up with a good CFD set-up (several hours while the PCA is running), that would leave about four hours to complete the job within a single day of turn-around time.

5. Results

The following section presents the results obtained from exploration and optimisation runs performed with SHIPFLOW and Neptuno, respectively, and analyses the possible gains from the PCA.

5.1. Optimization Studies Using Potential Flow Code

In order to check the process improvement, which can be achieved by using a PCA compared to the traditional approach using design variables in CAD space, a high number of simulations were performed using XPAN.

Figure 17 shows the total resistance made non-dimensional by the resistance of the baseline from two Sobol runs. The first one was performed in CAD space with 200 samples (green) and the second one using the PCA with 100 samples (orange). As can be seen, the design space covered by both runs is fairly similar. There are three design variables which are not really correlating to any of the chosen principal parameters and which are not covered when running the optimisation with a PCA, namely, keepCPCfor, startAsymmetry and deltaXCB. Apparently, the same geometry variation is also achievable with a suitable combination of other design variables, making these CAD variables rather superfluous. The longitudinal shift of the centre of buoyancy, which was achieved using a Lackenby transformation, could for example also be achieved by moving the parallel midship section (moveParallel). Most importantly, however, it can be seen that the range of the objective function is very similar for both the CAD space and the PCA space, the maximum being approximately 1.05 and the minimum being around 0.9. This means, that the optimisation potential was not reduced by reducing the dimensionality of the problem by a factor of two and evaluating 100 instead of 200 variants.

In the next step, six optimisation runs are performed starting from the baseline design. Figure 18 shows the various histories and associated improvement in non-dimensional resistance over the iteration count. All three deterministic optimisation algorithms are applied for both the traditional CAD approach and the PCA. The first thing to observe is, that the T-Search finds the highest improvement, namely, a reduction in total resistance by 10.5% for the CAD based approach and 9.9% when using the PCA. However, the first significant improvement when using the T-Search with the PCA is achieved after 16 steps compared to 24 steps when using the traditional approach, which leads to a reduction in computational time of 33%. The best result in terms of computational time was achieved by the one-stop steepest descent, which lead to a reduction of 8.5% in 11 steps when using the PCA compared to 6.8% in 16 steps for the optimisation in CAD space. The performance of the Nelder–Mead simplex was worse under both aspects for the present application; however, the PCA still performed better than the optimisation in CAD space.

5.2. Validation of the Results Using the Viscous Flow Code

In order to validate the results obtained by the potential flow optimisation, computations with the viscous flow code Neptuno are run and the results are compared to the ones obtained with XPAN. It is a popular practice when performing optimisations with a lot of variants to perform the optimisation with a potential flow solver, which is cheap by means of computational costs, and only check the final result using the viscous flow solver.

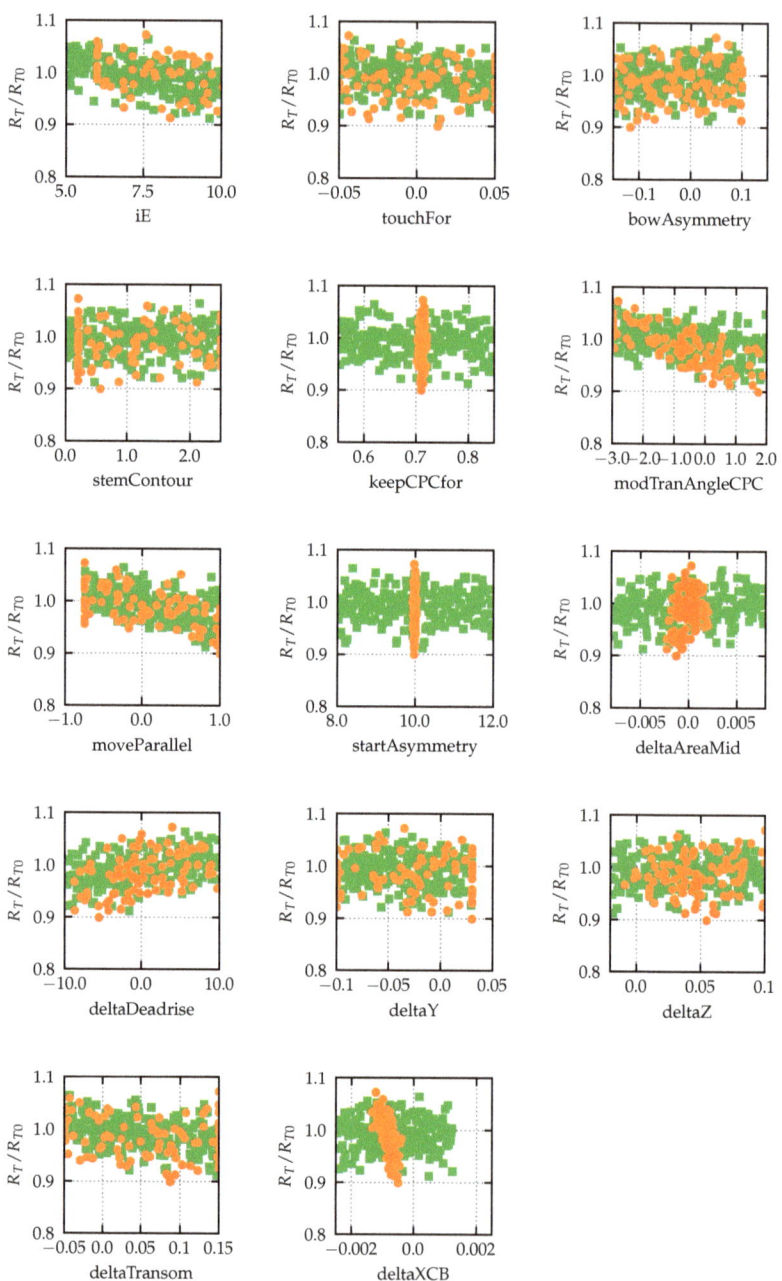

Figure 17. Dependency of the objective function evaluated with XPAN on CAD parameters (green) and on PCA parameters (orange).

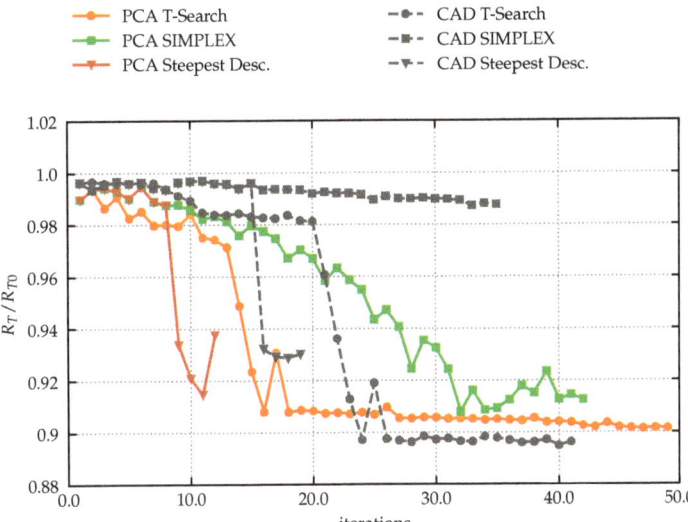

Figure 18. Evaluation of objective function with XPAN for three optimisation algorithms applied in CAD space and PCA space.

At first, the results for the baseline are compared. The results for the total resistance R_T, the trim angle θ and the change in sinkage ζ_O are summarised in Table 6. As can be seen, the agreement for the baseline is rather good, with a difference of 5.9% in the absolute value. It should be noted, however, that the coarse grid underestimates the resistance, so for the fine grid the discrepancy would be slightly larger. Interestingly, for the optimised hull form the differences significantly increase. The viscous flow solver did not confirm the optimisation potential found by the potential flow code. A possible explanation for this behaviour can be found by looking at the free surface elevation predicted by both codes. For the baseline, see Figure 19, a similar wave pattern can be observed for SHIPFLOW (upper half of the cat's wave pattern) and Neptuno (lower half of the cat's wave pattern), although the wave at the bow seems to be a little bit underestimated. The optimised design, see Figure 20, features a bow that is a little bit more blunt, increasing the bow wave which turns out to be problematic for the potential flow code.

Table 6. Comparison of total resistance, trim angle and sinkage computed with SHIPFLOW (SHF) and Neptuno (NEP).

		NEP	SHF	(SHF-NEP)/NEP
Baseline	R_T	10.92 kN	10.28 kN	−5.9%
	θ	1.4°	1.5°	+6.4%
	ζ_O	0.072 m	0.065 m	−8.9%
SHF-OPT	R_T	10.91 kN	9.62 kN	−11.8%
	R_T/R_{T0}	0.999	0.936	−11.8%
	θ	1.12°	1.18°	+5.3%
	ζ_O	0.085 m	0.083 m	−2.4%

5.3. Final Optimisation Using a Viscous Flow Code

Although the optimisation potential seems overestimated by the potential flow code, the reduction of the degrees-of-freedom worked well. Thus, it is possible to perform the optimisation directly with Neptuno instead of using it only to confirm the results of the potential flow solver. Figure 21 displays the results of the optimisation run carried out with Neptuno. In this case, due to the increased numerical effort required to compute the resistance, the optimisation was only performed in the PCA space and only for two of the three considered optimisation algorithms, namely, the steepest descend and the SIMPLEX algorithm. The behaviour of the RANS code is similar to the potential flow code optimisation. The performance of the steepest descend is rather good, whereas the SIMPLEX takes a little bit longer to achieve the same improvement.

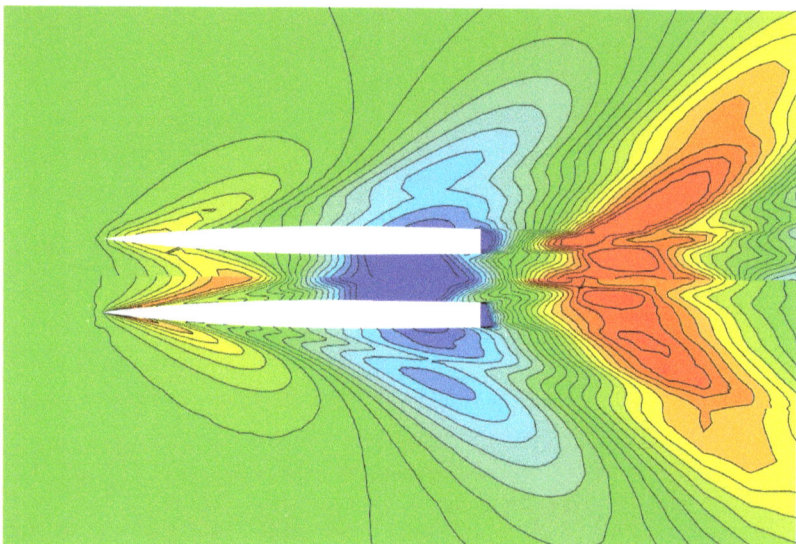

Figure 19. Free surface elevation computed for the baseline computed by SHIPFLOW (upper half of the cat's wave pattern) and Neptuno (lower half of the cat's wave pattern).

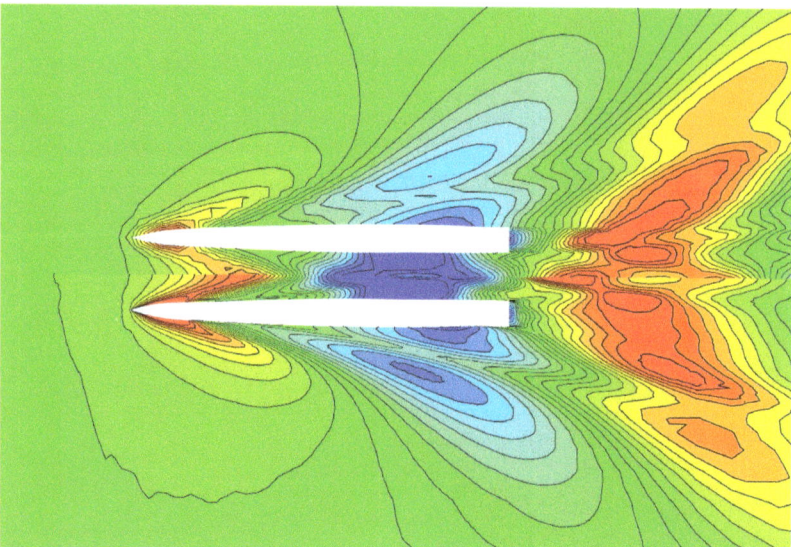

Figure 20. Free surface elevation computed for the design optimised with Shipflow (SHF-OPT) computed by SHIPFLOW (upper half of the cat's wave pattern) and Neptuno (lower half of the cat's wave pattern).

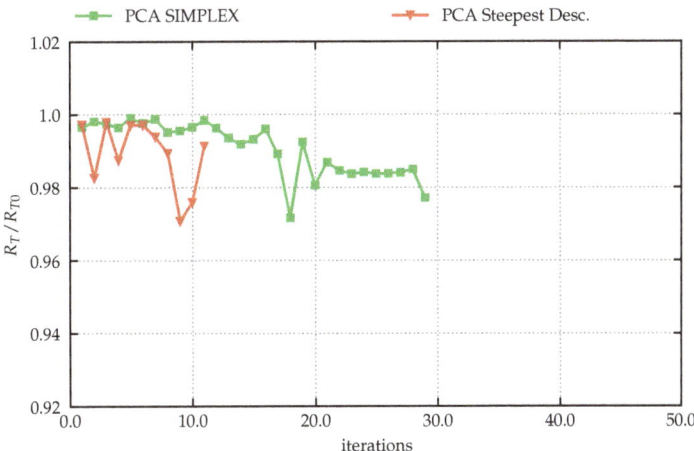

Figure 21. Evaluation of objective function with Neptuno for both optimisation algorithms using PCA.

The steepest descend shows an improvement of 3% in total resistance at model scale (and 10.54 kN total resistance, corresponding to 3.6% improvement at full scale) relative to the baseline after only 11 steps, whereas the SIMPLEX algorithm yields a slightly lower value after 18 iterations. The best design (NEP-OPT) is cross-checked with XPAN as well, which yields an improvement of 7.2%. Figure 22 shows the corresponding free surface elevation for the optimised design as computed with XPAN.

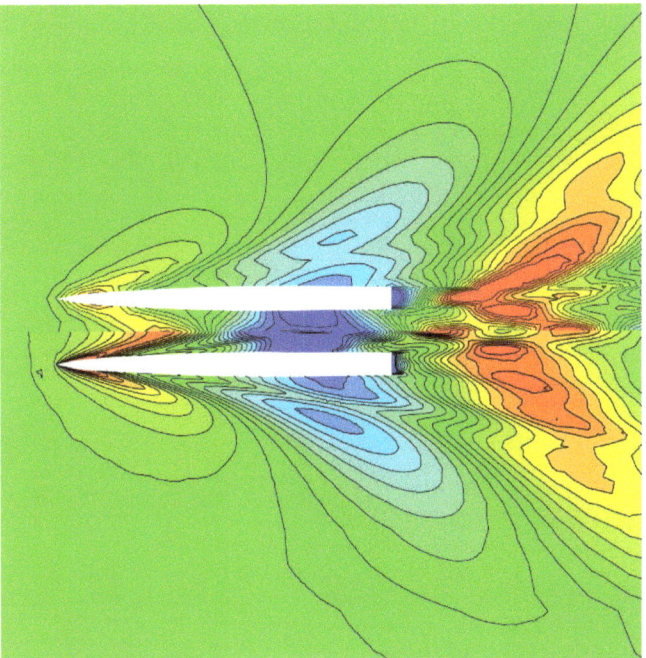

Figure 22. Free surface elevation computed for the design optimised with Neptuno (NEP-OPT) by SHIPFLOW (upper half of the cat's wave pattern) and Neptuno (lower half of the cat's wave pattern).

6. Conclusions and Future Work

With an approximate decrease of 3.6% in resistance at design draft and design speed when compared to the total resistance of the baseline, the improvements found for the catamaran are relatively small. The baseline's performance can therefore be considered pretty good already, at least when looking at the proposed and acceptable geometric variations. This is representative of practical design work where baselines are often quite mature already.

The performed optimisation shows that dimensionality reduction by using a PCA offers high potential in removing complexity from a model without giving away too much optimisation potential. In the present study, the number of free variables is reduced from 14 to 7, which lead to a reduction in computational time of 33%, without sacrificing much of the improvements.

It turns out that the limits of the potential code are reached during the course of the optimisation, requiring higher fidelity tools to be used. Due to the reduction in the degrees-of-freedom by performing a PCA, it is still feasible to run the whole optimisation using a viscous flow solver for practical application cases. This is an even bigger advantage for projects with even more free variables, where gradient-based algorithms take a very long time until they find the direction in which to improve the objective.

More research is needed, naturally. The one-stop steepest descent would be worthwhile to test within other optimisation campaigns. In addition, the PCA and its influence on both speed-up and optimisation gains requires more application cases to gather evidence. In addition, further fine-tuning should be undertaken with regard to questions of sample size for the PCA and the number of points collected. While a large set of samples and many points to consider certainly increase the accuracy of the dimensionality reduction, complex parametric models require a few seconds to a minute of time to update. The fewer samples are acceptable, the less time spent on the PCA.

Author Contributions: This work was jointly planned and carried out by both authors. Both authors have read and agreed to the published version of the manuscript.

Funding: This research was funded by the German Ministry of Economic Affairs through the research project AutoPLOP under grant number 03SX523A.

Data Availability Statement: No data are made available in public. Interested parties can contact the authors.

Acknowledgments: The authors would like to express their gratitude towards Andrés Cura Hochbaum for providing the RANS code Neptuno, FLOWTECH International AB—especially Michal Orych—for the support with the XPAN computations, Samuel James from GridPro for the support regarding the automated meshing, as well as Erik Bergmann for the implementation of the RBF morphing and dimensionality reduction in CAESES.

Conflicts of Interest: The authors declare no conflict of interest.

References

1. Harries, S. Practical Shape Optimization Using CFD: State-Of-The-Art in Industry and Selected Trends. In Proceedings of the Conference on Computer Applications and Information Technology in the Maritime Industries (COMPIT 2020), Pontignano, Italy, 17–19 August 2020.
2. Harries, S.; Abt, C. Faster Turn-around Times for the Design and Optimization of Functional Surfaces. *Ocean Eng.* **2019**, *193*, 106470. [CrossRef]
3. Harries, S.; Abt, C.; Brenner, M. *Upfront CAD—Parametric Modeling Techniques for Shape Optimization*; Springer: Heidelberg, Germany, 2018.
4. Sobol, I.M.; Levitan, Y.L. *The Production of Points Uniformly Distributed in a Multidimensional Cube*; Technical Report; Institute of Applied Mathematics, USSR Academy of Sciences: Moscow, Russia, 1976. (In Russian)
5. Harries, S.; Abt, C. Integration of Tools for Application Case Studies. In *A Holistic Approach to Ship Design. Volume 2: Application Case Studies*; Springer: Heidelberg, Germany, 2021.
6. Botsch, M.; Kobbelt, L. *Real-Time Shape Editing Using Radial Basis Functions*; Blackwell Publishing, Inc.: Oxford, UK; Boston, MA, USA, 2005; Volume 24.

7. Albert, S.; Hildebrandt, T.; Harries, S.; Bergmann, E.; Kovacic, M. Parametric Modeling and Hydrodynamic Optimization of an Electric Catamaran Ferry based on Radial Basis Function for an Intuitive Set-Up. In Proceedings of the Conference on Computer Applications and Information Technology in the Maritime Industries (COMPIT 2021), Muehlheim an der Ruhr, Germany, 2021.
8. Abt, C.; Harries, S. Hull Variation and Improvement using the Generalised Lackenby Method of the FRIENDSHIP-Framework. *Nav. Archit.* **2007**, 166–167.
9. Diez, M.; Campagna, E.F.; Stern, F. Design-space dimensionality reduction in shape optimization by Karhunen–Loève expansion. *Comput. Methods Appl. Mech. Engrg.* **2015**, *283*, 1525–1544. [CrossRef]
10. Pellegrini, R.; Serani, A.; Harries, S.; Diez, M. Multi-Objective Hull-Form Optimization of a SWATH Configuration via Design-Space Dimensionality Reduction, Multi-Fidelity Metamodels, and Swarm Intelligence. In Proceedings of the VII International Conference on Computational Methods in Marine Engineering (MARINE 2017), Nantes, France, 15–17 May 2017.
11. Bergmann, E.; Fütterer, C.; Harries, S.; Palluch, J. Massive Parameter Reduction for faster Fluid-dynamic Optimization of Shapes. In Proceedings of the International CAE Conference and Exhibition, Vicenza, Italy, 8–9 October 2018.
12. Uharek, S.; Cura Hochbaum, A. The influence of inertial effects on the mean forces and moments on a ship sailing in oblique waves Part B: Numerical prediction using a RANS code. *Ocean Eng.* **2018**, *165*, 264–276. [CrossRef]
13. Cura Hochbaum, A. Ein Finite-Volumen-Verfahren zur Berechnung Turbulenter Schiffsumströmungen. Ph.D. Thesis, Technische Universität Hamburg, Hamburg, Germany, 1993. (In Germany)
14. Cura Hochbaum, A.; Vogt, M. Towards the simulation of seakeeping and maneuvering based on the computation of the free surface viscous ship flow. In Proceedings of the 24th ONR Symposium on Naval Hydrodynamics, Fukuoka, Japan, 8–13 July 2002.
15. Patankar, S.V.; Spalding, D.B. A Calculation Procedure for Heat, Mass and Momentum Transfer in Three-Dimensional Parabolic Flows. *Int. J. Heat Mass Transf.* **1972**, *15*. [CrossRef]
16. Wilcox, D.C. *Turbulence Modeling for CFD*, 1st ed.; DCW Industries: Hong Kong, China, 1993.
17. Osher, S.; Sethian, J.A. Fronts Propagating with Curvature-Dependent Speed: Algorithms Based on Hamilton-Jacobi Formulations. *J. Comput. Phys.* **1988**, *79*, 12–49. [CrossRef]
18. Sussman, M.; Smereka, P.; Osher, S. A level set approach for computing solutions to incompressible two-phase flow. *J. Comput. Phys.* **1994**, *114*, 146–159. [CrossRef]
19. Cura Hochbaum, A.; Pierzynski, M. Flow Simulation for a Combatant in Head Waves. In Proceedings of the CFD Workshop Tokyo, Tokyo, Japan, 13–16 November 2005; pp. 1–5.
20. Cura Hochbaum, A.; Vogt, M.; Gatchell, S. Maneuvering prediction for two tankers based on RANS simulations. In Proceedings of the SIMMAN 2008 Workshop, Lyngby, Denmark, 14–16 April 2008; p. F22.
21. Cura Hochbaum, A.; Uharek, S. Prediction of the manoeuvring behaviour of the KCS based on virtual captive tests. In Proceedings of the SIMMAN 2014 Workshop, Copenhagen, Denmark, 2014.
22. ITTC. 1978 ITTC Performance Prediction Method. Recommended Procedures and Guidelines, Number 7.5-02-03-01.4. 2017.
23. Uharek, S. Numerical Prediction of Ship Manoeuvring Performance in Waves. Ph.D. Thesis, Technical University Berlin, Berlin, Germany, 2019.
24. ITTC. Uncertainty Analysis in CFD. Recommended Procedures and Guidelines. 2002, pp. 7.5–03.
25. Hilleary, R. *The Tangent Search Method of Constrained Minimization*; Technical Report, Resarch Paper 59; United States Naval Postgraduate School: Monterey, CA, USA, 1966.
26. Nelder, J.A.; Mead, R. A Simplex method for function minimization. *Comput. J.* **1965**, *7*, 308–313. [CrossRef]

Article

Adaptive Polynomials for the Vibration Analysis of an L-Type Beam Structure with a Free End

Duck Young Yoon [1] and Jeong Hee Park [2,*]

[1] Department of Naval Architecture and Ocean Engineering, Chosun University, Gwangju 61452, Korea; dyyun@chosun.ac.kr
[2] Structural Design Departament, Hyundai Samho Heavy Industries, Jeollanam-do 58462, Korea
* Correspondence: _for_dream@hanmail.net

Abstract: Vibration analysis using the component mode method has been less popular than before, since computers are powerful enough to solve complicated structures by a single large finite model. However, many structural engineers designing local structures on a ship still need simple tools to check anticipated vibration problems during their design work. Since most of local structures on a ship are simple enough to consist of several substructures, the component mode method could be of use as long as good, natural mode functions can be provided so that reasonable natural frequencies can be yielded. In this study, since mode polynomials based on static deflection of cantilever beams fail to work to cover the various configurations of L-type beams with a free end, two alternatives are suggested. One is based on more flexible mode functions—we call them adaptive polynomials. The other is a purely mathematical approach, which makes realistic mode functions unnecessary. Suggested alternatives yield very good numerical results.

Keywords: L-type beam structure; adaptive polynomials; pure mathematical functions

1. Introduction

Two component mode methods are well-known: component mode synthesis, suggested by Hurty [1,2] and Craig [3,4], and the branch mode method, suggested by Hunn [5] and Gladwell [6].

The L-type beam structure with a free end studied here has been used for the explanation of the branch mode method, where normal modes of cantilevers, together with rigid mode, are suggested for generating branch modes [7].

Bhat suggested higher mode functions using orthogonal polynomials, and applied these mode functions for the free vibration analysis of a single plate with different boundary conditions [8].

Bourquin [9], Hou [10], Hintz [11], and Benfield [12] have studied the constraints at the junction of the connected structures.

Recent studies have focused on nonlinear mechanics. Pagani et al. explained that the natural frequency and mode shape can be changed significantly when the metal structure is subjected to large displacement and rotation under geometrical nonlinear conditions [13]. Carrera et al. developed the Lagrange formula, including cross-sectional deformation, in order to implement the vibration mode of the composite beam structure in the nonlinear region [14].

Furthermore, Carrera et al. developed a theory that can be solved by converting a three-dimensional model for large deformation of a structure into one dimension, which was developed and applied to the calculation [15].

Pagani and Carrera [16] introduced the unified formulation for geometric, nonlinear analysis of metal structures, and explained the formula for handling large displacements and rotations. In order to solve the geometric nonlinear problem of the plate, Pagani et al. [17]

explained various nonlinear theories, and how these theories affect the nonlinear static behavior of thin-walled structures in the large displacement and rotation.

Alessandro et al. [18] introduced application of the fundamental model reduction techniques used in structural dynamics to flexible, multibody systems.

Park [19] suggested mode functions for L-type beam structures with fixed ends, where constraints at a junction are described using fixed and simple supported boundary conditions. An application of this mode function for plate structure has also been found [20].

As mentioned, if suitable mode functions are available, the component mode method can be a powerful tool for the free vibration analysis of simple local structures.

The purpose of this study is to provide powerful mode functions for the free vibration analysis of L-type beam structures, as shown in Figure 1, which can work for various length ratios ($0 \leq L_B/L_A \leq \infty$). L_A and L_B are the lengths of the connected L-type beam in Figure 1.

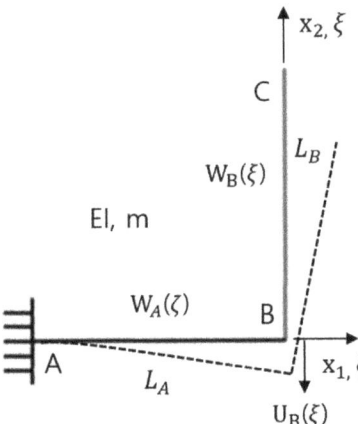

Figure 1. Simplified model of one end of a supported structure.

Fundamental mode function, based on a fourth-order polynomial that satisfies four boundary conditions of a cantilever beam, failed to work for free vibration analysis of an L-type beam structure with a free end, although it describes deflections of a cantilever beam reasonably well. It is a reasonable guess that any mode functions that can describe well the deflections of substructures, which are cantilever beams, may not be able to describe deflections of the L-type beam structure with a free end.

New fundamental mode functions, using a second-order polynomial together with higher orthogonal mode functions, are suggested. These new mode functions have been found to be suitable for the free vibration analysis of L-type beam structures for various length ratios ($0 \leq L_B/L_A \leq \infty$). This good performance is because of the fact that new mode functions based on lower-order polynomials are flexible enough to describe various shapes of deflections of varying length ratios. In this sense, these new polynomials can be named as adaptive polynomials.

In addition, a purely mathematical approach is suggested, where no efforts to describe meaningful mode functions are necessary. Instead, pure mathematical polynomials that only satisfy geometrical boundary conditions at a free end are used.

2. Problem Description and Mathematical Model

An L-type beam structure with a free end is shown in Figure 1:

Where $m_A = m_B = m$, $EI_A = EI_B = EI$ are assumed to be same for notational simplicity; m is mass per unit length of beam, and E and I are the Young's modulus and moment of inertia, respectively.

x_1 and x_2 are coordinates of substructrues of L-type structures in Figure 1.

In addition, x_1 and x_2 are non-dimensionalized, such that $\zeta = \frac{x_1}{L_A}$, $\xi = \frac{x_2}{L_B}$.

$W_A(\zeta)$ and $W_B(\xi)$ are lateral deflections of the horizontal and vertical beam, respectively:

$$W_A(\zeta, t) = \sum_{i=1}^{m} \phi_i(\zeta) p_i(t) \tag{1}$$

$$W_B(\xi, t) = \sum_{j=1}^{n} \psi_j(\xi) q_j(t) \tag{2}$$

$$U_B(\xi, t) = r_1(t) \tag{3}$$

where $p_i(t), q_j(t)$, and $r_1(t)$ are the generalized coordinates, and $\phi_i(\zeta)$ and $\psi_j(\xi)$ are corresponding mode functions to describe lateral deflections of beams A and B. $U_B(t)$ is vertical displacement of beam B.

The method proposed by Bhat to generate higher orthogonal polynomials is as follows:

$$\phi_2(\zeta) = (\zeta - B_1)\phi_1(\zeta) \tag{4}$$

$$\phi_k(\zeta) = (\zeta - B_k)\phi_{k-1}(\zeta) - C_k \phi_{k-2}(\zeta) \tag{5}$$

$$B_k = \int_0^1 \zeta \cdot \phi_{k-1}^2(\zeta) d\zeta \bigg/ \int_0^1 \phi_{k-1}^2(\zeta) d\zeta \tag{6}$$

$$C_k = \int_0^1 \zeta \cdot \phi_{k-1}(\zeta) \phi_{k-2}(\zeta) d\zeta \bigg/ \int_0^1 \phi_{k-2}^2(\zeta) d\zeta \tag{7}$$

It can be shown that the polynomial $\phi_k(\zeta)$ satisfies the orthogonality condition:

The coefficients B_k and C_k are implemented using the orthogonal formula of the beam function:

$$\int_0^1 \phi_k(\zeta) \phi_l(\zeta) d\zeta = \left\{ \begin{array}{l} 0 \; if \; k \neq l \\ 1 \; if \; k = l \end{array} \right\} \tag{8}$$

Note that this polynomial $\phi_k(\zeta)$ only satisfies geometrical boundary conditions, although a fundamental polynomial can be chosen to satisfy natural boundary conditions.

Given mode functions, generalized mass is

$$m_{Aij} = mL_A \int_0^1 \phi_i \phi_j d\zeta \tag{9}$$

$$m_{Bij} = mL_B \int_0^1 \psi_i \psi_j d\xi \tag{10}$$

$$m_{Brr} = mL_B \int_0^1 d\xi \tag{11}$$

and generalized stiffness (k) is

$$k_{Aij} = \frac{EI}{L_A^3} \int_0^1 \phi''_i \phi''_j d\zeta \tag{12}$$

$$k_{Bij} = \frac{EI}{L_B^3} \int_0^1 \psi''_i \psi''_j d\xi \tag{13}$$

where, $m_{Aij}, m_{Bij}, m_{Brr}$ and k_{Aij}, k_{Bij} are generalized mass and generalized stiffness of substructures of L-type structure.

Applying displacement continuity (r_1) at the junction results in the following:

$$r_1 = \sum \phi_i(1) p_i = (\phi_1(1) p_1 + \phi_2(1) p_2 + \cdots + \phi_m(1) p_m) \tag{14}$$

Similarly, applying slope continuity (q_n) yields,

$$q_n = \frac{L_B}{L_A}\left(\frac{\phi'_1(1)}{\psi'_n(0)} p_1 + \frac{\phi'_2(1)}{\psi'_n(0)} p_2 + \cdots + \frac{\phi'_m(1)}{\psi'_n(0)} p_m - \frac{\psi'_1(0)}{\psi'_n(0)} q_1 - \frac{\psi'_2(0)}{\psi'_n(0)} q_2 - \cdots - \frac{\psi'_{n-1}(0)}{\psi'_n(0)} q_{n-1}\right) \tag{15}$$

The mass and stiffness matrices (M_A and K_A, respectively) using the suggested polynomials are shown in Equations (16) and (17):

$$[M_A] = mL_A \begin{vmatrix} \phi_1\phi_1 & \cdots & \phi_1\phi_m & \phi_1\psi_1 & \cdots & \phi_1\psi_n \\ \vdots & \ddots & \vdots & \vdots & \ddots & \vdots \\ \phi_m\phi_1 & \cdots & \phi_m\phi_m & \phi_m\psi_1 & \cdots & \phi_m\psi_n \\ \psi_1\phi_1 & \cdots & \psi_1\phi_m & \psi_1\psi_1 & \cdots & \psi_1\psi_n \\ \vdots & \ddots & \vdots & \vdots & \ddots & \vdots \\ \psi_n\phi_1 & \cdots & \psi_n\phi_m & \psi_n\psi_1 & \cdots & \psi_n\psi_n \end{vmatrix} \tag{16}$$

$$[K_A] = \frac{8EI}{L_A^3} \begin{vmatrix} \phi''_1\phi''_1 & \cdots & \phi''_1\phi''_m & \phi''_1\psi''_1 & \cdots & \phi''_1\psi''_n \\ \vdots & \ddots & \vdots & \vdots & \ddots & \vdots \\ \phi''_m\phi''_1 & \cdots & \phi''_m\phi''_m & \phi''_m\psi''_1 & \cdots & \phi''_m\psi''_n \\ \psi''_1\phi''_1 & \cdots & \psi''_1\phi''_m & \psi''_1\psi''_1 & \cdots & \psi''_1\psi''_n \\ \vdots & \ddots & \vdots & \vdots & \ddots & \vdots \\ \psi''_n\phi''_1 & \cdots & \psi''_n\phi''_m & \psi''_n\psi''_1 & \cdots & \psi''_n\psi''_n \end{vmatrix} \tag{17}$$

and M_B and K_B can be expressed in a similar manner. Using the displacement and slope continuity in Equations (14) and (15) yields Equation (18):

$$\begin{Bmatrix} p_1 \\ \vdots \\ p_m \\ q_1 \\ \vdots \\ q_{n-1} \\ q_n \\ r_1 \end{Bmatrix} = \begin{vmatrix} \phi_1\phi_1 & \cdots & \phi_1\phi_m & \phi_1\psi_1 & \cdots & \phi_1\psi_n \\ \vdots & \ddots & \vdots & \vdots & \ddots & \vdots \\ \phi_m\phi_1 & \cdots & \phi_m\phi_m & \phi_m\psi_1 & \cdots & \phi_m\psi_n \\ \psi_1\phi_1 & \cdots & \psi_1\phi_m & \psi_1\psi_1 & \cdots & \psi_1\psi_n \\ \vdots & \ddots & \vdots & \vdots & \ddots & \vdots \\ \psi_n\phi_1 & \cdots & \psi_n\phi_m & \psi_n\psi_1 & \cdots & \psi_n\psi_n \\ \alpha\frac{\psi'_1(1)}{\phi'_n(0)} & \cdots & \alpha\frac{\psi'_m(1)}{\phi'_n(0)} & -\alpha\frac{\psi'_1(1)}{\phi'_n(0)} & \cdots & -\alpha\frac{\psi'_{n-1}(1)}{\phi'_n(0)} \\ \phi_1(1) & \cdots & \phi_m(1) & 0 & \cdots & 0 \end{vmatrix} \begin{Bmatrix} p_1 \\ \cdots \\ p_m \\ q_1 \\ \cdots \\ q_{n-1} \end{Bmatrix} \tag{18}$$

where α is the ratio of length for the subcomponents ($\alpha = L_B/L_A$):

3. FEM (Finite Element Method) Analysis

For comparison, FEM analysis was performed first. The eam properties used are shown in Table 1, and Figure 2 shows the L-type finite element method (FEM) model and geometric boundary conditions.

Table 1. Properties of the finite element method (FEM) model.

Property	W_A	Cross Section
Density (kg/m^3)	7850	
Total Length ($L_A + L_B$) (m)	10	0.2m
Young's modulus (N/mm^2)	2.1×10^5	
Moment of inertia (mm^4)	3.33×10^8	0.5m

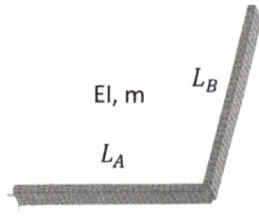

Figure 2. L-type beam structure with a free end.

In Figure 2, m is mass per unit length of beam, E and I are Young's modulus and moment of inertia, respectively.

Natural frequencies are shown in Figure 3, Figure 4, Figure 5.

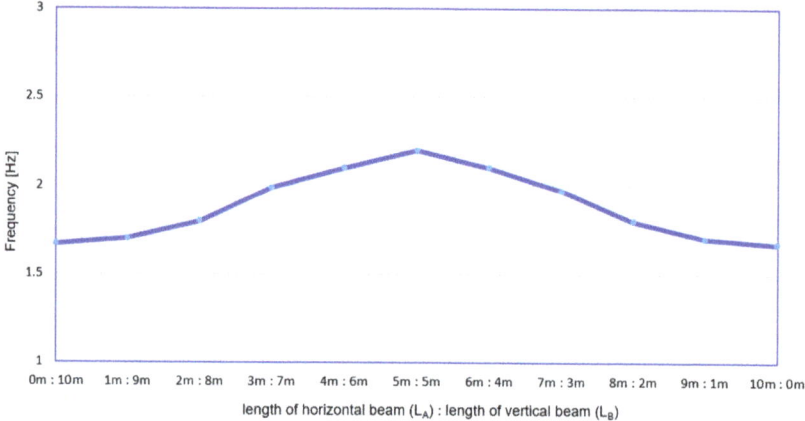

Figure 3. Natural frequency of the first mode for an L-type beam structure.

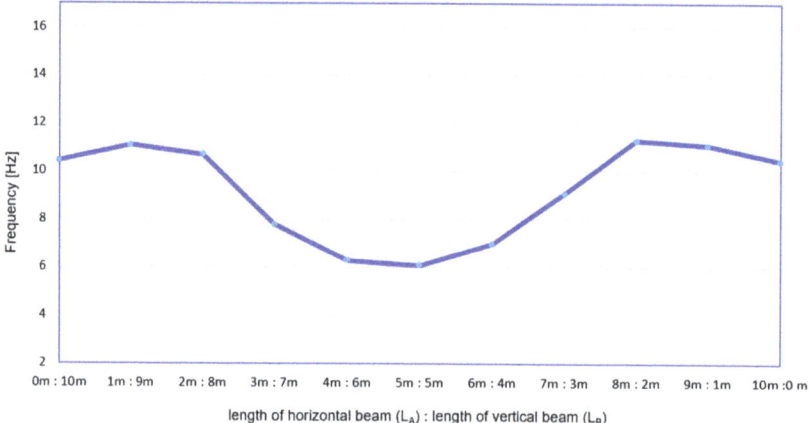

Figure 4. Natural frequency of the second mode for an L-type beam structure.

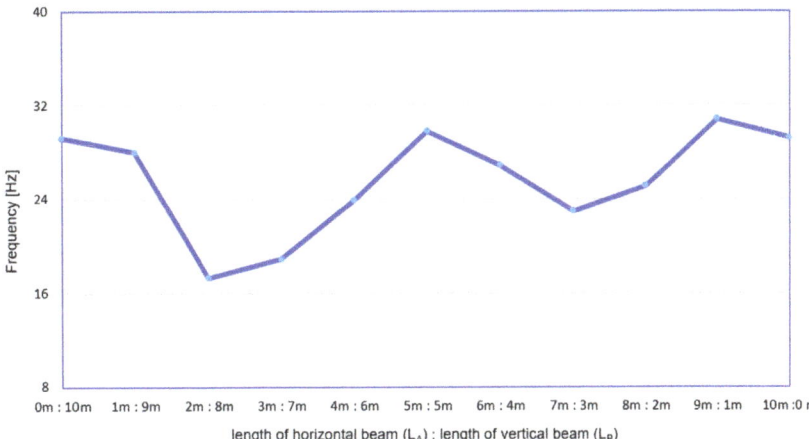

Figure 5. Natural frequency of the third mode for an L-type beam structure.

It is worth noting that the natural frequencies for the length ratio L_B/L_A are relatively similar to those for the length ratio L_A/L_B as like Figure 6, although mode shapes are different; this is somewhat interesting. However, it can be understood because this structure becomes a cantilever beam as L_A or L_B approaches zero.

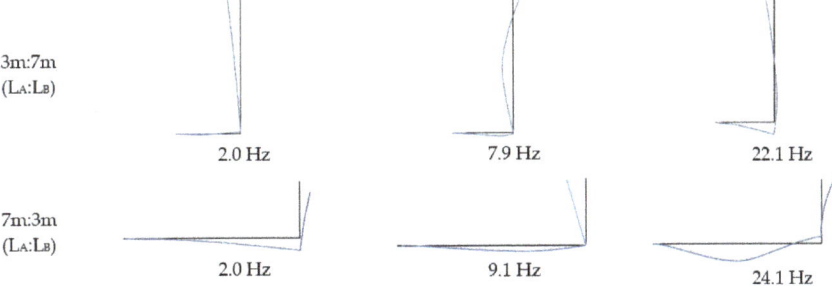

Figure 6. Typical mode shape of FEM result: first, second, and third mode.

4. Fundamental Mode Function Using Fourth-Order Polynomial and Numerical Results

The fourth-order polynomial for fundamental mode function $\varnothing_1(\zeta)$ can be easily obtained from four boundary conditions of a cantilever beam:

$$w(0) = w'(0) = 0, w''(1) = w'''(1) = 0$$

The lower four polynomials are shown in Table 2.

Table 2. The mode function using fourth-order polynomial.

i	Mode Functions (ϕ_i)
1	$\phi_1(\zeta) = \zeta^4 - 4\zeta^3 + 6\zeta^2$
2	$\phi_2(\zeta) = \zeta^5 - 4.8022\zeta^4 + 9.2088\zeta^3 - 4.8132\zeta^2$
3	$\phi_3(\zeta) = \zeta^6 - 5.4477\zeta^5 + 12.2838\zeta^4 - 10.6580\zeta^3 + 2.6575\zeta^2$
4	$\phi_4(\zeta) = \zeta^7 - 6.0363\zeta^6 + 15.4515\zeta^5 - 17.7016\zeta^4 + 8.8724\zeta^3 - 1.5534\zeta^2$

Free vibration analysis using these mode functions for a cantilever beam was performed.

The numerical results was compared with those of the analytical solution of Euler's beam [21] and are shown in Table 3. For reference, the comparison result in Table 3 is the calculated value of the relationship between the natural frequency of the beam and the properties.

Table 3. Comparison of FEM result and using a fourth-order polynomial.

	Fundamental $(\beta_n l)^2$	Second $(\beta_n l)^2$	Third $(\beta_n l)^2$	Remark
FEM result	3.51	21.99	61.44	$(\beta_n l)^2 = w_n / \sqrt{\frac{EI}{\rho l^4}}$
Using fourth-order polynomial	3.51	22.00	61.70	

However, free vibration analysis for the L-type structure with a free end using these mode functions was not satisfactory, as shown in Figures 7 and 8.

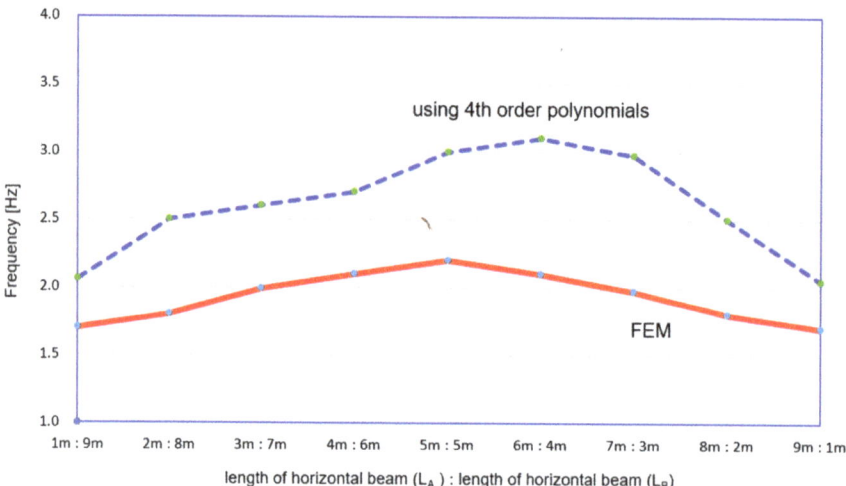

Figure 7. The comparison result of FEM and an L-type connected beam (using a fourth-order polynomial): First mode.

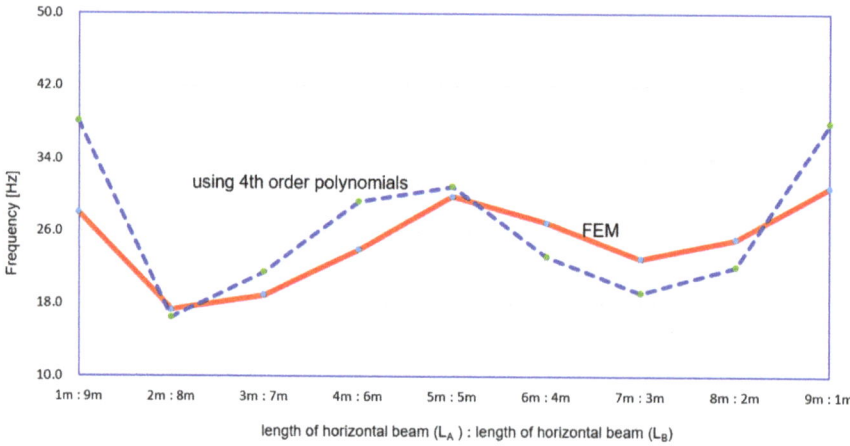

Figure 8. The comparison result of FEM and an L-type connected beam (using a fourth-order polynomial): Third mode.

5. Fundamental Mode Function Using Second-Order Polynomial (Adaptive Mode Function)

In order to make mode functions as flexible as possible, a second-order polynomial was chosen as a fundamental mode function, and higher orthogonal polynomials were generated, as suggested before by Bhat. In order to consider the rigid rotation of a vertical beam, $\psi_1(\xi) = \xi$ is added.

The lower four mode functions are shown Table 4.

Table 4. The mode function using adaptive polynomials.

i and j	Horizontal Component (ϕ_i)	Vertical Component (ψ_j)
1	ζ^2	ξ
2	$\zeta^3 - 0.8333\zeta^2$	ξ^2
3	$\zeta^4 - 1.5\zeta^3 + 0.5357\zeta^2$	$\xi^3 - 0.8333\xi^2$
4	$\zeta^5 - 2.1\zeta^4 + 1.4\zeta^3 - 0.2917\zeta^2$	$\xi^4 - 1.5\xi^3 + 0.5357\xi^2$

The numerical results are shown in Table 5, Table 6, Table 7. Thirteen mode functions, seven for ϕ_i and five for ψ_j, were used for this numerical analysis.

Table 5. Comparison of FEM results and using adaptive polynomials: First mode.

Length (L_A:L_B)	0:10	1:9	2:8	3:7	4:6	5:5	6:4	7:3	8:2	9:1	10:0
FEM	1.67	1.70	1.80	1.99	2.10	2.20	2.10	1.97	1.80	1.70	1.67
Adaptive polynomial	1.67	1.71	1.82	2.00	2.10	2.20	2.10	1.97	1.80	1.71	1.67

Table 6. Comparison of FEM results and using adaptive polynomials: second mode.

Length (L_A:L_B)	0:10	1:9	2:8	3:7	4:6	5:5	6:4	7:3	8:2	9:1	10:0
FEM	10.45	11.10	10.70	7.80	6.30	6.10	7.00	9.10	11.30	11.10	10.45
Adaptive polynomial	10.45	11.10	10.75	7.87	6.30	6.10	6.96	9.10	11.30	11.10	10.45

Table 7. Comparison of FEM results and using adaptive polynomials: third mode.

Length (L_A:L_B)	0:10	1:9	2:8	3:7	4:6	5:5	6:4	7:3	8:2	9:1	10:0
FEM	29.20	28.00	17.30	18.90	23.90	29.80	26.90	23.00	25.10	30.80	29.20
Adaptive polynomial	29.20	28.90	17.40	19.00	24.12	29.80	27.00	23.20	25.40	31.50	29.20

Typical corresponding natural modes are shown in Figure 9.

The numerical results showed very good agreement with the FEM results. This good agreement is due to the fact that suggested mode functions are flexible enough to follow anticipated deflections of an L-type beam with a free end. In that sense, we named these mode polynomials as having "adaptive mode function".

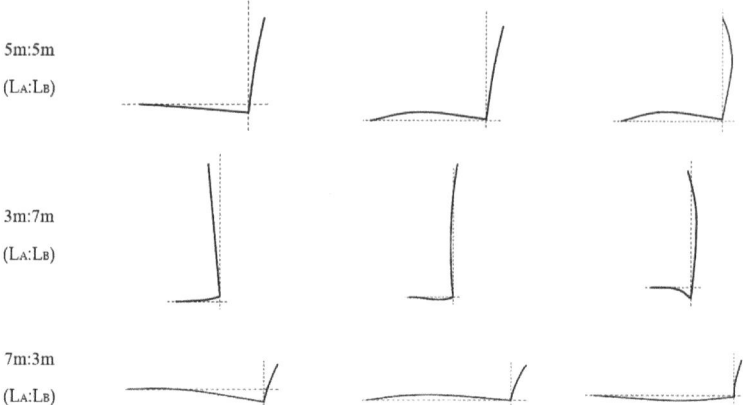

Figure 9. The mode shape of using adaptive polynomials.

6. Pure Mathematical Method

Mode functions for typical structures have been proposed, including this study [22,23]. There may be some structures where suitable mode functions may not be easy to generate.

In this case, a purely mathematical approach is suggested, where no meaningful higher-order mode functions are necessary. We may take mode functions in the following form:

$$W_A(\zeta) = \sum_{i=1}^{m} \phi_i(\zeta) p_i = \sum_{i=1}^{m} \zeta^{i+1} p_i \quad (19)$$

$$W_B(\xi) = \sum_{j=1}^{n} \psi_j(\xi) q_j = \sum_{j=1}^{n} \xi^j q_j \quad (20)$$

$$U_B(\xi) = r_1 \quad (21)$$

Note that no higher-order mode functions are assumed. Most accurate natural frequencies are obtained using the how approach, although the eigenvectors obtained have no physical meaning.

This is due to the fact that no assumption for higher mode functions has been made. The numerical results were compared with those obtained from FEM analysis. Figure 1 was used for the calculation model and the beam properties mentioned in Table 1. The calculation results are shown in Table 8, Table 9, Table 10.

Table 8. Comparison of FEM results and using mathematical function: First mode.

Length ($L_A:L_B$)	0:10	1:9	2:8	3:7	4:6	5:5	6:4	7:3	8:2	9:1	10:0
FEM	1.67	1.70	1.80	1.99	2.10	2.20	2.10	1.97	1.80	1.70	1.67
Mathematical function	1.67	1.71	1.82	1.99	2.10	2.20	2.10	1.97	1.82	1.71	1.67

Table 9. Comparison of FEM results and using mathematical function: Second mode.

Length ($L_A:L_B$)	0:10	1:9	2:8	3:7	4:6	5:5	6:4	7:3	8:2	9:1	10:0
FEM	10.45	11.10	10.70	7.80	6.30	6.10	7.00	9.10	11.30	11.10	10.45
Mathematical function	10.45	11.16	10.75	7.87	6.34	6.08	6.96	9.10	11.33	11.15	10.45

Table 10. Comparison of FEM results and using mathematical function: Third mode.

Length ($L_A:L_B$)	0:10	1:9	2:8	3:7	4:6	5:5	6:4	7:3	8:2	9:1	10:0
FEM	29.20	28.00	17.30	18.90	23.90	29.80	26.90	23.00	25.10	30.80	29.20
Mathematical function	29.20	28.40	17.48	19.11	24.12	30.00	27.19	23.31	25.47	31.50	29.20

To better understand how good results can be obtained, use mode shapes together with eigenvectors. The mode shape is shown in Figure 10, while the eigenvectors are shown in Table 11.

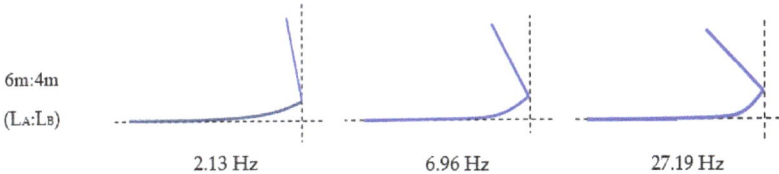

6m:4m

($L_A:L_B$)

2.13 Hz   6.96 Hz   27.19 Hz

Figure 10. Mode shapes using mathematical function.

Table 11. Eigenvectors of the mode shape.

Coordinate	2.1 Hz	6.96 Hz	27.2 Hz	Coordinate	2.1 Hz	6.96 Hz	27.2 Hz
P1 (ζ^2)	1.00	1.00	1.00	P11 (ζ^{12})	2.23	26.60	−15.65
P2 (ζ^3)	1.10	3.78	2.65	P12 (ζ^{13})	2.36	29.07	−18.33
P3 (ζ^4)	1.22	6.45	2.36	P13 (ζ^{14})	2.49	31.54	−21.04
P4 (ζ^5)	1.34	9.05	1.08	P14 (ζ^{15})	2.62	34.00	−23.75
P5 (ζ^6)	1.46	11.61	−0.75	P15 (ζ^{16})	2.76	36.47	−26.48
P6 (ζ^7)	1.58	14.14	−2.91	P16 (ζ^{17})	2.89	38.93	−29.23
P7 (ζ^8)	1.71	16.65	−5.28	P17 (ζ^{18})	3.02	41.39	−31.99
P8 (ζ^9)	1.84	19.15	−7.78	P18 (ζ^{19})	3.16	43.84	−34.77
P9 (ζ^{10})	1.97	21.64	−10.36	P19 (ζ^{20})	3.29	46.30	−37.57
P10 (ζ^{11})	2.10	24.12	−12.99	P20 (ζ^{21})	3.42	48.75	−40.39

7. Discussion

Our work deals with a very classic subject, and little research based on the assumed mode method has been found in last 20 years. Furthermore, free vibration analysis of an L-type beam with a free end is a typical example, even in the textbooks, for explaining component mode synthesis.

However, we believe that our work can renew appreciation of the usefulness of component mode method for free vibration analysis, by providing powerful mode functions.

As you can see. it will not be an easy task to find mode functions that can work on various configurations of L-type beam structures (length ratio L_B/L_A varies from 0 to ∞). Certain mode functions that can work for one specific value of L_B/L_A may not work for different values of L_B/L_A.

Although we do not include it in the paper, the suggested mode function comes from dozens of candidates. If a component is divided into subcomponents which may have geometrical boundary conditions only at one end, like a cantilever, then free vibration solutions may be very sensitive to the choice of mode functions. Most methods based on the Rayleigh–Ritz method use assumed modes.

However, we suggest using pure mathematical functions instead of using an assumed mode function. Although mode shapes using mathematical functions have nothing to do with real mode shapes, the results of the proposed method are compared with the FEM results and shown in Table 5, Table 6, Table 7, Table 8, Table 9, Table 10.

As a result, the function is accurate enough to show an error rate of less than 2% in all sections, regardless of the length ratio of the connected structure.

8. Conclusions

A second-order fundamental polynomial, together with higher orthogonal polynomials, is suggested as the most suitable assumed mode functions for an L-type beam structure with a free end.

The robustness of the suggested polynomials is proven through numerical analysis for an L-type beam structure with a free end against varying length ratios.

A purely numerical approach has been suggested for the structures where substructures have geometrical conditions only at one end, like a cantilever beam.

The most accurate natural frequencies are obtained this way, since any assumptions for higher-mode functions are unnecessary. Once natural frequencies are obtained, the way to find corresponding natural modes is worth studing.

Author Contributions: Conceptualization, D.Y.Y., J.H.P.; methodology, D.Y.Y., J.H.P.; software, J.H.P.; validation, J.H.P., D.Y.Y.; formal analysis, J.H.P., D.Y.Y.; investigation, J.H.P.; writing—original draft preparation, J.H.P., D.Y.Y.; writing—review and editing, J.H.P., D.Y.Y.; funding acquisition. All authors have read and agreed to the published version of the manuscript.

Funding: This research has been supported by the Chosun University research fund.

Conflicts of Interest: The authors declare no conflict of interest.

References

1. Hurty, W.C. Vibrations of structural systems by component mode synthesis. *J. Eng. Mech. Div.* **1960**, *86*, 51–70. [CrossRef]
2. Hurty, W.C. Dynamic analysis of structural systems using component modes. *AIAA J.* **1965**, *3*, 678–685. [CrossRef]
3. Craig, R.R.; Bampton, M.C.C. Coupling of sub-structures for dynamic analysis. *AIAA J.* **1968**, *6*, 1313–1319. [CrossRef]
4. Craig, R.R.; Chang, C.J. A review of substructure coupling methods for dynamic analysis. *Adv. Eng. Sci.* **1976**, *2*, 393–408.
5. Hunn, B.A. A method of calculating space free resonant modes of an aircraft. *J. Roy. Aero. Soc.* **1953**, *57*, 420. [CrossRef]
6. Gladwell, G.M.L. Branch mode analysis of vibrating systems. *J. Sound Vib.* **1964**, *1*, 41–59. [CrossRef]
7. Thorby, D. *Structural Dynamics and Vibration in Practice*, 1st ed.; Elsevier: London, UK, 2008; pp. 194–213.
8. Bhat, R.B. Natural Frequencies of Rectangular Plates Using Characteristic Orthogonal Polynomials in Rayleigh-Ritz Method. *J. Sound Vib.* **1985**, *102*, 493–499. [CrossRef]
9. Bourquin, F. Component mode synthesis and eigenvalues of second order operators: Discretization and algorithm. *ESAIM Math. Model. Numer. Anal.* **1992**, *26*, 385–423. [CrossRef]
10. Hou, S. Review of modal synthesis techniques and a new approach. *Shock Vib. Bull.* **1969**, *40*, 25–39.
11. Hintz, R.M. Analytical Methods in Component Modal Synthesis. *AIAA J.* **1975**, *13*, 1007–1016. [CrossRef]
12. Benfield, W.A.; Hruda, R.F. Vibration Analysis of Structures by Component Mode Substitution. *AIAA J.* **1971**, *9*, 1255–1261. [CrossRef]
13. Pagani, A.; Augello, R.; Carrera, E. Frequency and mode change in the large deflection and post-buckling of compact and thin-walled beams. *J. Sound Vib.* **2018**, *432*, 88–104. [CrossRef]
14. Carrera, E.; Pagani, A.; Augello, R. Effect of large displacements on the linearized vibration of composite beams. *Int. J. Non-linear Mech.* **2020**, *120*, 103390. [CrossRef]
15. Carrera, E.; Pagani, A.; Giusa, D.; Augello, R. Nonlinear analysis of thin-walled beams with highly deformable sections. *Int. J. Non-Linear Mech.* **2021**, *128*, 103613. [CrossRef]
16. Carrera, E.; Pagani, A.; Augello, R. Evaluation of geometrically nonlinear effects due to large cross-sectional deformations of compact and shell-like structures. *Mech. Adv. Mater. Struct.* **2018**, *27*, 1269–1277. [CrossRef]
17. Pagani, A.; Daneshkhah, E.; Xu, X.; Carrera, E. Evaluation of geometrically nonlinear terms in the large-deflection and post-buckling analysis of isotropic rectangular plates. *Int. J. Non-Linear Mech.* **2020**, *121*, 103461. [CrossRef]
18. Alessandro, C.; Pappalardo, C. On the use of component mode synthesis methods for the model reduction of flexible multibody systems within the floating frame of reference formulation. *Mech. Syst. Signal Process* **2020**, *142*, 106745.
19. Park, J.H.; Yoon, D.Y. A Proposal of Mode Polynomials for Efficient use of Component mode synthesis and Methodology to simplify the calculation of the connecting beams. *J. Mar. Sci. Eng.* **2021**, *9*, 20. [CrossRef]
20. Park, J.-H.; Yang, J.-H. Normal Mode Analysis for Connected Plate Structure Using Efficient Mode Polynomials with Component Mode Synthesis. *Appl. Sci.* **2020**, *10*, 7717. [CrossRef]
21. Thomson, W.T. *Theory of Vibration with Applications*, 4th ed.; Prentice Hall: Upper Saddle River, NJ, USA, 1993; pp. 360–365.
22. Han, S.Y.; Huh, Y.C. Vibration Analysis of Quadrangular plate having attachments by the Assumed Mode Method. *Trans. Soc. Nav. Archit. Korea* **1995**, *32*, 116–125. (In Korean)
23. Rubin, S. Improved component-mode representation for structural dynamic analysis. *AIAA J.* **1975**, *13*, 995–1006. [CrossRef]

MDPI AG
Grosspeteranlage 5
4052 Basel
Switzerland
Tel.: +41 61 683 77 34

Journal of Marine Science and Engineering Editorial Office
E-mail: jmse@mdpi.com
www.mdpi.com/journal/jmse

Disclaimer/Publisher's Note: The statements, opinions and data contained in all publications are solely those of the individual author(s) and contributor(s) and not of MDPI and/or the editor(s). MDPI and/or the editor(s) disclaim responsibility for any injury to people or property resulting from any ideas, methods, instructions or products referred to in the content.

www.ingramcontent.com/pod-product-compliance
Lightning Source LLC
LaVergne TN
LVHW070157120526
838202LV00013BA/1308